Plant Community History

PLANT COMMUNITY HISTORY

Long-term changes in plant distribution and diversity

J. H. Tallis
Senior Lecturer
Department of Environmental Biology
University of Manchester

CHAPMAN AND HALL
London • New York • Tokyo • Melbourne • Madras

UK	Chapman and Hall, 2–6 Boundary Row, London SE1 8HN
USA	Chapman and Hall, 29 West 35th Street, New York NY10001
JAPAN	Chapman and Hall Japan, Thomson Publishing Japan, Hirakawacho Nemoto Building, 7F, 1–7–11 Hirakawa-cho, Chiyoda-ku, Tokyo 102
AUSTRALIA	Chapman and Hall Australia, Thomas Nelson Australia, 102 Dodds Street, South Melbourne, Victoria 3205
INDIA	Chapman and Hall India, R. Seshadri, 32 Second Main Road, CIT East, Madras 600 035

First edition 1991

Typeset in 10/12pt Meridien by
Rowland Phototypesetting Ltd,
Bury St Edmunds, Suffolk
Printed in Great Britain by
St Edmundsbury Press Ltd,
Bury St Edmunds, Suffolk

ISBN 0 412 30320 5

British Library Cataloguing in Publication Data
Tallis, John H.
Plant community history.
1. Plants. Evolution
I. Title
581.38

ISBN 0-412-30320-5

Library of Congress Cataloging-in-Publication Data available

Contents

Preface

At a time of increasing concern over the future of the global environment, a book about plant communities of the past might appear an academic irrelevance, an ostrich-in-the-sand avoidance of the crucial issues. Yet a knowledge of how plants have responded to environmental changes in the past is our principal source of evidence for predicting how they might respond to the new environments of the future, whilst a knowledge of past climates may enable us to assess more clearly the human contribution to the climatic changes of the future. Moreover, with the natural plant cover of the world relentlessly diminishing in extent the identification of areas of prime conservation status becomes increasingly more vital. That too requires information about their history – as I hope this book will show.

Although the content of this book thus has relevance to the future, the main concern is nevertheless for the present. *Plant Community History* sets out to demonstrate that the composition and distribution of modern plant communities can be understood only in the light of their past history. In the developed world we are accustomed to affording human manipulation pride of place in that history. Yet the moulding of the modern plant cover has been proceeding much longer than Man's presence on the earth – and has taken place in the context of major changes in climate and the global configurations of land, sea and ice. Our understanding of events contributing to that moulding has increased enormously in the last 30 years, with a proliferation of studies in such varied disciplines as geophysics, oceanography, climatology and palynology. The relevance of much of the work to plant community history at the global level is often difficult for the non-specialist to comprehend, as published synthetic accounts tend to be parochial or circumscribed in outlook, concentrating either on a particular country (the British Isles, for example) or on a particular time-period (the last 10 000 years, for example), and dealing with certain aspects only (methods of investigation, or vegetation history, for example). Only one book (the recently-published multi-author *Vegetation History**) attempts a near-global synthesis of long-term environmental and vegetational change, and its treatment is aimed at readers already familiar with the

* Huntley, B. and Webb, T., III (eds) (1988), *Vegetation History*, Kluwer Dordrecht.

subject matter. *Plant Community History* can thus claim to be the first single-author book aimed primarily at the non-specialist, and designed to form a baseline to which new discoveries can be related. Because of the breadth of coverage demanded, the uneven quality of the data available and the inevitable simplifications and generalizations that must result from compressing that material into 300 pages of text, I recognize that the synthesis must be an imperfect one. My justification for attempting it at this point in time is that any account is better than none at all.

I hope that *Plant Community History* will appeal to all biologists with an informed interest in the world's plant cover. Its aim is to present the results of investigations into the history of this plant cover over the last 65 million years, not the methods by which these results have been achieved, and to document a selection of the more accessible published accounts, both general and specific, for further reading. The book is organized into four main parts, dealing successively with the tenet that modern plant distributions and communities have been moulded by long-term environmental change (Chapters 1–3), the nature of global environmental change over the last 65 million years (Chapters 4–6), the ways in which plants and plant communities have responded to environmental change (Chapters 7 and 8), and the global history of forest, dryland and coldland communities over the last 65 million years (Chapters 9–14). Finally, Chapter 15 looks to the future. The attempted breadth of coverage is bound to lead to unevenness of treatment – and to omissions and deficiencies that will be obvious to the specialist. However, I hope that this same breadth of coverage will compensate for any shortcomings, and that the reader will be introduced to new perspectives in the study of plant communities that may help towards a better understanding of the status and management of modern landscapes.

The final form of this book owes much to the helpful comments of two anonymous referees and Dr E. A. Livett. I thank especially all those authors whose published work provided the basis for the illustrations.

PART ONE

The Global Kaleidoscope of Vegetation and Climate

CHAPTER ONE

The reality of long-term change

1.1 INTRODUCTION

Change is an integral feature of the world we live in. The changes we perceive most strongly are those centred on the diurnal and seasonal patterns of climate, the unpredictable daily vagaries of weather, and the processes of growing older. The natural plant cover clearly responds to these changes also, but there is widespread implicit, if not explicit, belief in its overall stability if left undisturbed or with minimal management. Major vegetation changes at the present day are seen as the product of human mismanagement rather than of natural environmental change. The fossil record suggests otherwise. Even if we ignore the grand sweep of evolutionary change over geological time that produced, *inter alia*, landscapes dominated by dinosaurs and tree ferns, there is abundant evidence from more recent times of the remains of plants and animals preserved in situations where they could not exist today. Some examples are considered in the following pages. The picture revealed, of major shifts in the distributions of plants and the communities that they formed in response to equally striking environmental changes, is then elaborated in subsequent chapters.

The scenario covers the last 65 million years of geological time (the Cenozoic Era). Throughout that period of time mammals and flowering plants have been the dominant groups of organisms. The customary subdivisions of the Cenozoic era are shown in Table 1.1, together with brief characterizations of earlier geological time. The abbreviation 'Ma' (million years) used in that table is used subsequently throughout the text for convenience, together with 'ka' (thousand years) for shorter time-periods.

1.2 SPITSBERGEN CONIFERS

At the present day some 50% of the available land surface of the world is reckoned to be climatically suitable for the growth of trees (Table 1.2 and

Table 1.1 The background timescale

(*a*) SUBDIVISIONS OF GEOLOGICAL TIME

Time-period	Duration (Ma ago)*	Characteristic fossils
Precambrian Era	4 600–640	Sparse microscopic organisms
Palaeozoic Era	640–245	Predominant invertebrate faunas (e.g. trilobites and graptolites), together with early land plants and land vertebrates (fish and amphibians)
Mesozoic Era	245–65	Now-extinct reptiles (e.g., dinosaurs), together with ferns, seed-ferns and gymnosperms
Cenozoic Era	65–present	Mammals and flowering plants

(*b*) SUBDIVISIONS OF THE CENOZOIC ERA [18]

Era	Period		Epoch		Date (Ma ago)
Cenozoic	Tertiary	Palaeogene	Palaeocene		
				—	53.5
			Eocene		
				—	37.5
			Oligocene		
					23.7
		Neogene	Miocene		
				—	5.3
			Pliocene		
					1.6
	Quaternary		Pleistocene		
				—	0.01
			Holocene		

* Ma = million years.

Figure 3.1), although the actual extent of forest and woodland is considerably less (see Chapter 13). The other 50% of the earth's surface, however, is too cold, too wet, or too dry for trees to grow. In the northern hemisphere trees do not reach nearer to the North Pole than latitude 70°N – some 2200 km to the south. The land masses that extend further north (Figure 1.1) are treeless. One of the most thoroughly investigated of these arctic lands is Spitsbergen (or Svalbard), lying between latitudes 74° and 80°30′N, and separated from the North Pole by about 1000 km of ice-laden Arctic Ocean. The nearest trees to Spitsbergen are in northern Norway, 700–800 km to the south (pine and birch), and only a few hardy dwarf shrubs (notably *Betula nana, Cassiope tetragona, Empetrum nigrum* ssp. *hermaphroditum* and *Salix polaris*) occur on Spitsbergen today, as a fringe around sheltered fjords on the western sides of the islands as far north as 78°N [1]. Spitsbergen was

Table 1.2 The approximate global extent (percentage of total land surface) of various vegetation types [19]

Potentially forested land		**31.7**
Tropical forest	11.4	
Temperate broadleaf forest	6.4	
Boreal and montane conifer forest	13.9	
Land marginal for tree growth		**19.4**
Tropical savanna and dry woodland	16.1	
Temperate dry woodland and scrub	3.3	
Land too cold for tree growth		**20.7**
Permanent ice and snow	10.7	
Tundra	8.2	
High-altitude grassland	1.8	
Land too dry for tree growth		**25.9**
Desert and semi-desert	19.6	
Temperate grassland	6.3	
Land too wet for tree growth		**1.6**
Bog and fen	1.6	

visited by a series of Swedish arctic expeditions in the latter part of the last century. These expeditions brought back with them a large quantity of excellently preserved plant material, from rocks in the southwest of Spitsbergen of Early Tertiary age (50–65 Ma old). The most abundant plant remains were leaf impressions of broadleaved deciduous trees, most belonging to species long extinct and not readily assignable even to modern families. Remains of coniferous trees, on the other hand – leaves, whole shoots, male and female cones, pollen grains – though less common, could more easily be related to modern taxa (genera and even species). From these collections, Heer (in his *Flora Fossilis Arctica* of 1877), and later Schweitzer [2], recorded fourteen genera of conifers, eleven of which still have living representatives (Table 1.3). Five of these genera (*Glyptostrobus*, Chinese cypress; *Metasequoia*, dawn redwood; *Pseudolarix*, golden larch; *Sciadopitys*, umbrella pine; and *Taiwania*) have living representatives that are frost-sensitive and confined to eastern Asia between about latitudes 24° and 30°N, where mean annual temperatures of 15–18°C are to be expected. Current mean annual temperatures on Spitsbergen are 3–6°C below freezing point [3]. If the climatic tolerances of the extinct Tertiary conifers are assumed to have been not markedly different from those of their living near relatives, then massive and spectacular changes in climate and vegetation must have taken place within the last 50 Ma: a lowering of temperature by 18–24°C and a displacement of plant ranges by 5000–6000 km. (The possibility of continental displacements as well cannot be ignored – Chapter 4.)

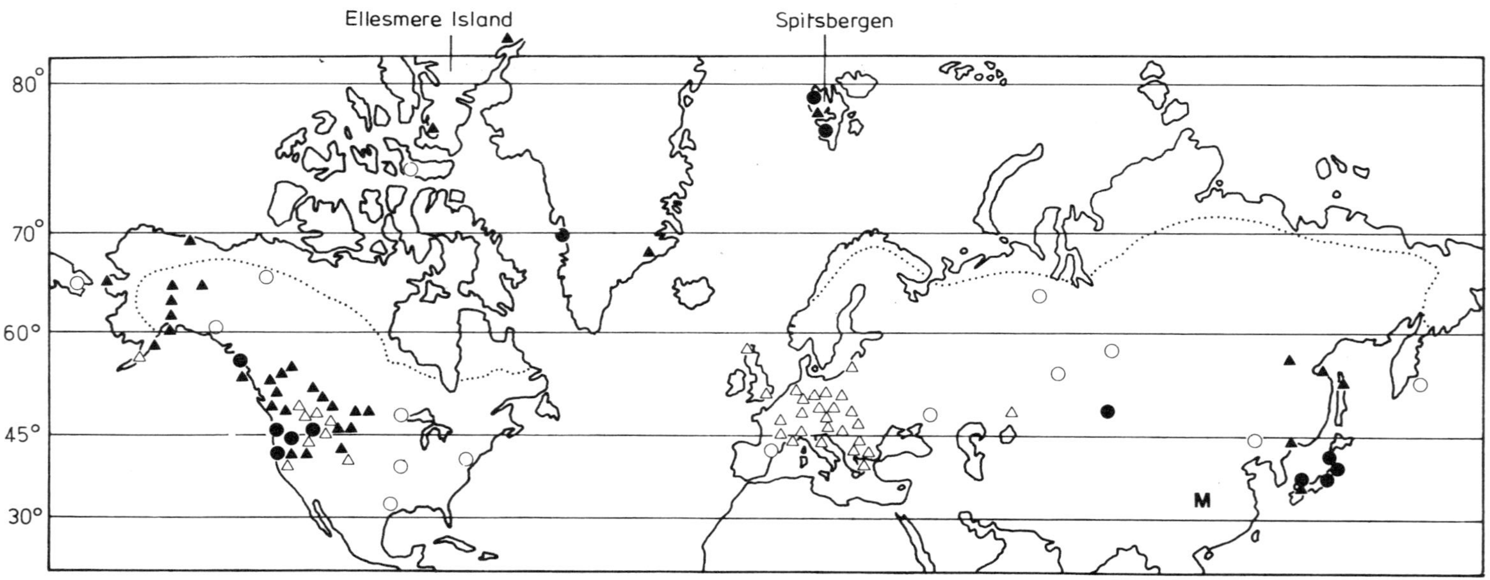

Figure 1.1 The fossil redwood record. Sites with fossil remains of *Sequoia* ▲, *Metasequoia* △, and both genera ● from the Tertiary Period, based on the maps in Schloemer-Jäger [21]; ○, additional sites from Chaney [11] which could be either genus. M, present-day locality of *Metasequoia*; ....., approximate northern limit of present-day forest communities (the arctic timberline).

Table 1.3 Early Tertiary conifers from Spitsbergen, and the distribution of their extant relatives [2,20]

	Distribution of modern relatives
A. Macroscopic remains	
Glyptostrobus cf. *ungeri*; *Glyptostrobus nordenskioeldi*	1 sp., south-east China
Metasequoia occidentalis	1 sp., Hupeh and Szechuan, south-west China
Pseudolarix septentrionalis	1 sp., eastern China
Sequoia langsdorfii	1 sp., western United States
Taiwania schaeferi	1 sp., Taiwan; 1 sp., south-west China and northern Burma
Taxodium dubium	3 spp., south-east USA and Mexico
Thuja ehrenswaerdi	5 spp., eastern and western USA, eastern Asia
Cupressinocladus interruptus; *Elatocladus olriki*; *Parataxodium wigginsii*	Modern affinities uncertain
B. Pollen	
Picea	Throughout northern hemisphere
Pinus	Throughout northern hemisphere
Sciadopitys	1 sp., Japan
Tsuga	c. 10 spp., eastern and western USA, eastern Asia

The commonest conifer in the Spitsbergen deposits is *Metasequoia occidentalis*. The only surviving stands of its extant relative *M. glyptostroboides* occur in the mountains of western China, in a climate resembling that of Georgia, USA, with hot, wet summers and mild winters [4,5]. Other analogies between Tertiary Spitsbergen and the southeastern United States also exist. Thus Heer and Schweitzer both suggest that *Taxodium dubium* occurred in Tertiary Spitsbergen as a component of swamp forest similar to that found today on the coastal plains of the United States from Virginia southwards to Louisiana, where *Taxodium* (bald cypress) and *Nyssa* (sweet gum) are the major trees [6]. Fossil faunas of the same age as the Spitsbergen floras, and from the same latitude – from northern Ellesmere Island in the Canadian Arctic (Figure 1.1) – include a tortoise (*Geochelone*) and an alligator (*Allognathosuchus*) whose closest modern relatives likewise occur in the coastal plains of the southeastern United States [7]. This evidence fortifies the conclusion that the environment of Spitsbergen in the Early Tertiary must have been warm–temperate.

At the same time as the Spitsbergen floras were being described, palaeobotanists were speculating about the affinities of the large, single-

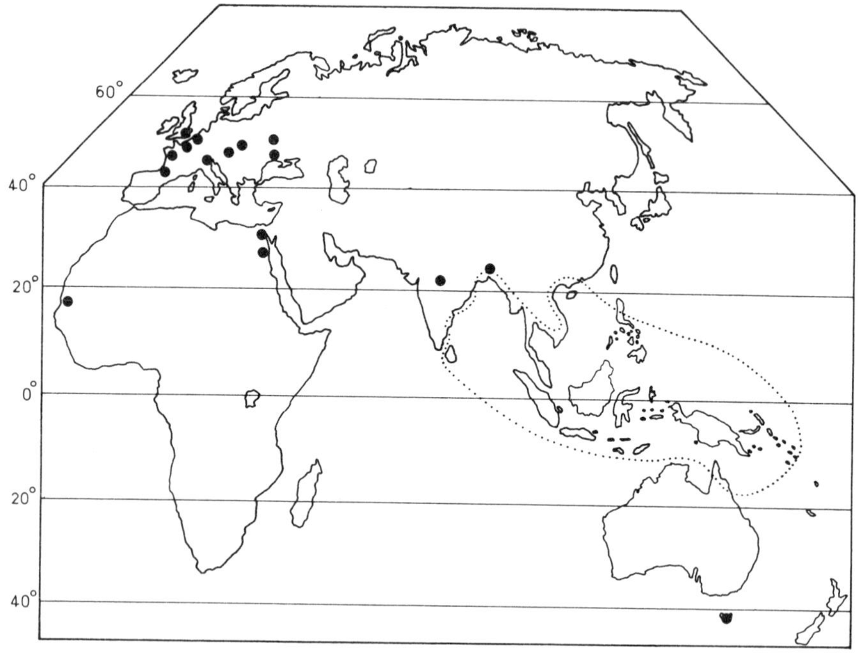

Figure 1.2 Fossil finds of the palm genus *Nypa* •. The present-day range is enclosed by the dotted line. Re-drawn from Tralau [10].

seeded, thick-walled fruits that were being recovered from sediments of similar age (i.e., Early Tertiary) in south-east England, Belgium and Italy. These fossil fruits were eventually recognized to be similar to those of the modern *Nypa fruticans*, a prostrate creeping palm confined to the tropics of South-east Asia, where it is often the dominant plant in brackish-water coastal and estuarine mangrove swamps (Figure 1.2). Associated with the *Nypa* fruits in the London Clay beds of the London and Hampshire basins of south-east England was a rich flora of evergreen angiosperms, with modern relatives characteristic of the rainforests and mangrove swamps of the Indo-Malayan region. A tropical environment, with temperatures 8–10°C higher than at present, was reconstructed for south-east England 50 Ma ago [8,9,10].

1.3 FOSSIL REDWOODS

The abundant remains of *Metasequoia* in the Spitsbergen deposits were originally ascribed by Oswald Heer to *Sequoia*. In the second half of last century and the first half of the present century, fossilized remains of

Sequoia were reported from Tertiary rocks stretching in a broad belt across North America and Eurasia (Figure 1.1). At many sites *Sequoia* was a major component of the flora. From these finds at least twelve species of *Sequoia* were described, of which *S. langsdorfii* was the most widespread. With the *Sequoia* fossils at these sites occurred a wide range of other woody plants. As a common core of accompanying tree taxa was recorded repeatedly at site after site, similarities between the various floras appeared so marked as to suggest a similar age for all of them. The geological evidence, however, did not support this view, indicating rather that the most northerly records were the oldest (50–65 Ma old) and the most southerly ones the youngest (10–20 Ma old). A picture thus began to emerge of a wide-ranging, mixed conifer–broadleaf forest, developed initially at high latitudes in e.g., Greenland, Alaska, Spitsbergen and Siberia, that moved gradually southwards over North America and Eurasia with little change in floristic composition as climate worsened through the Tertiary Period [11]. In these mixed conifer–broadleaf forests occurred tree taxa ancestral to a large number of present-day species characteristic of a variety of different modern temperate broadleaf and coniferous forests. The modern forests, it was suggested, had resulted from the slow fragmentation of this floristically-rich 'Arctotertiary Forest' into various less species-rich assemblages with separated geographical ranges; in the process many taxa, including *Sequoia*, were eliminated over large areas.

Among the best-worked of these *Sequoia* fossil floras were two from nearby sites in north-eastern Oregon – the Bridge Creek flora (about 25 Ma old) and the later and richer Blue Mountains flora (about 15 Ma old). Chaney [12,13] emphasized the similarities between these two floras and the flora of the modern coastal redwood forests of California dominated by *Sequoia sempervirens* (Table 1.4). Subsequent workers drew even closer parallels. Thus Cain [14] asserted that 'the redwood forest relics along the California coast are very similar . . . to the redwood forests of Oregon in the Early Miocene, or of Alaska during the Eocene. To walk into these forests today is to walk back forty to sixty million of years in vegetational history to what lived in far-flung places those long ages ago.'

The coastal redwoods are restricted naturally today to a narrow coastal belt, seldom more than 40 km in width, stretching from Monterey in California northwards for 725 km to the Oregon border [15]. The tallest trees (60–90 m or more in height) form nearly pure, densely-packed stands on moist coastal plains, alluvial valley bottoms and river deltas, but on the backing hillslopes of the western half of the Coast Ranges more mixed redwood forests occur up to altitudes of 750 m (and locally to over 900 m). Some common associated tree species of these hillslope forests are listed in Table 1.4.

To view these impressive *Sequoia* forests, now so localized in distribution,

Table 1.4 Woody taxa recorded as fossils from the Bridge Creek (A) and Blue Mountains (B) Floras of north-eastern Oregon, and as living plants from the modern redwood forests of California (C) and the upland forests of Hupeh, China (D). Sources of data [4, 12; 13, 15].

	A	B	C	D
Abies		+	+	+
**Acer*	+	+	+	+
**Alnus*	+	+	+	+
Cornus	+		+	+
Corylus	+	+	+	+
**Crataegus*	+	+	+	+
**Mahonia/Berberis*	+	+	+	+
Picea		+	+	+
**Pinus*	+	+	+	+
Pseudotsuga		+	+	+
**Quercus*	+	+	+	+
Rhamnus	+	+	+	+
Rhus		+	+	+
**Salix*		+	+	+
Sequoia/Metasequoia	+	+	+	+
Tsuga	+	+	+	+
**Ailanthus*		+		+
Celtis		+		+
Cercidiphyllum		+		+
**Fagus*	+	+		+
**Hydrangea*		+		+
Ilex		+		+
Juglans		+		+
Liquidambar		+		+
**Nyssa*		+		+
**Ostrya/Carpinus*	+	+		+
**Populus*		+		+
Prunus		+		+
**Pterocarya*		+		+
**Sassafras*		+		+
**Ulmus/Zelkova*	+	+		+
Fraxinus	+		+	
Myrica	+		+	
Philadelphus	+		+	
Torreya	+		+	
Umbellularia	+		+	

* Taxa marked with an asterisk are referred to in section 2.2.3.

as direct descendants of the wide-ranging Tertiary forests is to highlight the magnitude of the changes in vegetation that must have occurred in the last 50 Ma. The comparison may not be a valid one, however, for in 1941 a Japanese palaeobotanist, Shigeru Miki, reported significant differences between fossil *Sequoia* material from Japan and Korea and living material of *S. sempervirens*. Miki proposed the creation of a new genus, *Metasequoia*, to accommodate his fossil material, and suggested that some other fossil records of *Sequoia* might also be *Metasequoia*. Accordingly, Chaney [16] undertook a revision of North American records of fossil *Sequoia*, and found that true *Sequoia* probably had a rather restricted Tertiary distribution in the subcontinent (Figure 1.1). Most of the more northerly records (from present-day arctic regions) were of *Metasequoia*. In Eurasia the status of the two genera is still uncertain, as no systematic revision of pre-1950 material has been carried out. Even some post-1950 records are suspect, as it is not always clear whether separation of *Sequoia* and *Metasequoia* pollen has been attempted [17]. The lack of European records of *Metasequoia* in Figure 1.1 probably reflects this uncertainty rather than a real absence.

In the same year that Miki discovered fossil *Metasequoia*, three large living individuals of a hitherto undescribed conifer were found in the mountainous interior of China. This conifer was placed in Miki's genus *Metasequoia*, and named *M. glyptostroboides* [5]. More than one thousand mature trees were subsequently found, in a totally enclosed valley in the rugged northern part of Hupeh Province. The *Metasequoia* occurred in a species-rich mixed conifer–broadleaf forest containing many genera recorded from Tertiary sites in North America and Eurasia. The resemblances of this forest with the arctotertiary forest are probably even closer than those shown by the modern redwood forests [4] (Table 1.4).

1.4 PERSPECTIVES AND IMPLICATIONS

The fossil sequences we have been describing show that 50 Ma ago, when tropical forest occupied southern Britain, mixed conifer–broadleaf forest occurred far to the north, in latitudes now occupied by sparse tundra communities. Those northerly forests apparently moved progressively southwards through the Tertiary, changing gradually in composition in the process; at the same time the London Clay tropical forest also moved southwards, and as a result many of its derivative plants are now widespread in the forests of south-east Asia. Many taxa must have migrated over distances of hundreds or thousands of kilometres. The changes did not just affect the distributions of individual taxa, however: the whole character of the world's forests changed in tandem. We have already noted the break-up of the complex arctotertiary forest into less species-rich assemblages as it moved southwards. Similar changes may be inferred for the tropical forests further south. Thus more than one-tenth of the identified genera in the London Clay floras have modern relatives that occur today outside the tropics, in quite different forest types. Almost certainly the London Clay forests grew in a climate of a more seasonal character than that associated with tropical rainforest today. In these forests, as in the arctotertiary forest further north, an assemblage of species may well have existed together which has the modern descendants segregated among a range of different modern forest types.

Some of the fossil redwood records from the Early Tertiary Period are from areas at low latitudes that are now covered by non-forest vegetation – steppe, and even desert. The inference must be that those areas were formerly forested, and that many millions of years ago trees were able to grow at high latitudes and in the interiors of continents where conditions are now too cold or too dry for forest growth. If that is so, when and why did tundra, steppe and desert originate, and from what sources were their floras derived? Indeed, do these modern vegetation types have fossil analogues, or are they modern phenomena? Did tundra vegetation, in particular, originate only during the Pleistocene Ice Age that succeeded the Tertiary Period, when high-latitude ice developed extensively over areas that were formerly forested, and large parts of the world's land surfaces became unsuitable for tree growth? Where did trees survive then, and in what sorts of forest communities? These are crucial questions that this book attempts to answer.

In the scenarios we have been discussing, the observed changes in the ranges of certain plant taxa over the last 65 Ma have been used to deduce the possible magnitude of the climatic changes that could have caused them. Such a procedure is unsatisfactory for a number of reasons. Thus if the same biological evidence is used to determine both range changes and the likely causal factors, then circular arguments may ensue. Moreover, the underlying assumption is made that the tolerances of modern species are similar to those of their fossil ancestors, and that these ancestral taxa occupied similar climatic regimes to those of their modern relatives. Such an assumption ignores possible evolutionary changes in tolerance, and changes in tolerance are inherently no less likely than morphological changes to have occurred over geological time. Where a genus has undergone marked contraction or expansion of range through time, then it is particularly likely that certain genotypes will have been lost or will have evolved concomitantly with the range changes. Accordingly, the physiological behaviour of the modern species cannot reliably be used as a guide to environmental changes in the past. A further assumption made with this form of enquiry is that the ranges of modern species are in equilibrium with their environment – and there is no *a priori* reason why that should be so. Independent assessment of environmental change, rather than inferences based on modern species-ranges, should be a necessary and integral part of the palaeoecological* data-set. The evidence for environmental change over the last 65 Ma that is presented in Chapters 4 to 6 is accordingly divorced as far as possible from the floristic/vegetational database.

REFERENCES

1. Summerhayes, V. S. and Elton, C. S. (1928) Further contributions to the ecology of Spitsbergen. *J. Ecol.*, **16**, 193–268.
2. Schweitzer, H.-J. (1980) Environment and climate in the early Tertiary of Spitsbergen. *Palaeogeogr. Palaeoclimatol. Palaeoecol.*, **30**, 297–311.
3. Høeg, O. A. (1956) The present and past vegetation of Spitsbergen. *Proc. Linn. Soc. Lond.*, **166**, 144–9.
4. Chu, K.-L. and Cooper, W. S. (1950) An ecological reconnaissance in the native home of *Metasequoia glyptostroboides. Ecology*, **31**, 260–78.
5. Fulling, E. H. (1976) Metasequoia, living and fossil. *Bot. Rev.*, **42**, 215–315.
6. Penfound, W. T. (1952) Southern swamps and marshes. *Bot. Rev.*, **18**, 413–46.
7. Estes, R. and Hutchinson, J. H. (1980) Eocene lower vertebrates from Ellesmere Island, Canadian Arctic Archipelago. *Palaeogeogr. Palaeoclimatol. Palaeoecol.*, **30**, 325–47.
8. Collinson, M. E. and Hooker, J. J. (1987) Vegetational and mammalian faunal changes in the Early Tertiary of southern England, in *The Origins of Angiosperms and their Biological Consequences* (eds M. E. Friis, W. G. Chaloner and P. R. Crane), Cambridge University Press, Cambridge, pp. 259–304.
9. Pennington, W. (1969) *The History of British Vegetation*, English Universities Press, London.
10. Tralau, H. (1964) The genus *Nypa* Van Wurmb. *K. Svenska Vetensk. Akad. Handl.*, **10**, 1–29.
11. Chaney, R. W. (1940) Tertiary forests and continental history. *Geol. Soc. Am. Bull.*, **51**, 469–88.
12. Chaney, R. W. (1925) A comparative study of the Bridge Creek flora and the modern redwood forest. *Publs Carnegie Instn*, **349**, 1–22.
13. Chaney, R. W. (1959) Miocene floras of the Columbia Plateau. Part I. Composition and interpretation. *Publs Carnegie Instn*, **617**, 1–134.
14. Cain, S. A. (1944) *Foundations of Plant Geography*, Harper, New York.
15. Wright, H. A. and Bailey, A. W. (1982) *Fire Ecology, United States and Southern Canada*, Wiley, New York.
16. Chaney, R. W. (1950) A revision of fossil *Sequoia* and *Taxodium* in western North America based on the recent discovery of *Metasequoia. Trans. Am. Phil. Soc.*, New Series, **40**, 171–263.

* Palaeoecology: the study and understanding of the relationships between past organisms and the environment in which they lived.

17. Simpson, J. B. (1949) Fossil pollen of *Metasequoia* type. *Nature*, **163**, 771–2.
18. Saarnisto, M. (1988) Time-scales and dating, in *Vegetation History* (eds B. Huntley and T. Webb III), Kluwer Academic Publishers, Dordrecht, pp. 77–112.
19. *The New Encyclopaedia Britannica* (1984) Encyclopaedia Britannica, Chicago, vol. 18, p. 145.
20. Li, H.-L. (1953) Present distribution and habitats of the conifers and taxads. *Evolution*, **7**, 245–61.
21. Schloemer-Jäger, A. (1958) Alttertiäre Pflanzen aus Flözen der Brögger-Halbinsel, Spitzbergens. *Palaeontographica* B, **104**, 39–103.

CHAPTER TWO

Diversity – the global plant cover

2.1 THE CONTEMPORARY PLANT COVER

The world's plant cover is composed of an immense number of different plant species – about a quarter of a million angiosperm species alone. Not one of these species occurs worldwide – all are restricted in distribution, so that each occupies part only of the total available land surface; some indeed have very localized geographical ranges. The world's plant cover thus differs considerably in composition from one region to another, and consequently displays marked heterogeneity. Within this floristic heterogeneity, marked **patterns of diversity** can be recognized. ('Diversity' = the state of being diverse; difference; unlikeness; variety – *Chambers Twentieth Century Dictionary*.) These patterns are of three main types:

1. Differences in species-richness (numbers of species), or some other measure of species-diversity, between regions of comparable size. There are, for example, more species generally at low latitudes or altitudes than at high latitudes or altitudes.
2. Differences in range patterns. Because many species (and also higher taxa such as genera and families) have rather similar geographical ranges, it is possible to subdivide the world's land surface into a series of **floristic regions** (Figure 2.1), each of which is characterized by a distinctive suite of taxa that includes some that are **endemic** (i.e., exclusive) to that particular region. The more important regions (kingdoms and subkingdoms) are characterized by suites of endemic families; lesser-ranked regions (provinces) show differences principally at the generic and species levels.
3. Differences in vegetation. The plants that occur within a circumscribed area, whether it be a patch of a few square metres, or a major land area of thousands of square kilometres, form social groupings or communities that are widely believed to represent integrated functional units (especially if the associated animals are also included). Conventional definitions of **plant communities** include: subdivisions of a vegetation cover; sociological units of every degree from the simplest (as an

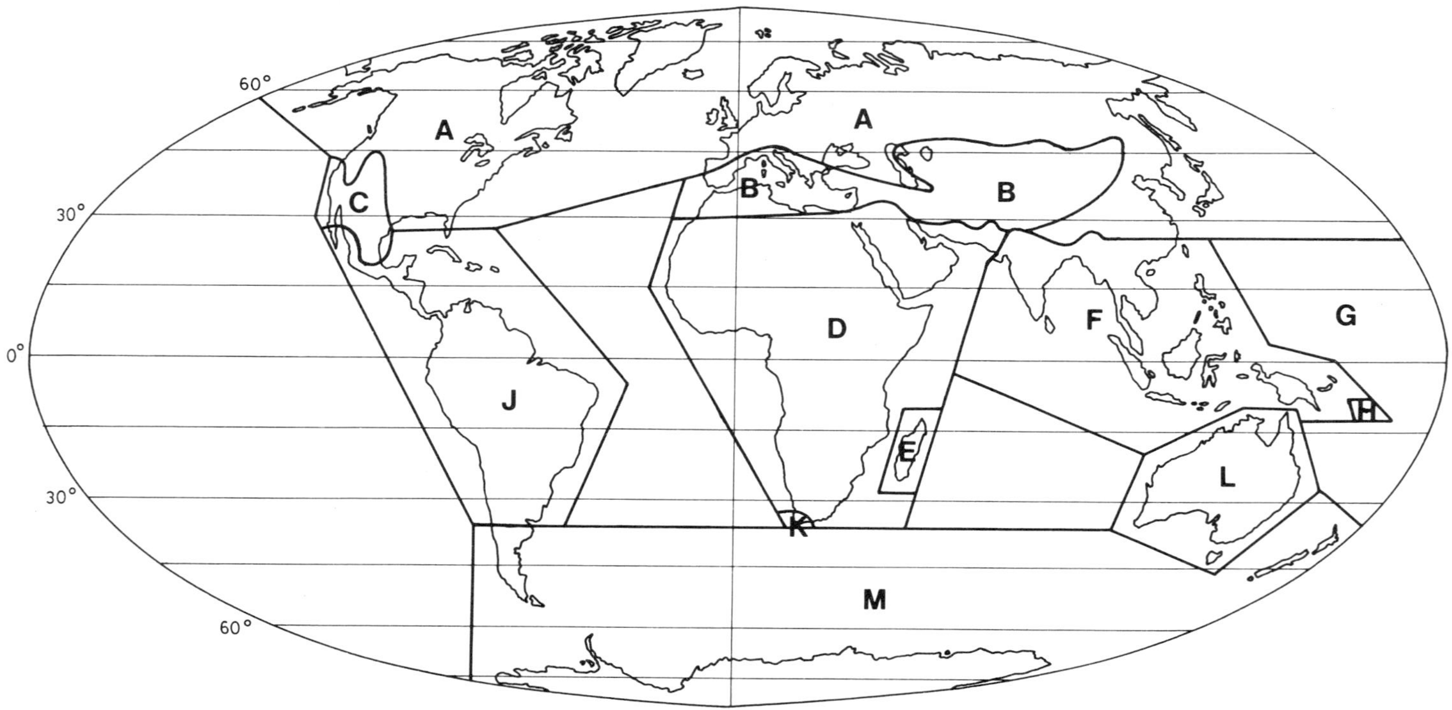

Figure 2.1 Floristic regions of the world (after Takhtajan [31]). A–C, Holarctic kingdom: A = Boreal subkingdom; B = Tethyan subkingdom; C = Madrean subkingdom. D–H, Palaeotropical kingdom: D = African subkingdom; E = Madagascan subkingdom; F = Indo-Malesian subkingdom; G = Polynesian subkingdom; H = Neocaledonian subkingdom. J = Neotropical kingdom; K = Cape kingdom; L = Australian kingdom; M = Antarctic kingdom.

unrooted mat of algae) to the most complex (as a multistoried rainforest); assemblages of different species which interact with one another; and assemblages of plants and/or animals found living together with some degree of permanence [1]. Social groupings on an intercontinental scale that generate large-size patterning of the global plant cover are called **biomes**. Each biome has a characteristic set of taxa and also a characteristic overall physiognomy, comprising features such as the stature and growth habit of the major plant taxa; the range of leaf size, shape and texture among these taxa; the proportions of evergreen and deciduous taxa; the diversity of plant taxa present; and any other features which impart a distinctive visual signature and architecture to large portions of the global plant cover.

Table 2.1 lists the world biomes recognized by five recent ecology textbooks. Common to all five are: evergreen rainforest, tropical seasonal forest, savanna and dry woodland, desert, mediterranean scrub and woodland, temperate forest, steppe, boreal forest, and tundra. (The terms used are a subjective choice of those considered to be the most appropriate.) A generalized scheme, or model, of the distribution of these biomes (Figure 2.2) shows them occupying nearly transverse belts in particular regions of each major land mass, and arranged mirror-image-wise in the two hemispheres.

Conventional small-scale vegetation maps of the world that show the distribution of these biomes may appear unrealistically simple – representing Amazonia, for example, as continuous tropical jungle, the American Midwest as an uninterrupted expanse of prairie, and England entirely forested. Most of us would recognize that the real world is not like this – that large parts of Amazonia are no longer forested, that much of the American Midwest is under cultivation, and that England has less than 5% of its native woodland left. Thus the world vegetation map is clearly an abstraction, a model of an ideal world without humans and without environmental instability. Yet it is not a fantasy world. The components of the picture (the biomes) have reality. They exist as surviving stands of vegetation, and as written descriptions from the recent past; many of their participant taxa persist today in derivative plant communities. Biomes such as rainforest have undoubtedly been more widespread in the recent past; and from their surviving relics a reconstruction of their former extent can be made. The vegetation map thus models their potential distribution under climatic conditions similar to those of the present day – and depicts a world with a markedly non-uniform plant cover. There is considerable disagreement as to the numbers of biomes recognizable on a world scale, rather widespread agreement that in many parts of the world the biomes that should ideally be present have been destroyed or modified by human

Table 2.1 World biomes according to different authors

Evergreen rainforest	Tropical rainforest	Tropical rainforest Subtropical rainforest	Tropical rainforest	Tropical forest
Semi-evergreen and deciduous tropical forest	Tropical seasonal forest	Monsoon forest	Tropical deciduous forest	
Dry woodland and savanna	Tropical savanna	Tropical savanna	Tropical grassland and savanna	Savanna and open forest
	Tropical thorn scrub and woodland	Thorn forest and scrub	Tropical scrub forest	
Desert and semidesert	Desert	Dry desert	Desert	Desert
Sclerophyllous woodland	Mediterranean scrub and woodland	Temperate evergreen woodland	Chaparral	Maquis and chaparral
Warm temperate woodland	Temperate forest	Temperate rainforest	Temperate deciduous and rain forest	Temperate forest
Temperate deciduous forest		Temperate deciduous forest		
Steppe	Temperate grassland	Steppe and half-desert	Temperate grassland	Steppe and prairie
Boreal coniferous forest	Boreal forest (taiga)	Boreal conifer forest	Northern conifer forest	Boreal forest (taiga)
Tundra	Tundra	Tundra and cold woodland	Tundra	Tundra
Mountains	Mountain Polar ice cap	Cold desert	Mountains	High mountain
(according to Walter [2])	(according to Whittaker [27])	(according to Clapham [28])	(according to Simmons [29])	(according to Dajoz [30])

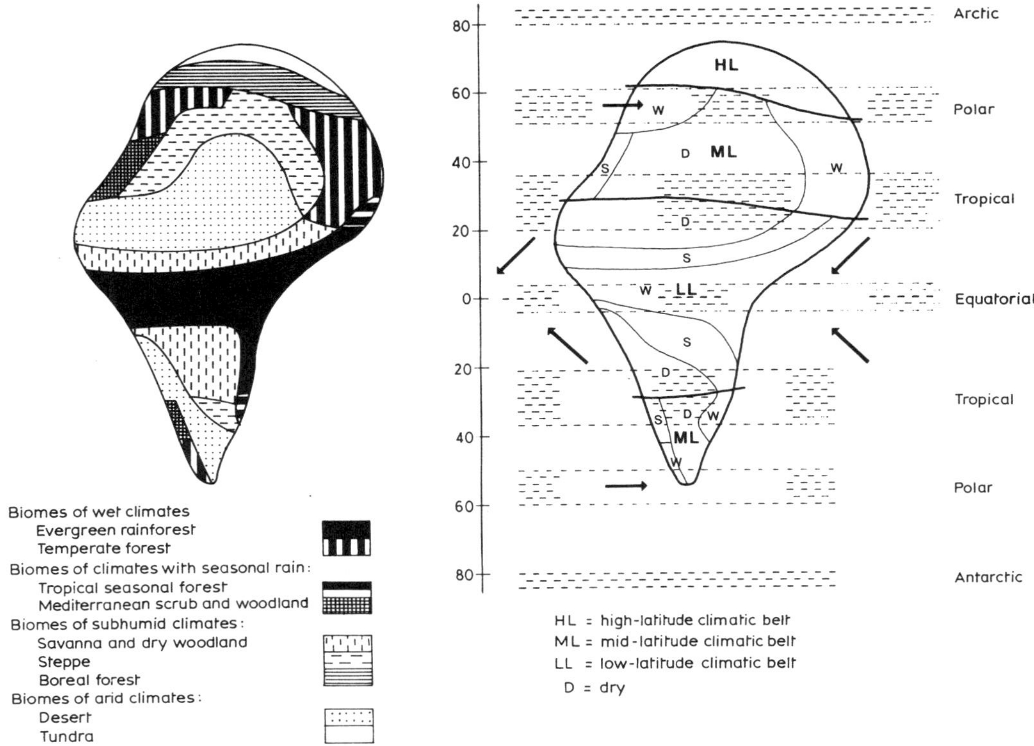

Figure 2.2 Distribution of biomes (left) and source regions and air masses (right) in relation to a generalized land mass (based on diagrams by Strahler and Strahler [32]). Source regions are shown shaded; arrows show direction of prevailing winds. HL = high-latitude climatic belt; ML = mid-latitude climatic belt; LL = low-latitude climatic belt; D = dry; W = wet; S = seasonal.

settlement, and a general consensus that broad ecotones (transitions) exist between neighbouring biomes [2]. The essential validity of the concept that the global plant cover exhibits large-scale patterning has nevertheless never seriously been questioned.

Comparable patterning has also existed in the past. It is clear from the examples discussed in the previous chapter that the patterning must at times have been very different. Reconstructing that former patterning and linking it to the present provides a **palaeoecological perspective** on modern plant distributions and communities. Differences in past patterning have involved changes in the genotypes of the plant taxa themselves (evolutionary change), changes in the geographical ranges of the taxa (distributional change), and changes in the nature of the plant communities that they formed (vegetational change). Recognition and analysis of these differences requires evidence from a wide range of sites about the plants and plant communities that grew there at various times in the past. How that evidence may be obtained and interpreted is considered briefly below.

2.2 THE PLANT COVER OF THE PAST

2.2.1 The nature of the evidence

The primary sources of evidence about where plants grew in the past, and the communities they formed, are the preserved remains of the plants themselves. Some of these remains, **microfossils**, are visible only under the microscope: pollen grains, spores and diatoms, for example. Others, **macrofossils**, are visible to the naked eye, though microscopic examination may be necessary for their identification: seeds, fruits and leaves, for example. The processes leading to the preservation of plant material, especially of higher plants, as fossils almost always involves the break-up (loss of integrity) of the individual plant, however, so that separate plant organs and fragments rather than whole plants are preserved. These separate remains, moreover, are rarely preserved in the same proportions as they were present in the parent community; indeed, some species may not be preserved at all (those that were rare or whose remains decay readily, for example). Furthermore, the collection of plant remains that is preserved at a given site (the **fossil assemblage**) may not be derived solely from the on-site community; additional plant material may be transported to the site of preservation from other adjacent, or more distant, communities, by wind, water, or animals. Because preservation occurs only in specialized environments (characterized by waterlogging, or by extremes of cold, acidity, or dryness), certain plant communities, as well as certain plant components (the more resistant ones), are more likely than others to be preserved. All these features of the preservation process raise problems with the interpretation of fossil assemblages. The following are some of the problems. Can the various fossils be identified with certainty, and, if so, how specific are the identifications? Do all the plant fossils preserved at a particular site come from a single plant community, or are they derived from two or more adjoining, or more distant, communities? If that problem can be resolved, and the fossils originating from one community can be specified, does the fossil data-set carry enough information for the analogous modern plant community to be recognized? Recent work [3,4] shows that there are no simple answers to these problems. The two principal techniques of investigation, pollen analysis and macrofossil analysis, differ in the clarity of the responses that they supply to the questions. However, the clear lack of equivalence of plant fossil assemblages to modern plant communities has led to the development of special methods for collecting and evaluating the data. Because of the large size of the data-sets involved, computer-based methods are increasingly being used. Details of such methods are outside the scope of this book – the interested reader should consult Birks and Birks [3], Birks and Gordon [5], Delcourt and Delcourt [6] and Grimm [7].

Whatever methods of data-analysis are used, all palaeoecological reconstruction ultimately relies on precise knowledge of the taxa involved. The degree of precision with which a particular plant fossil can be identified depends on its state of preservation and the type of organ involved. Poor preservation alone can preclude positive identification, but more commonly it is the lack of distinctive morphological features of a particular plant organ that is the constraint. Seeds and fruits offer the best possibilities among macrofossils for specific identification; for most other macrofossils, and for most pollen grains, identification to generic level is the best that can be hoped for [8,9,10]. Leaves are rather generally regarded by plant systematists as the most plastic and variable of all plant organs, and of least diagnostic value [11,12]. Because of the resulting 'blind spots' in the identification of fossil plant remains, both macrofossil and microfossil records are desirable.

The complementary palaeoecological information that can be obtained from microfossil and macrofossil identifications from the same fossil assemblage is illustrated by the work of Birks and Mathewes [13] on Early Holocene sediments from Abernethy Forest in north-east Scotland. Of a total of 53 taxa recorded, only 13 were present as both microfossils and macrofossils; 32 taxa (including locally important ones such as *Corylus*, *Juniperus* and *Lycopodium*), were present only as pollen grains, and 8 (including *Juncus* and *Narthecium*) were present only as macrofossils.

Because of selective preservation, inadequate information, and possible adulteration by non-local plant material, the information content of a fossil assemblage thus falls far short of what could be obtained from a related modern community. In trying to identify the local vegetation cover that such an assemblage represents, the palaeoecologist seeks to pinpoint within the total fossil data-set either individual taxa or groups of taxa that are characteristic of one specific modern vegetation type ('indicator taxa'), or (in the case of preserved pollen) levels of abundance of an individual taxon that signify either local presence or local dominance. Obtaining that information almost always requires the complementary examination of the assemblages of dead plant material (both pollen grains and macro-remains) that are collecting at the present day below particular modern vegetation types – so-called 'surface samples' [3,14]. Good examples of how that information may then be processed are given by Davis and Webb [15], Ritchie and Yarranton [16], Birks and Birks [3], Webb *et al.* [17] and Delcourt and Delcourt [6].

In one particular palaeoecological system – that of **packrat middens** – the preserved plant remains conform closely in their representation and abundance with those of the contemporary parent plants growing in the neighbourhood of the sampling site (within a radius of about 100 m). An excellent account of this unique system and its significance for palaeo-ecological reconstructions is given by Wells [18].

2.2.2 The form of the fossil record

The form of the plant-fossil record is represented schematically by the chart in Figure 2.3. Each open rectangle on the chart represents a fossil record located in time and space, and comprising one or more identified taxa. Some of the records consist of long fossil sequences (A, B on the chart), usually in the form of preserved pollen, that may extend to the present day (rectangle A), or may cease at some point back in time (B). Other evidence is discontinuous: isolated samples, often of plant macrofossils, each related to one short time-interval in the past. Some of the samples (C, D) may constitute the only evidence of any age for a particular location; others (E, F) may be from the same location, but separated in time. The hatched rectangles (1–6) represent present-day records of dead plant remains about to be preserved (i.e., surface samples), of either single species or assemblages of species at a given location. On the chart, successive records in a vertical direction represent vegetational changes at a single site, driven by environmental change. These changes can be regarded as the appearance or disappearance of individual taxa or groups of taxa, or variations in their abundance, at a site. The disappearance of an individual taxon from one site, in response to adverse environmental conditions, may signify total extinction of that taxon, but more usually it is associated with

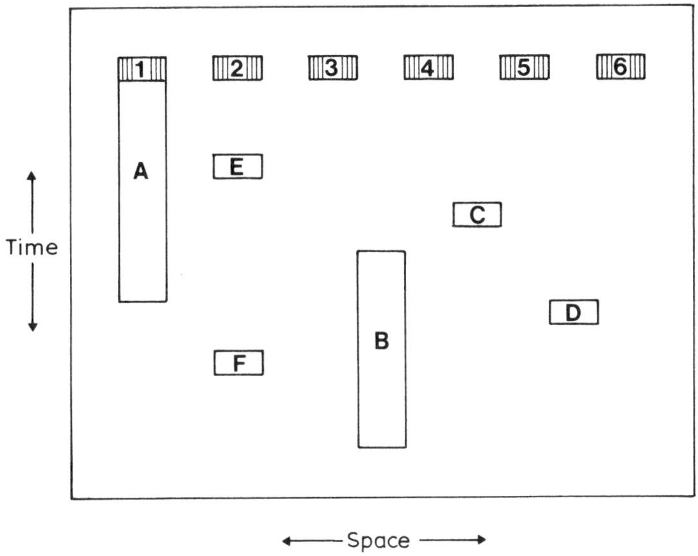

Figure 2.3 Schematic representation of the palaeoecological database. For explanation, see section 2.2.2.

migration to another, more favourable, site (i.e., 'appearance' in the fossil record at that new site). Migration is thus represented on the chart by 'cross-linkages' in a predominantly horizontal direction. In order to follow such migration patterns in the palaeoecological record an independent timescale is essential – a timescale that will establish unequivocally, for example, that 'C' and 'E' on the chart are of different age.

Absolute dating of fossil plant material or of particular events in the past can be achieved most easily by radiometric methods – methods such as radiocarbon-dating that utilize the radioactive-decay processes of certain naturally-occurring isotopes (e.g., C-14, Pb-210). In more limited situations the annual growth patterns of laminated sediments (varves) or woody tissues of trees (growth rings) can be used. Other dating methods merely place an object or event in its correct relative position within a chronological framework [19]. Different dating methods are applicable to different time-intervals of the Cenozoic era, and to different biological and non-biological materials. The precision of the dating estimates given by the different methods varies, and generally decreases with increasing age of the material. Only for the last 20 ka, for the most part, when radiocarbon-dating can be used, is a fairly precise and well-documented dating framework available. Beyond that it is often not possible to date events even to within 20 ka, and the reliability may be much less for events in the Early Tertiary Period. For further details of dating methods the reader should consult the comprehensive accounts by West [20], Lowe and Walker [21] and Saarnisto [22].

2.2.3 Limitations of the fossil record

From a dated data-set of the type shown in Figure 2.3, reconstructions of global patterning at specified times in the past should be possible. The reconstructions may be for single taxa only, or for assemblages of taxa. Three examples are shown in Figure 2.4: (a) sites in eastern North America with *Fagus* pollen present (presumed presence of the tree nearby) 14 ka and 10 ka ago; (b) sites in western North America with plant assemblages of similar composition and age to the Blue Mountains flora (Chapter 1, section 1.3) – sites are categorized according to the numbers of the taxa asterisked in Table 1.4 that are present as fossils; (c) sites in eastern North America 10 ka ago with pollen spectra similar to those in surface samples from different modern forest types.

In these reconstructions the problems inherent in the palaeoecological record that were discussed in the previous section have been largely overcome, and data from a sufficient number of sites were available to provide good spatial coverage. This is not always the case. Research effort over the last 40 years particularly has been concentrated into certain geographical regions, so that for large areas of the earth's surface, and also for certain time-periods, palaeoecological evidence is as yet scanty. The uneven distribution of study sites with good fossil plant evidence is illustrated in Figure 2.5, which shows the coverage of information for 18 ka ago. That map, though now somewhat out-of-date, nevertheless highlights the large areas in middle and low latitudes that are essentially *terra incognita* for 18 ka ago (much land at high latitudes was covered by permanent snow and ice then). A greater number of data points is available for the last 10 ka, but the coverage of sites is still very uneven; for earlier time-periods the inadequacies of the data are even more apparent. These inadequacies necessarily limit the extent to which the spatial variability of the global plant cover can be reconstructed for different times in the past.

2.3 THE REALITY OF THE PLANT COMMUNITY

A further problem exists: can plant communities really be traced through time with some degree of continuity, in the same way as individual plant taxa can? If so, then plant communities must possess a unity of behaviour and a degree of permanence similar to that of individual taxa. Such characteristics were indeed ascribed to plant communities by Frederic

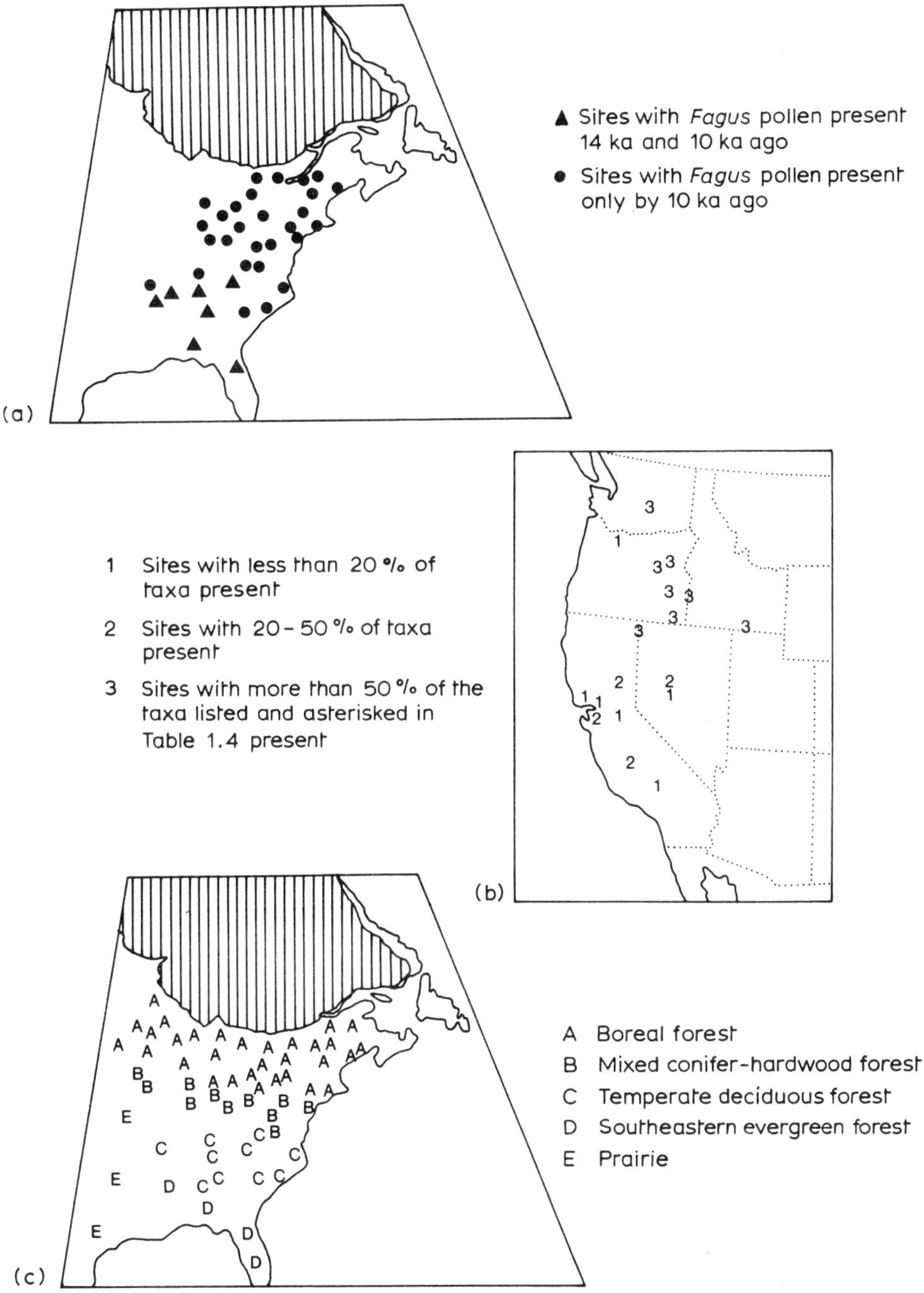

Figure 2.4 Examples of former patterning of the global plant cover: (a) Sites in eastern North America with *Fagus* pollen present 14 ka and 10 ka ago (re-drawn from Delcourt and Delcourt [6]; (b) Sites in western North America with fossil assemblages of similar composition to the Miocene-age Blue Mountains flora (data of Robichaux and Taylor [33]); (c) Sites in eastern North America 10 ka ago with pollen spectra similar in composition to those from a range of different modern vegetation types (re-drawn from Delcourt and Delcourt [6]). Permanent ice at 10 ka ago is shaded.

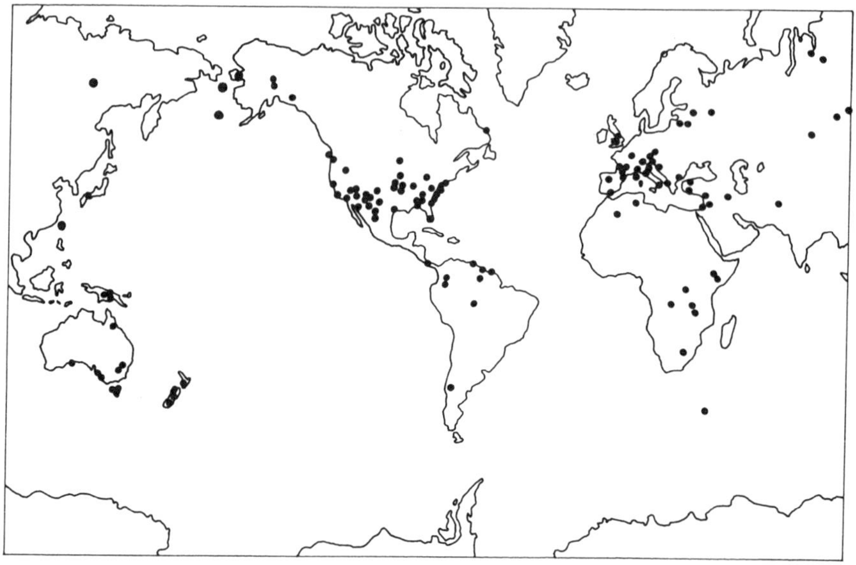

Figure 2.5 Location of sites with fossil plant evidence from 18 ka ago. Re-drawn from Webb [34].

Clements in the early 1900s, and his view of the climax community as a self-perpetuating living organism in which all its members are interdependent had considerable influence during the formative years of plant ecology as a scientific discipline [23,24]. The interpretation of the history of the fossil redwood assemblages given in Chapter 1, section 1.3, for example, was strongly influenced by Clementsian doctrine. Nowadays, few ecologists would accept the doctrine in its extreme form [25], and there is growing acceptance, especially among palaeoecologists, of the opposing viewpoint originally proposed by Gleason [26] that plant communities are chance assemblages of species in space and time, and that each species has unique requirements and behaviour. The Gleasonian viewpoint would not accept climax communities as self-perpetuating assemblages over long periods of time.

Ultimately, resolution of this dispute relies on palaeoecological information. Some of that information is presented in this book, and some tentative conclusions are formulated in the final chapter. In the meantime, it is sufficient to point out that interpretation of the palaeoecological record is bound to be affected by the theoretical viewpoint of the nature of the plant community that is adopted. The crucial role of the Pleistocene Ice Age in shaping modern plant distributions, for example, is widely recognized; yet it may not be profitable to seek vegetation reconstructions for that period that involve merely a re-location of modern biomes over the earth's surface (as the Clementsian viewpoint would demand). The alternative Gleasonian viewpoint raises the possibility that some plant communities may be dissolved completely under stress, and that their component species may then become re-sorted into quite different assemblages. Much of the vegetation of the Pleistocene Ice Age, accordingly, could have no modern analogues. Such a possibility makes interpretation of the palaeoecological record even more difficult.

REFERENCES

1. Anderson, D. J. and Kikkawa, J. (1986) Development of concepts, in *Community Ecology: Pattern and Process* (eds J. Kikkawa and D. J. Anderson), Blackwell, Oxford, pp. 3–16.
2. Walter, H. (1973) *Vegetation of the Earth*, 2nd edn (transl. by J. Wieser), Springer Verlag, Berlin.
3. Birks, H. J. B. and Birks, H. H. (1980) *Quaternary Palaeoecology*, Arnold, London.
4. Krasilov, V. A. (1975) *Paleoecology of Terrestrial Plants. Basic Principles and Techniques* (transl. by H. Hardin), Wiley, New York.
5. Birks, H. J. B. and Gordon, A. D. (1985) *Numerical Methods in Quaternary Pollen Analysis*, Academic Press, London.
6. Delcourt, P. A. and Delcourt, H. R. (1987) *Long Term Forest Dynamics of the Temperate Zone*, Springer Verlag, New York.
7. Grimm, E. C. (1988) Data analysis and display, in *Vegetation History* (eds B. Huntley and T. Webb III), Kluwer Academic Publishers, Dordrecht, pp. 43–76.
8. Dickson, C. A. (1970) The study of plant macrofossils in British Quaternary deposits, in *Studies in the Vegetational History of the British Isles* (eds D. Walker and R. G. West), Cambridge University Press, Cambridge, pp. 235–54.
9. Dilcher, D. L. (1974) Approaches to the identification of angiosperm leaf remains. *Bot. Rev.*, **40**, 1–157.
10. Moore, P. D. and Webb, J. A. (1978) *An Illustrated Guide to Pollen Analysis*, Hodder and Stoughton, London.
11. Dilcher, D. L. (1973) A palaeoclimatic interpretation of the Eocene floras of southeastern North America, in *Vegetation and Vegetational History of Northern Latin America* (ed. A. Graham), Elsevier, Amsterdam, pp. 39–59.
12. Graham, A. (ed.) (1972) *Floristics and Palaeofloristics of Asia and Eastern North America*, Elsevier, Amsterdam.
13. Birks, H. H. and Mathewes, R. W. (1978) Studies in the vegetational history of Scotland. V. Late Devensian and Early Flandrian pollen and macrofossil stratigraphy at Abernethy Forest, Inverness-shire. *New Phytol.*, **80**, 455–84.
14. Spicer, R. A. (1980) The importance of depositional settling to the biostratigraphy of plant macrofossils, in *Biostratigraphy of Fossil Plants. Successional and Paleoecological Analyses* (eds D. L. Dilcher and T. N. Taylor), Dowden, Hutchinson and Ross, Stroudsburg, PA, pp. 171–83.
15. Davis, R. B. and Webb, T. (1975) The contemporary distribution of pollen in

eastern North America: a comparison with the vegetation. *Quat. Res.*, **5**, 395–434.

16. Ritchie, J. C. and Yarranton, G. A. (1978) The Late-Quaternary history of the boreal forest of central Canada, based on standard pollen stratigraphy and principal components analysis. *J. Ecol.*, **66**, 199–212.
17. Webb, T., Cushing, E. J. and Wright, H. E., Jr (1984) Holocene changes in the vegetation of the Midwest, in *Late-Quaternary Environments of the United States*, Vol. 2, *The Holocene* (ed. H. E. Wright Jr), Longman, London, pp. 142–65.
18. Wells, P. V. (1976) Macrofossil analysis of wood rat (*Neotoma*) middens as a key to the Quaternary vegetational history of arid America. *Quat. Res.*, **6**, 223–48.
19. Colman, S. M. and Birkeland, P. W. (1987) Suggested terminology for Quaternary dating methods. *Quat. Res.*, **28**, 314–9.
20. West, R. G. (1977) *Pleistocene Geology and Biology*, 2nd edn, Longman, London.
21. Lowe, J. J. and Walker, M. J. C. (1984) *Reconstructing Quaternary Environments*, Longman, London.
22. Saarnisto, M. (1988) Time-scales and dating, in *Vegetation History* (eds B. Huntley and T. Webb III), Kluwer Academic Publishers, Dordrecht, pp. 77–112.
23. Clements, F. E. (1936) Nature and structure of the climax. *J. Ecol.*, **24**, 252–84.
24. Tobey, R. C. (1981) *Saving the Prairies*, University of California Press, Berkeley, CA.
25. Tansley, A. G. (1935) The use and abuse of vegetational concepts and terms. *J. Ecol.*, **16**, 284–307.
26. Gleason, H. A. (1917) The structure and development of the plant association. *Bull. Torrey Bot. Club*, **44**, 463–81.
27. Whittaker, R. H. (1970) *Communities and Ecosystems*, Collier-Macmillan, Toronto.
28. Clapham, W. B., Jr (1983) *Natural Ecosystems*, 2nd edn, Macmillan, New York.
29. Simmons, I. (1982) *Biogeographical Processes*, Allen and Unwin, London.
30. Dajoz, R. (1977) *Introduction to Ecology*, Hodder and Stoughton, London.
31. Takhtajan, A. (1969) *Flowering Plants. Origin and Dispersal* (transl. by C. Jeffrey), Oliver and Boyd, Edinburgh.
32. Strahler, A. N. and Strahler, A. H. (1984) *Elements of Physical Geography*, 3rd edn, Wiley, New York.
33. Robichaux, R. H. and Taylor, D. W. (1977) Vegetation-analysis techniques applied to Late Tertiary fossil floras from the western United States. *J. Ecol.*, **65**, 643–60.
34. Webb, T., III (1985) Holocene palynology and climate, in *Paleoclimate Analysis and Modeling* (ed. A. D. Hecht), Wiley, New York, pp. 163–95.

CHAPTER THREE

Diversity – global climate

3.1 CLIMATIC CONTROL OF DIVERSITY

Climate is generally believed to exert a controlling influence on global patterns of plant diversity. Species-richness has been shown to vary regionally with both temperature and precipitation [1,2], and even at the family level a clear diversity gradient with temperature can be demonstrated [3]. The geographical ranges of individual plant taxa likewise indicate an underlying climatic control, so that floristic regions can be seen as a segregation of the world's flora into different subsets, each exposed to a particular set of climatic conditions. Biomes too are segregated subsets of the world's flora that express through their physiognomy a basic adaptation to climate; because they are constrained in composition by the taxa available to them in particular areas, they also reflect the climatic characteristics of the floristic regions where they occur.

Ultimately, the climatic control of diversity is exerted through the physiological tolerances of the individual plant taxa. Woodward [3] has shown that the poleward limits of biomes in both hemispheres can be successfully predicted from a knowledge of the responses of the component plant species to low temperature (their sensitivity to chilling and frost drought), drought stress (their sensitivity to soil water deficit), and the length of the growing season. Soil water deficit is determined by the balance between precipitation and evapotranspiration, and length of the growing season by the total accumulated heat sum above a threshold temperature. Both of these factors govern the complexity of the leaf canopy that can be maintained. Sensitivity to chilling and frost drought differentially limits the distribution of broadleaved evergreen, broadleaved deciduous, and needle-leaved tree taxa. Some critical climatic thresholds suggested by Woodward [3] are given in Table 3.1.

Approaches of this sort, incorporating the results of experimental and modelling studies, demonstrate the close links between plant diversity and climate at the global level. In the current state of our knowledge, however, the critical climatic thresholds identified cannot yet be estimated for times

Table 3.1 Some critical climatic thresholds [3]

	Minimum temperature (°C)	Total heat sum (day-degrees above 0°C)
Broadleaved evergreen forest		
———	−15	
Broadleaved deciduous forest		
———	−40	
Coniferous forest		
———		600
Shrub tundra		
———		200
Herbaceous tundra		

prior to the present; and in order to investigate the effects of past climatic changes on the global plant cover, more empirical approaches need to be adopted – approaches that pinpoint climatic features which can be reconstructed for different times in the past. Two such approaches are considered below.

3.2 MODERN GLOBAL CLIMATIC PATTERNS

3.2.1 Climatic belts

Global climate exhibits large-scale patterning comparable to that of the global plant cover. Global climatic patterns are more difficult to appreciate, however, because their configurations are constantly changing. The short-term flux of climate from day to day that constitutes weather gives rise in a global context to diversity of almost unmanageable proportions. Worldwide climatic patterns in space are readily apparent only when that flux is simplified, and the component parameters of temperature, precipitation, atmospheric pressure, wind strength and direction, etc., are expressed as seasonal or annual mean values, preferably averaged over several years. Climatic values presented in this way show clear patterning – and that patterning may show similarities to global plant diversity patterns. The global temperature/precipitation mosaic shown in simplified form in Figure 3.1, for example, broadly resembles the distribution pattern of forest and woodland communities on a world scale.

An overall highly simplified picture of climatic patterns can be achieved if the values of the various parameters are averaged worldwide for particular latitudinal zones. Figure 3.2 shows the picture thus obtained for mean annual temperature and precipitation, mean annual irradiance (northern hemisphere only), mean January and July air pressure, and mean annual cloud cover, averaged at 10° latitudinal intervals for land and sea surfaces

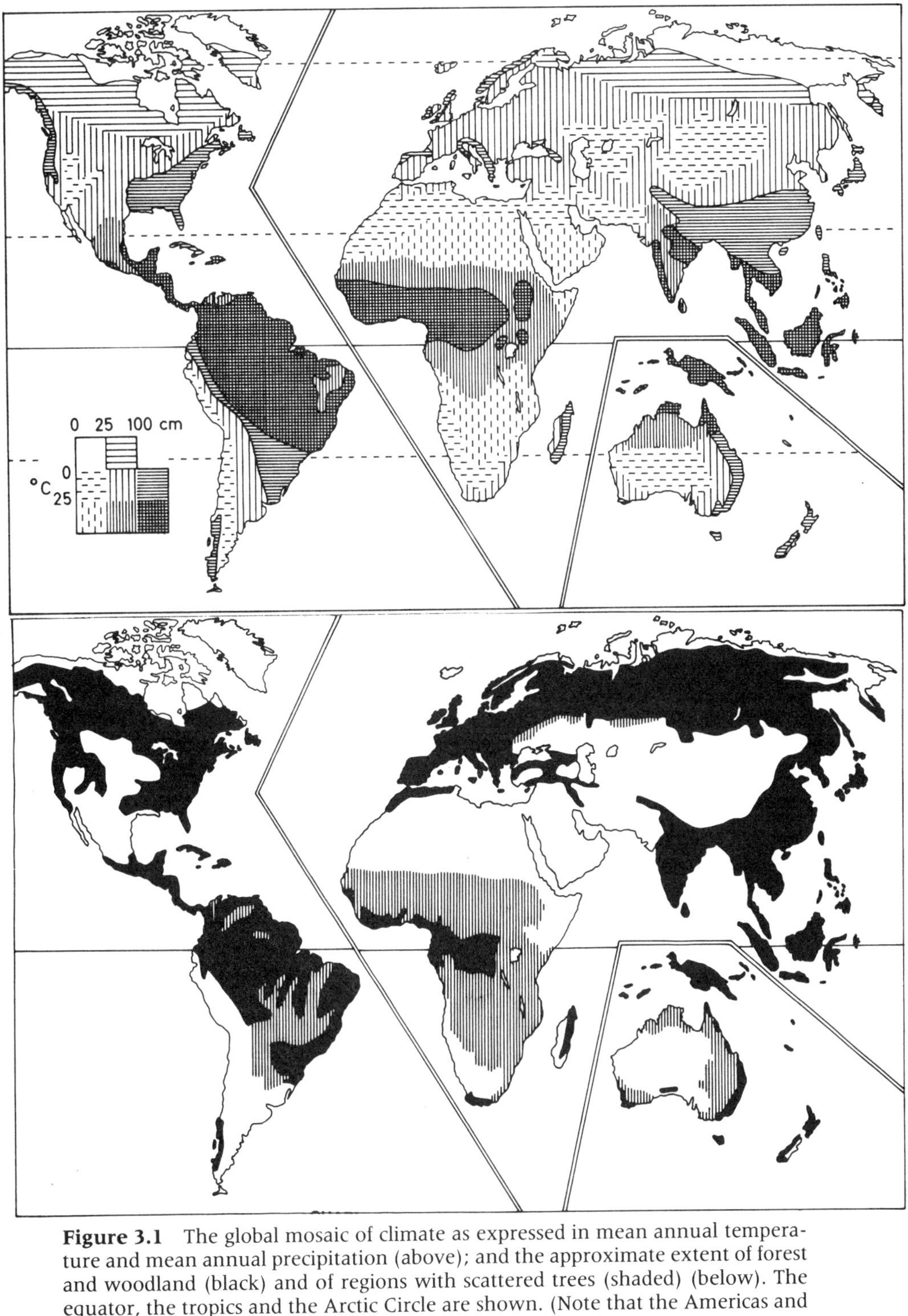

Figure 3.1 The global mosaic of climate as expressed in mean annual temperature and mean annual precipitation (above); and the approximate extent of forest and woodland (black) and of regions with scattered trees (shaded) (below). The equator, the tropics and the Arctic Circle are shown. (Note that the Americas and Australasia are displaced laterally from their true positions.)

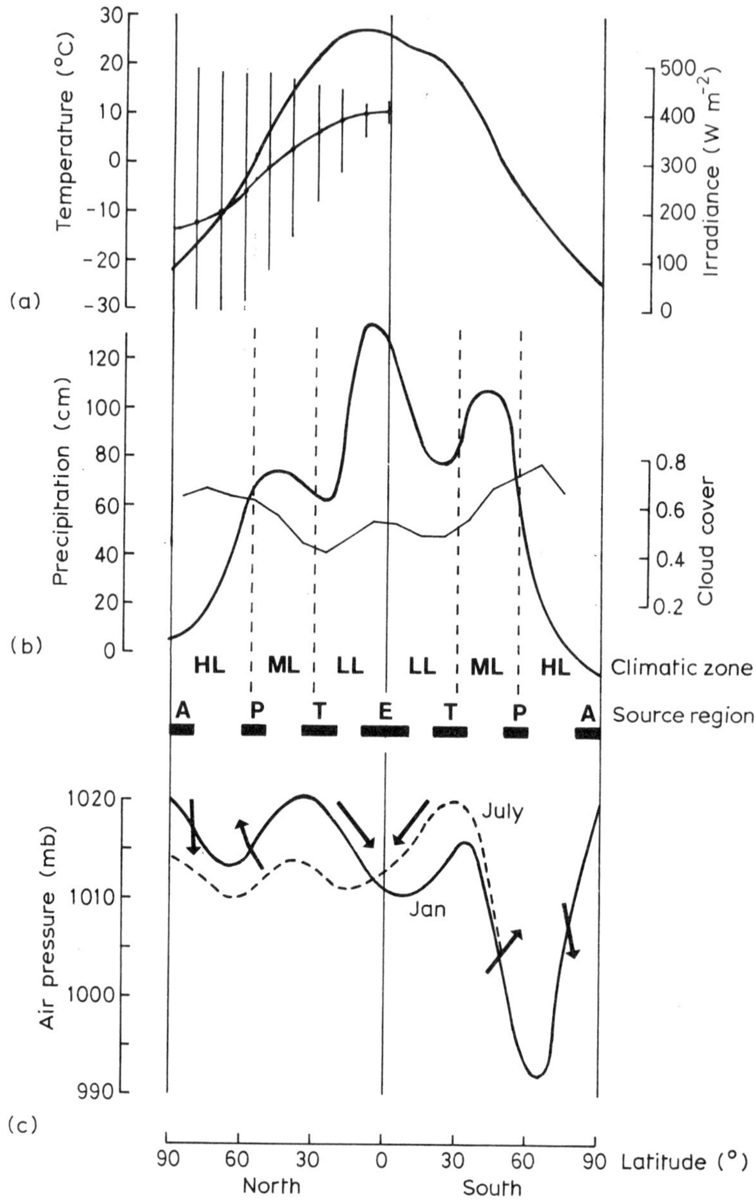

Figure 3.2 Worldwide averages at 10° latitudinal intervals (for land and sea surfaces combined) of (a) mean annual temperature (heavy line), and mean annual irradiance and range of mean monthly irradiance (line and vertical bars); (b) mean annual precipitation (heavy line) and mean annual cloud cover; (c) mean January and July air pressure, and approximate latitudinal positions of the major wind systems (arrows). Also shown on (b) are boundaries of main climatic zones and positions of source areas: HL = high-latitude, ML = mid-latitude, LL = low-latitude, A = arctic/antarctic, P = polar, T = tropical, E = equatorial. Redrawn from Lamb [4], Frakes [5] and Woodward [3].

combined (and expressed conventionally as values at sea level), together with the range of mean monthly irradiance (which broadly parallels the monthly temperature range). At this extreme of simplification, overall patterns are very clear. Particularly striking are the precipitation peaks centred on latitudes 40–50°N, 5–10°N and 40–50°S, and the alternating regions of high and low pressure in the two hemispheres. On the basis of these patterns, a series of broad **climatic belts** can be recognized [6]:

1. An equatorial belt of somewhat lower-than-average air pressure (1008–1011 mb), high temperature and high rainfall (the **equatorial trough**).
2. Belts of higher pressure (typically in excess of 1020 mb) and lower precipitation to north and south of the equatorial trough, and centred on latitudes 30°N and 30°S (the **subtropical high-pressure belts**).
3. A broad belt of low pressure and higher rainfall in the southern hemisphere centred on about latitude 65°S (the **subantarctic low-pressure belt**), and a corresponding less clearly defined belt in the northern hemisphere.
4. High pressure over both poles, and particularly over Antarctica (the **polar highs**).

In between these pressure belts are rather narrow zones with relatively steep gradients of barometric pressure. These zones are characterized by strong and persistent winds, often blowing from one prevailing direction: the **Trade Winds** at low latitudes (blowing from the northeast in the northern hemisphere and from the southeast in the southern hemisphere), the **Westerlies** in the middle latitudes of both hemispheres, and more variable but often easterly winds at high latitudes blowing around the polar highs (Figure 3.2).

The pressure belts in this much-simplified picture are not static within fixed latitudinal zones, but vary their position meridionally throughout the year as a result of seasonal changes in the **thermal gradients** (i.e. the temperature difference) between equator and pole in the two hemispheres. The thermal gradient of either hemisphere is least in summer, when solar radiation receipts (and hence the surface temperatures) at the pole and equator are most similar, and greatest in winter when the receipts are most different. The magnitude of the thermal gradient determines the strength of the air circulation in each hemisphere, which is most vigorous when the gradient is large (i.e., during the winter months). Periods of vigorous circulation in one hemisphere necessarily coincide with periods of weakened circulation in the other, and are accompanied by an expansion of the pressure belts in the 'winter' hemisphere, so that the belts are displaced towards the 'summer' hemisphere. The scale of this displacement is illustrated by the position of the **Intertropical Convergence Zone** (ITCZ), where the northern-hemisphere and southern-hemisphere Trade Winds converge within the equatorial trough; the extreme locations of the ITCZ south (January) and north (July) of the equator each year differ by 20–30° of latitude over the continents, but only by 5–10° over the oceans (Figure 3.3).

Such seasonal migrations of the pressure belts brought about by changes in the thermal gradients of the two hemispheres through the year produce climates at certain latitudes with strongly contrasting summer and winter rainfall patterns. Thus regions along the poleward margins of the subtropical high-pressure belt are affected for part of each year by the wetter conditions of the mid-latitude westerly-wind belt as it is displaced equatorwards in winter, and are thus characterized by climates with wet winters and dry summers. Concomitant winter displacement of the subtropical high-pressure belt equatorwards produces climates along the poleward margins of the equatorial trough with wet summers and dry winters.

Other more localized climates with similar strongly seasonal rainfall are characteristic of low latitudes in Asia, and to a lesser extent in Australia, Africa, and Central and South America, particularly on the eastern sides of these land masses influenced by warm moist winds off the ocean (Figure 3.3). Because land surfaces heat up to a greater extent than ocean surfaces during the summer months, and cool down more in winter, areas of low and high pressure, respectively, develop successively over the land through the year. As a result of the pressure differences set up, rain-bearing winds are drawn in over the land during the summer months (the 'monsoon season'), but not during the winter – which is accordingly dry, with prevailing offshore winds.

All water precipitated over land derives ultimately from the world's oceans, and depends on transport of moist air from ocean to land areas, so that climates with high precipitation (falling either throughout the year, or at one season only) are associated with land areas adjacent to the major oceans of the world. The interiors of large land masses, by contrast, receive relatively little precipitation, as the water content of air is depleted by long passage over land. Heightened precipitation also occurs wherever mountain ranges lie across the paths of moisture-bearing winds, but because air currents that have passed over such ranges will be dry, the gradient of diminishing precipitation inland is accordingly accentuated.

Other localized climatic diversity is present along coastal fringes influenced by the current systems of the world's oceans. Warm readily-evaporated water is carried by these currents to high latitudes along the eastern sides of continents, and cold less-easily-evaporated water is carried to lower latitudes along the western sides (Figure 3.3). In the southern oceans there is a single water flow at mid-latitudes, the Circum-Antartic

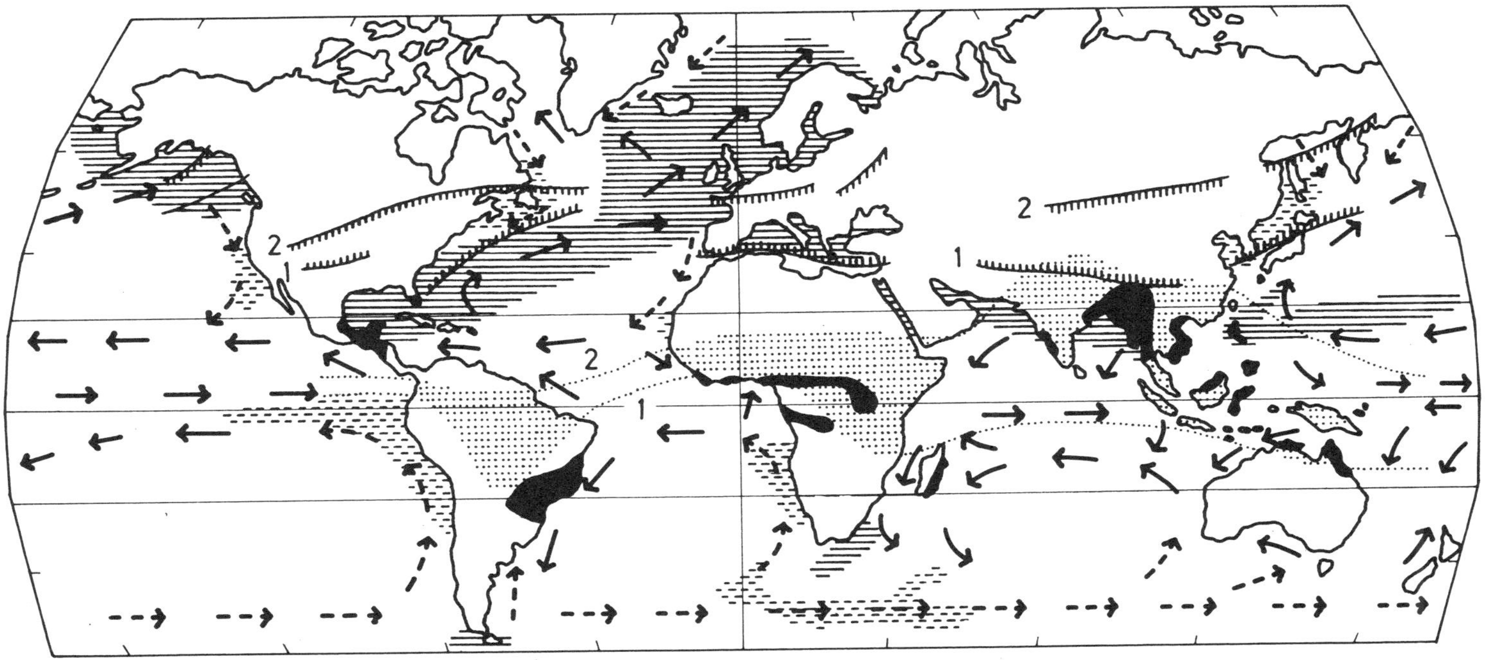

Figure 3.3 Spatial and temporal heterogeneity of climate. (a) Positions of the polar [symbol] and tropical fronts [symbol] in January (1) and July (2); land areas affected by the seasonal movements of the tropical front (the ITCZ) are stippled. (b) Principal regions with monsoon or related climate [symbol]. (c) Warm (→) and cold (-→) ocean currents. (d) Ocean regions where the surface temperatures are ⩾3°C higher [symbol] or lower [symbol] than the latitudinal average. The equator and tropics are shown. Adapted from Lockwood [7], Clapham [8], Strahler and Strahler [6] and Lydolph [9].

Current, encircling the hemisphere; this Circum-Antarctic Current effectively prevents the penetration of warm waters to high latitudes. In the northern hemisphere, by contrast, there are separate gyres in each of the oceans, and poleward extensions of warm mid-latitude currents extend mild climates far to the north. However, polar waters are also fed into the mid-latitude gyres by separate current systems in the arctic seas, and these gyres may carry icebergs south to latitude 45°N [5].

3.2.2 Climatic regimes

The picture presented above is an essentially static one that ignores the constant flux of climate, and takes no account of the processes that link the various components of climate into an integrated and dynamic global climate system. These processes include the input and loss of energy at the earth's surface (from solar radiation and longwave radiation, respectively), the redistribution of this energy over the earth's surface by air and ocean currents, and the cycling of water between the land and the oceans; for detailed consideration of these processes, standard texts such as Lamb [4], Frakes [5], Lockwood [7] and Strahler and Strahler [6] should be consulted. One simple picture of how the total system might function makes use of the concepts of 'source regions' and 'air masses'. **Source regions** relate to the large-scale mosaic of climate over the earth's surface that is summarized in Figures 3.1 and 3.2. That mosaic comprises a patchwork of regions with contrasting properties of moisture availability, temperature and overlying barometric pressure, each situated in one of the major climatic belts defined in section 3.2.1. Accordingly, a series of source regions linked to these belts can be recognized: **equatorial** (situated within the equatorial trough), **tropical** (within the subtropical high-pressure belt), **polar** (within the temperate low-pressure belt), and **arctic/antarctic**. Source regions can be further categorized as **maritime** or **continental** (i.e., situated over ocean and land surfaces, respectively). Each source region imparts a distinctive set of physical characteristics to the immediately overlying **air mass**. These characteristics include temperature and moisture content, and their lapse rates (i.e., the rate of decrease upwards in the air mass). An air mass is a body of air in which these characteristics are fairly uniform over a large area (Table 3.2). Air masses are rarely stationary over a source region for any length of time; normally they move from one geographical area to another along gradients of barometric pressure, gradually changing their physical properties as they do so – but nevertheless maintaining fairly stable weather conditions in these areas. They are continually recharged from their source areas, which are more nearly stationary. A given air mass typically has a sharply defined

Table 3.2 Characteristics of source areas and their associated air masses (data of Strahler and Strahler [6])

Source area	Latitude (°)	Type	Characteristics	Mean temperature (°C)
Equatorial	0–10	Maritime	Warm, very moist	27
Tropical	20–35	Continental	Warm, dry	24
		Maritime	Warm, moist	24
Polar	50–60	Continental	Cold, dry	−11
		Maritime	Cool, moist	4
Arctic/Antarctic	80–90	Continental	Very cold and dry	−46

boundary between itself and an adjoining air mass. Interactions between the two air masses are manifested along a broad **frontal zone** of unstable weather conditions, characterized by a concentration of storm tracks and precipitation, and highly variable winds and cloudiness.

The varying movements of air masses from their source regions, often to a distinct seasonal pattern, give rise to characteristic climatic regimes over the land (Figure 2.2). On the basis of the differing contributions from the various air masses, three major **zones** can be distinguished in each hemisphere [6]: **low-latitude**, influenced by equatorial and tropical air masses; **mid-latitude**, influenced by tropical and polar air masses; and **high-latitude**, influenced by arctic/antarctic and polar air masses (Figure 2.2). The boundaries of these climatic zones are formed by the tropical, polar and arctic/antarctic frontal zones, respectively. These frontal zones vary seasonally in their position, so that considerable areas of the earth's surface experience disturbed conditions for part at least of the year, and varying dominance of major air masses (Figure 3.3). Within each climatic zone, dominance of continental air masses from any source region typically leads to dry conditions, and of maritime air masses to wet conditions. Local or seasonal alternation of maritime and continental air masses within or between zones leads to seasonally dry climates.

3.2.3 Climate/vegetation correlations

Both climatic models that we have been discussing produce large-scale patterning that closely resembles that of the global plant cover. Thus each of the major biomes can be located within one of the major climate belts or the transition zones between them, and is influenced by a unique annual sequence of air masses. Certain climatic boundaries, moreover, appear to coincide with major vegetation boundaries. Thus in North America at least, the northern and southern limits of the boreal forest follow the summer and winter positions, respectively, of the arctic front [10]; the winter and summer positions of the polar front in the northern hemisphere enclose an area characterized by temperate forest and mediterranean woodland; and the winter and summer positions of the tropical front (i.e., the ITCZ) largely define the extent of tropical evergreen and seasonal forest. These correlations are understandable in the context of the physiological studies mentioned in section 3.1, and they form a basis for looking at the interactions between climate and vegetation in the past.

Such an approach assumes, however, that a similar climatic framework to the one we have been considering in previous pages also existed in the past. There are good theoretical reasons for believing that to be the case, and that global climatic diversity has probably always included the following components: (a) a broad latitudinal zonation expressed as zonal differences in temperature, precipitation, air pressure, day length, etc.; (b) more localized climates determined by topography and land/sea configurations; (c) predictable annual sequences of air masses in the various zones.

Nevertheless, the patterns in detail have been very different because of interactions between the various parameters of climate and topography.

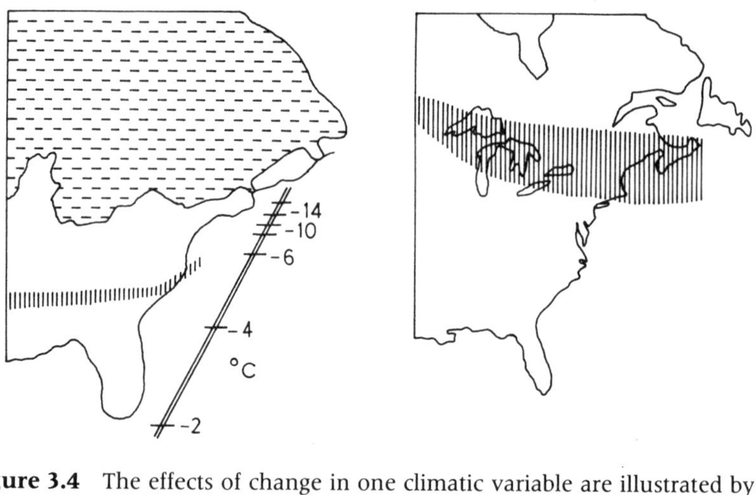

Figure 3.4 The effects of change in one climatic variable are illustrated by this reconstruction for the eastern United States (based on Delcourt and Delcourt [11] and Ruddiman [12]) for 18 ka (left) and 500 yr bp (right). (bp = before present, on a radiocarbon timescale; 'present' refers to the fixed baseline date 1950 AD.) Lowering of global temperature leads to an expansion of permanent ice -–-–-, a displacement and compression of the polar frontal zone ||||||||||, a steepening of the thermal gradient (the temperature scale shows lowering of August sea-surface temperatures relative to modern values), and a lowering of sea level.

Figure 3.4 illustrates the differences from near-present day for eastern North America under a temperature regime that allowed permanent ice to form over Canada. In the next two chapters we consider the global expression of that scenario 18–20 ka ago, and also the very dissimilar scenario for 50 Ma ago, as examples of the extremes of climate which have contributed to the development of the modern global plant cover.

REFERENCES

1. Gentry, A. H. (1982) Patterns of neotropical plant species diversity. *Evol. Biol.*, **15**, 1–84.
2. Whittaker, R. H. (1977) Evolution of species diversity in land communities. *Evol. Biol.*, **10**, 1–67.
3. Woodward, F. I. (1987) *Climate and Plant Distribution*, Cambridge University Press, Cambridge.
4. Lamb, H. H. (1972) *Climate: Present, Past and Future*, Vol. 1, *Fundamentals and Climate Now*, Methuen, London.
5. Frakes, L. A. (1979) *Climates Through Geological Time*, Elsevier, Amsterdam.
6. Strahler, A. N. and Strahler, A. H. (1984) *Elements of Physical Geography*, 3rd edn, Wiley, New York.
7. Lockwood, J. G. (1979) *Causes of Climate*, Arnold, London.
8. Clapham, W. B., Jr (1983) *Natural Ecosystems*, 2nd edn, Macmillan, New York.
9. Lydolph, P. E. (1985) *The Climate of the Earth*, Rowman and Allanheld, Totowa, N.J.
10. Nichols, H. (1967) Pollen diagrams from sub-arctic central Canada. *Science*, **155**, 1665–8.
11. Delcourt, P. A. and Delcourt, H. R. (1987) *Long Term Forest Dynamics of the Temperature Zone*, Springer Verlag, Berlin.
12. Ruddiman, W. F. (1985) Climate studies in ocean cores, in *Paleoclimate Analysis and Modeling* (ed. A. D. Hecht), Wiley, New York, pp. 197–257.

PART TWO

Environments of the Past

CHAPTER FOUR

Environmental change – the Tertiary period

4.1 INTRODUCTION

Details of the methods by which climatic scenarios of the past have been reconstructed are outside the scope of this book. Standard texts such as Frakes [1] and Bradley [2] should be consulted. Most of the methods have been developed since the 1950s, as a result of advances in techniques for studying the character and properties of the deep-ocean sediments of the world, for dating these and other sediments, and for handling large and complex data-sets. In summary, the principal methods used in climatic reconstructions are as follows:

1. Studies of the geophysical and geomorphological properties of sediments and landforms. These studies can give information both on past temperature regimes (the occurrence of glacial sediments and landforms) and on past precipitation regimes (the occurrence of wind-borne terrestrial material in offshore sediments, the presence of terrestrial sediments associated with either wet or arid environments of deposition, and the evidence of lake levels markedly different from present-day ones, for example). Both recent and older, traditional, techniques are included here.
2. Studies of the geochemical properties of sediments, particularly those from the deep oceans of the world. The most comprehensive studies here are those that yield temperature estimates based on the deviations from present-day values of the $^{18}O/^{16}O$ ratios in certain calcareous and siliceous components of the sediments (so-called $\delta^{18}O$ values).
3. Studies of preserved biological material that utilize information independent of any vegetation reconstructions. Included here are temperature estimates based on leaf-margin analysis of fossil land floras, the composition of planktonic assemblages in ocean cores, and the tree-ring analysis of woody remains; general climatic deductions based on the character of the associated faunas are also appropriate in some circumstances.

4. Simulations based on computer-modelling of global climatic regimes. The input data for these simulations are estimates of one or more climatic or topographic parameters at particular times in the past, derived by other methods.

The results obtained by these various methods, together with more specific references to published work, are considered briefly in the next three chapters. The presentation adopted is governed by the level of detail that is possible, both in the climatic reconstructions and in the available plant evidence (to be discussed in subsequent chapters). Moreover, different processes of environmental change have predominated at different times in the past. Accordingly, the material is presented separately for the Tertiary Period (Chapter 4), the Pleistocene Ice Age (Chapter 5) and the Holocene (Chapter 6).

4.2 LAND/SEA CONFIGURATIONS

From a land plant's point of view the global surface is a four-phase system: a complex mosaic of land, ice, deep water and shallow water. One of those phases, dry land, is clearly habitable, provided that climate and soil are suitable and that the plant's propagules can be carried there. Ice and water, on the other hand, are not habitable, and form potential barriers preventing the plant from passing from one land area to another – though both ice and shallow water can be transformed to habitable land if the climate improves or the level of the world's oceans falls. Deep water is a major, and often insurmountable, barrier. Each phase of the mosaic consists of numerous sub-units, some isolated, some interconnected. Currently the four phases occupy about 26%, 3%, 66% and 5%, respectively, of the total surface. Although these relative proportions have fluctuated somewhat over the last 100 Ma (the ice and shallow water phases, for example, each increasing at times to about 10%), much more significant has been the recurring re-assembly of the various sub-units of the mosaic into new patterns (as in a kaleidoscope), as a result of shifting positions of the land masses. A convincing mechanism for the process is provided by the modern theory of **plate tectonics**, itself an update of Alfred Wegener's earlier theory of continental drift [3]. The theory postulates a continuous carapace of juxtaposed plates over the earth's surface, their margins recognizable often as lines of intense seismicity and vulcanicity in the earth's crust (Figure 4.1). Most people delimit seven major plates, and a variable number of smaller ones. These plates are always in contact with one another, and there is no free space between; they are continually jostling one another, like ice-floes in a polar sea, their motive force provided by

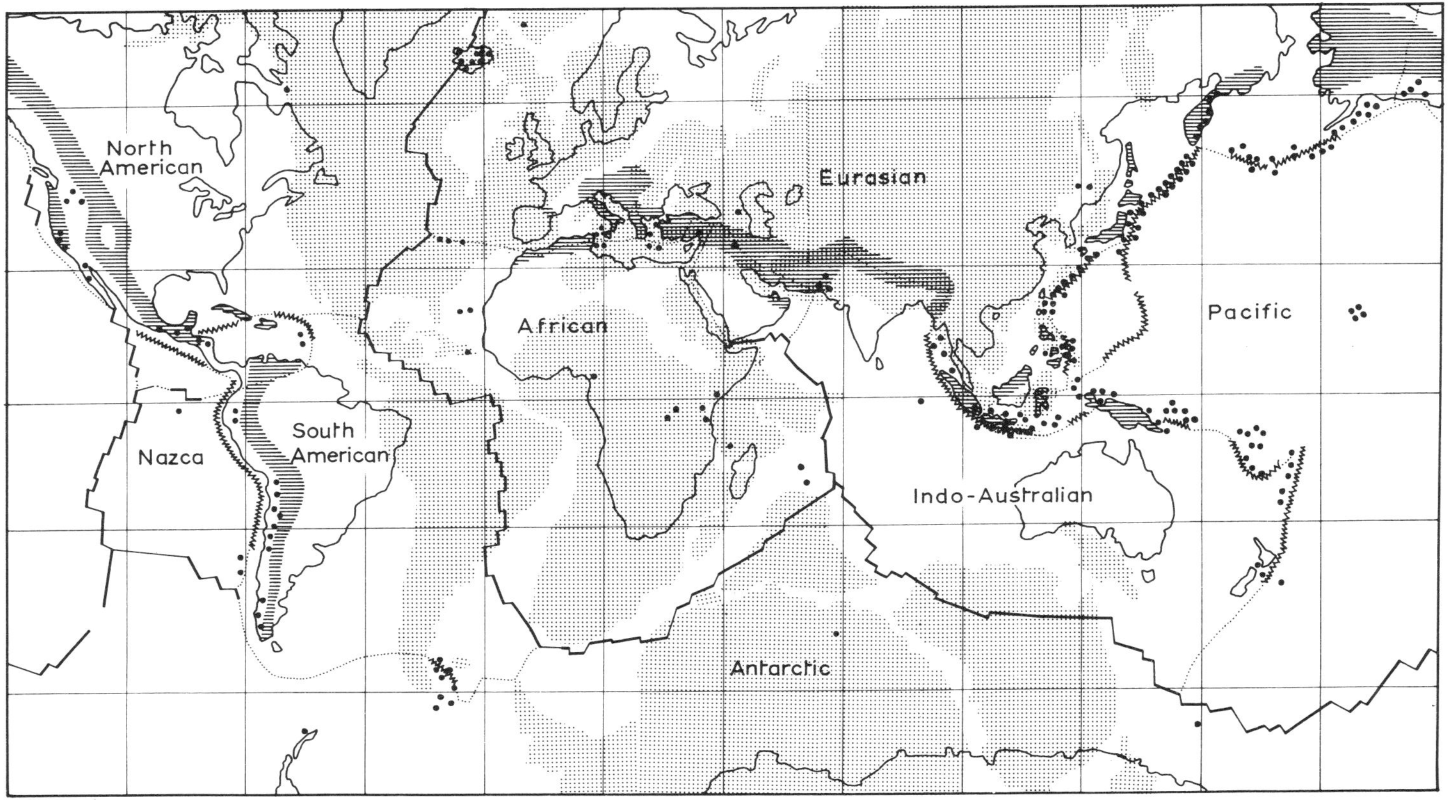

Figure 4.1 Plate tectonic features: the major plates forming the earth's crust, and continental positions 120 Ma ago (stippled). Plate boundaries are shown by a dotted line .... where they are not otherwise delimited by the line of the Mid-Ocean Ridge ⌐L or by ocean trenches ʌʌʌ. ●, active volcanoes; ≡, mountain ranges uplifted during the Cenozoic Era. Continental positions re-drawn from Smith and Bridèn [6].

upwelling of new ocean-floor material along mid-ocean ridges. Adjustment takes place either by subduction (one plate riding over another) or by rotational slipping (one plate slipping past another). For further details, see e.g., Heirtzler [4], Hurley [5] and Dietz and Holden [7].

About 200 Ma ago the global disposition of land and sea was much simpler than it is today. The land masses were connected into a single supercontinent, **Pangaea**, that was centred over the equator, and much of the rest of the world's surface was occupied by a greatly enlarged Pacific Ocean. Subsequent plate movements, however, led to the gradual break-up of Pangaea, and the development of new land masses and ocean basins. The break-up led initially to the formation of an equatorial seaway, the **Tethys Sea**, that separated off a northern land mass (**Laurasia**) from a southern one (**Gondwanaland**). These land masses then split up into smaller units, isolated from each other at times by deep-water barriers. From about 140 Ma onwards, India, Africa, South America, North America, Eurasia, Australia and Greenland were successively split off from what had been Pangaea, together with smaller land masses such as Madagascar and New Zealand. The situation 120 Ma ago (at an early stage in the break-up) is shown in Figure 4.1, superimposed onto the present situation, both drawn on a rectangular grid (Mercator's projection). Such a cartographic representation enables the scale of the configurational changes that have taken place over the last 120 Ma to be readily appreciated, but suffers from the disadvantage of distortion of land areas at higher latitudes.

Land connections between continental areas that are currently isolated survived to varying times into the Tertiary, the most recent major severances being those of Australia from Antarctica (37–55 Ma ago) and Greenland from Europe (35–50 Ma ago). However, continental displacements also brought some previously isolated land masses into contact: Alaska with Siberia across the Bering 'land bridge' about 70 Ma ago, India with Asia about 42 Ma ago, North Africa with Europe about 17 Ma ago, New Guinea with the South-east Asian archipelago 5–15 Ma ago, and North America with South America via the Panama Isthmus about 6 Ma ago. The consequences of these changing linkages on plant community development were profound, and are discussed in later chapters.

The continental displacements were accompanied by equally striking climatic changes that were in part a direct result of the changing global configurations of land and sea (section 4.7). They were also accompanied by fluctuating sea levels that exposed offshore continental shelf areas and created land bridges at times of low sea-level, but flooded over low-lying land within and around the continents and formed epicontinental seas at times of high sea-level. Moreover, the dry land-surfaces themselves were altering, as plate-margin contacts led to mountain uplift and volcanic outpourings. Mountain uplift (orogenesis) has been episodic through geological time. Many major mountain ranges were uplifted before the Cenozoic era, but by 65 Ma ago the older ranges had been substantially worn down by millennia of erosion, whilst the most recently uplifted ranges (the Rocky Mountains, Andes and Atlas Mountains) were still considerably lower than they are today. Uplift at the contact of the Indian and Asian Plates began 40–55 Ma ago, but even by the Pliocene the Himalayan Massif was still low (about 1500 m average elevation [8]); final uplift was delayed until the Quaternary Period. Contact of the northern edge of the African Plate and associated Arabian Plate with the Eurasian Plate led to massive upheaval in south and central Europe and the Middle East, with uplift first apparent 25–35 Ma ago in the Pyrenees, Carpathians, Caucasus and Zagros Mountains. The Alps were uplifted rather later (5–15 Ma ago). Uplift of the American Cordillera, begun perhaps 200 Ma ago, continued spasmodically from the Late Eocene onwards, with present-day elevations not being achieved until the Miocene or Pliocene [9]. The parallel Sierra–Cascade and Coast Ranges in the western United States were formed only late in the Pliocene (less than 5 Ma ago), as were the New Guinea and New Zealand Highlands and the mountains of Yakutia (northern Siberia).

At the start of this episode of mountain-building activity, about 50 Ma ago, the land surface was one of low relief, with no formidable mountain barriers to influence regional climates. As explained in Chapter 9, it is to this period of time, the Eocene Epoch, that we must look for the precursors of most modern plant taxa and plant communities. Accordingly, the environmental scenario for 50 Ma ago forms a convenient starting point from which to develop a palaeoecological perspective on the modern global plant cover. That scenario is considered in the next section.

4.3 THE EOCENE WORLD

The reconstruction of the Eocene world of 50 Ma ago in Figure 4.2 shows a very different disposition of the land masses from that at the present day, with the northern supercontinent, Laurasia, still largely intact, but with break-up of the southern supercontinent, Gondwanaland, well advanced. The relief of the land surfaces was considerably muted, with the uplift of many of the major mountain systems of the present day yet to occur (cf. Figure 4.1). Because of the prevailing low relief, coupled with high sea levels, rain-bearing winds were able to penetrate far into the interiors of all the main land masses. As a result, regions of arid climate were restricted in extent.

The Eocene world was characterized by a greatly-expanded Pacific

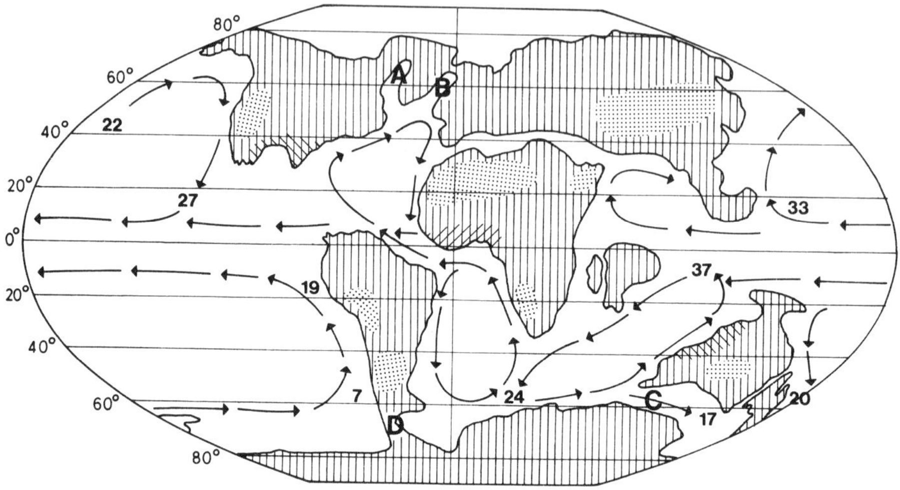

Figure 4.2 The Eocene world. Areas of drier climate and monsoonal climate are indicated; arrows show postulated ocean currents, and numerals are estimated sea surface temperatures (°C). A = Labrador Sea, B = Greenland Sea, C = Tasman Sea, D = Drake Passage. Re-drawn from Frakes and Kemp [10].

Ocean, with vast ocean gyres carrying heated tropical waters to high latitudes in both hemispheres. Moreover, because of the much greater uninterrupted expanse of ocean at low latitudes, a correspondingly greater absorption of solar energy must have occurred (section 4.7), to give warmer-than-present surface waters there. Direct estimates of ocean surface temperatures indicate that values 50 Ma ago were indeed higher than present-day ones, particularly at higher latitudes, where values of 18–20°C from shallow Antarctic waters south of New Zealand at latitude 50°S [11], for example, and about 15°C for waters around Ellesmere Island in the Canadian Arctic at 80°N [12], have been recorded. The close proximity of the South American and Australian land masses to Antarctica (Australia was separated by only a narrow sea passage 50 Ma ago) prevented the development of a cold transglobal current around Antarctica, such as is present today (Figure 3.3), and allowed warm waters to penetrate far to the south. Moreover, the Arctic Ocean was virtually landlocked, so that warm ocean currents reached far northwards also. As a result, permanent ice over the poles was absent, and the temperature gradient between equator and poles must have been considerably lower than it is today – only about half its present-day value [1].

The benign climatic conditions of the Eocene resulted in predominantly woody floras at all latitudes (Chapter 9). Fossil floras with a significant component of non-woody plants are unknown even in polar regions.

Nevertheless, because of the scarcity of fossil floras from continental interiors [13], it is not possible to state unequivocally that forest and woodland were universal in the Eocene.

4.4 THE TEMPERATURE RECORD OF THE TERTIARY

Global temperature changes over the last 50 Ma are now known with a degree of precision that would have been undreamt of even 20 years ago. Much of this new knowledge has come from oxygen-isotope analyses of a large number of deep-ocean cores, some with a record extending back more than 150 Ma. Examples are shown in Figure 4.3. The primary data from these analyses are $^{18}O/^{16}O$ ratios in the sediments expressed in relation to modern values ($\delta^{18}O$ values). The changing $\delta^{18}O$ values are in part a direct temperature signal, and the theoretically-predicted relationship between $\delta^{18}O$ and temperature can be used to calibrate the record [19]. This is the basis of the temperature scales in Figure 4.3. However, it is now clear that changes in global ice volume also affect the $\delta^{18}O$ values, and these changes, although temperature-dependent, cannot easily be calibrated in temperature terms. Once global ice is present, therefore, interpretation of $\delta^{18}O$ patterns is more difficult [20]. Most workers would probably agree that the overall patterns of change are still then valid, but that precise calibration is no longer possible. Thus as far as the data in Figure 4.3 are concerned, it is only for the early part of the record that translation into temperature terms is really appropriate.

Figure 4.3(a) shows separate records of change for high and low latitudes. In the Early Tertiary the two records followed a similar pattern, with an overall downward trend, and at low latitudes surface temperatures reached minimum values in the Late Oligocene that were appreciably below those of the present day. From the Early Miocene onwards, however, there took place what Savin *et al.* [14] refer to as 'uncoupling' – the development of quite different patterns of change in high-latitude and

Table 4.1 The age of glacially-deposited sediments

Glacial-marine sediments (ice-rafted debris)			
	Ross Ice Shelf, Antarctica [1]	26	Ma
	North Polar regions [21]	2.4	Ma
Glacial till	West Antarctica [1]	7–10	Ma
	Alaska [22]	9–10	Ma
	South America [23]	5–7	Ma
	Iceland [21]	3.1	Ma
	Sierra Nevada [1]	2.7–3.1	Ma
Glacial outwash	European Alps [1]	3.0	Ma

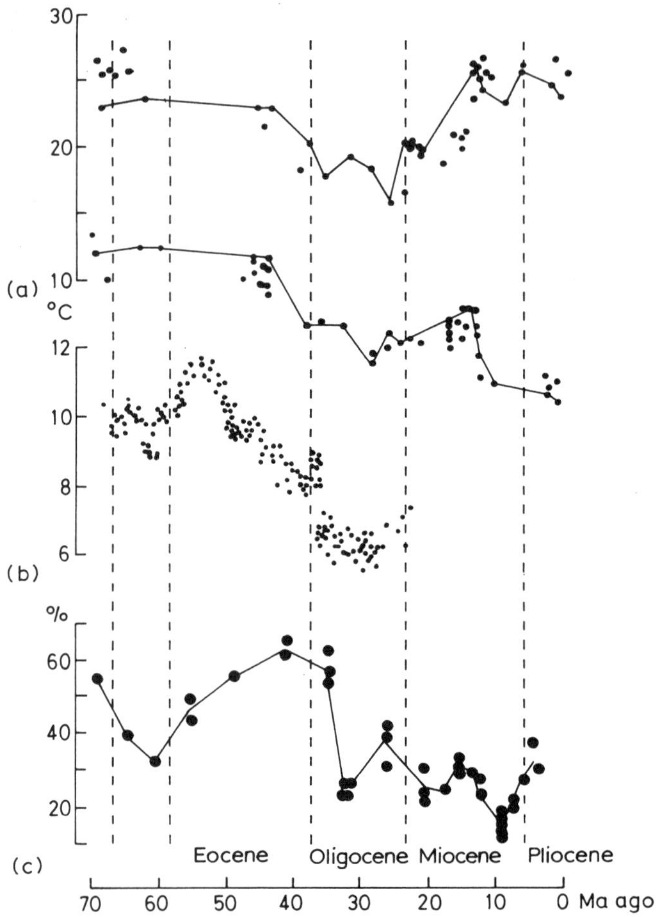

low-latitude waters. As a result of this uncoupling, the temperature gradient between equator and poles was progressively enhanced during the later part of the Tertiary Period.

At all latitudes, ocean-surface temperatures rose in the Early Miocene, to values that were not very different from those in the later Eocene. Thereafter, temperature declined again at higher latitudes; however, because glacial ice was undoubtedly present in Antarctica by the Middle Miocene (Table 4.1), the $\delta^{18}O$ changes cannot any longer be calibrated in precise temperature terms. Detailed oxygen-isotope analyses from ocean sediments that accumulated in the later Tertiary Period show characteristic closely-spaced fluctuations superimposed on the general downward trend.

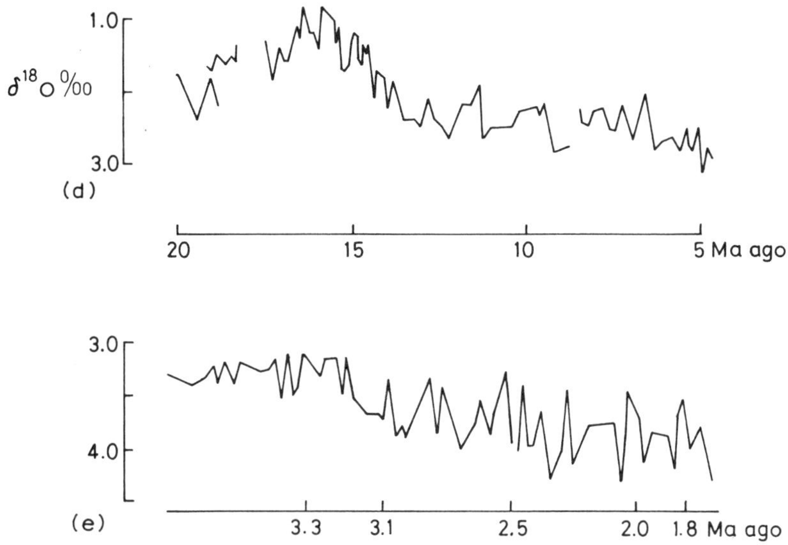

Figure 4.3 Temperature trends in the Tertiary Period. (a) Oxygen-isotope analyses of five low-latitude sites (near-surface Foraminifera, above) and twelve high-latitude sites (benthic Foraminifera, below); the continuous lines connect data from site 167 (mid-equatorial Pacific Ocean). Re-drawn from Savin *et al.* [14]. (b) Oxygen-isotope analyses of benthic Foraminifera from six South Atlantic sites. Re-drawn from Shackleton [15]. (c) Leaf-margin analyses (percentage of entire-margined woody taxa) of fossil floras from the north-western United States. Re-drawn from Axelrod and Bailey [16]. (d) Oxygen-isotope analyses of three deep-ocean cores, through sediments from the Miocene Epoch. Re-drawn from Keller and Barron [17]. (e) Oxygen-isotope analyses of benthic Foraminifera in Pliocene sediments of core V28-179 (equatorial Pacific Ocean). Re-drawn from Shackleton and Opdyke [18].

In some cases these fluctuations are apparent only in the later Pliocene (after 3.2 Ma ago – Figure 4.3(e)), but in others they are present also in Middle and Late Miocene sediments (Figure 4.3(d)). A marked accentuation in the amplitude of the fluctuations occurred 2.4–2.5 Ma ago [24], and provides direct continuity with the high-amplitude fluctuations of the Quaternary Period (Chapter 5).

The oxygen-isotope analyses yield information about ocean temperatures, and not about temperatures on land, and because of the uncertainties about the interpretation of $\delta^{18}O$ values, independent assessment of land temperatures is clearly desirable. Such assessment, albeit rather more fragmentary, can be derived from the proxy record of fossil angiosperm leaf floras – where a direct linear relationship between mean annual temperature and the proportion of woody taxa with entire-margined leaves is claimed [25,26]. The most detailed compilation of leaf-margin data is for a series of fossil floras from the north-western United States [27], but other less complete records are available for European, Japanese and Rocky Mountain floras [9,28,29]. Together they give a picture of temperature changes on land through the Tertiary Period which broadly corroborates that provided by the deep-ocean isotope record (Figure 4.3(c)).

4.5 THE ACCUMULATION OF PERMANENT ICE

Ice sheets are known to have expanded over East Antarctica in the Middle Miocene (or even as early as 36 Ma ago), and over West Antarctica at the close of the Miocene [1,30,31]. Ice accumulation can be seen as the direct consequence of the falling temperatures at higher latitudes through the Tertiary, coupled with continuing continental displacements. In the northern hemisphere the opening of the Labrador Sea and the Norwegian Sea (A and B in Figure 4.2) in the Late Tertiary allowed the establishment of ocean current systems bringing Arctic Sea waters southwards for the first time, and blocked off the northward penetration of warm currents [32]. In the southern hemisphere the Tasman Sea and Drake Passage (C and D in Figure 4.2) were formed rather early in the Tertiary, so that by 22 Ma ago a fully developed Circum-Antarctic Current was present; permanent ice started to accumulate over Antarctica around this time (Table 4.1) [1,31]. Crowley *et al.* [33] used a simple climatic model to demonstrate that the isolation of Greenland and Antarctica by sea channels would have resulted in reduced seasonal extremes of climate, and that, in particular, the onset of cooler summers would have hastened the accumulation of permanent snow and ice on these land masses.

Permanent ice formed later in the northern hemisphere than in the southern hemisphere, being delayed until the Middle Pliocene (Table 4.1). Changes in ocean current patterns caused by the opening of the Bering Straits and the union of the North and South American land masses across the Panama Isthmus 3.5–3.0 Ma ago [21,34], or in atmospheric circulation patterns caused by the continuing uplift of the Tibetan Plateau and the mountain ranges of the western United States [21], have been suggested as possible reasons for the icing-up of North Polar regions. Before the Greenland, Iceland and Baffin ice caps formed there was a period of about 10 Ma when permanent ice was present only in the southern hemisphere (and on some Alaskan mountains; Table 4.1).

4.6 CLIMATIC DRYNESS

Another signal – that of the moisture changes that accompanied the temperature changes through the Tertiary Period – is present in ocean sediments. About 75% of the non-biogenic component of these sediments (i.e., the component not formed from or by living organisms) may be of terrestrial origin, derived as wind-borne dust from arid and semi-arid regions. Considerable variability occurs with depth in the sediments, both in the overall quantities of dust and in particle size. Figure 4.4 shows the

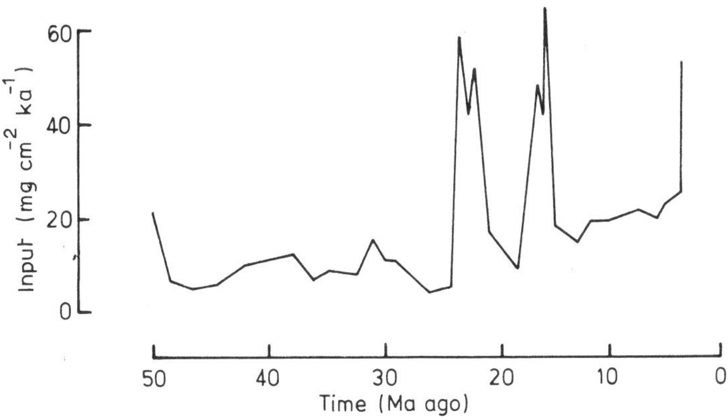

Figure 4.4 Aeolian dust in deep-ocean sediments. Mass accumulation rates of wind-borne dust in Tertiary sediments from a North Pacific core. Re-drawn from Rea *et al.* [35].

dust record in sediments from the North Pacific Ocean [35] over the last 50 Ma, expressed as mean accumulation rate over time-intervals of 1 Ma or more. Peak values at 0–3 Ma ago (Pleistocene/Pliocene), and 15–16 and 21–23 Ma ago (Miocene), are believed to represent increased aridity in the desert regions of central Asia at these times. More wide-ranging evidence from terrestrial sediments [1,36] indicates that by the Miocene seasonally dry or semi-arid climates were present over large areas of the continental interiors of Eurasia and Australia, and to the east of the Rocky Mountains and the Andes. Some grassland or desert vegetation was probably present in all the major land areas (Chapter 9).

The dryness was associated with falling sea levels (Figure 4.5), beginning about 14 Ma ago and accentuated after 10 Ma ago. By the end of the Miocene sea levels were perhaps 50 m below present-day positions [38]. The overall falling sea levels through the Tertiary Period shown in Figure

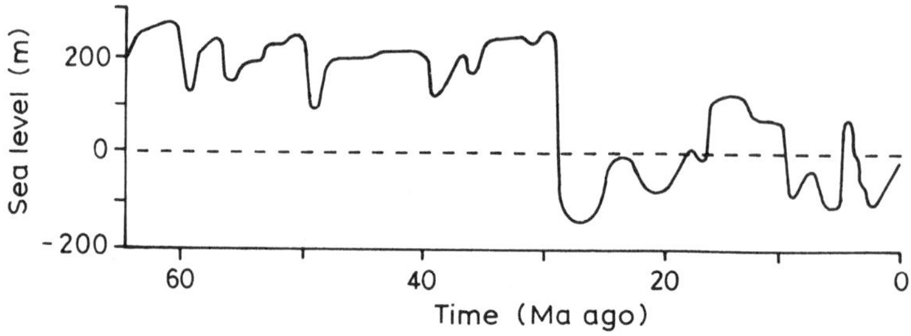

Figure 4.5 Generalized course of sea-level changes over the last 65 Ma, relative to present-day sea level (the interrupted line). Re-drawn from Summerhayes [37].

4.5 were the result of two processes: a gradual uplift of the continental land masses by plate movements (from a mean altitude of less than 300 m in the Early Miocene to about 800 m altitude average at the present day), and the removal of water from the world's oceans to form the permanent Antarctic and Greenland ice caps [39,40,41].

Increased aridity of climate in the Miocene was also associated with intensified mountain uplift (section 4.2). As a result, the later part of the Tertiary Period witnessed an enormous diversification in the range of terrestrial environments: the various montane and alpine environments associated with high-relief topography, together with a series of local and regional arid and semi-arid environments, developed in rainshadow areas. The consequences of this diversification for plant community history were far-reaching, and are discussed in Chapter 9.

4.7 MECHANISMS OF CLIMATIC CHANGE

Could the major changes in the global configurations of land and sea that are known to have occurred in the Tertiary Period explain the climatic trends recorded over the same time-interval? Various possible mechanisms have been suggested, and tested by a variety of models in recent years. Thus it is clear that the spatial distribution of solar radiation actually absorbed by the earth's surface (and hence capable of heating it) will be affected by the relative positions of land and sea because of their different albedos (the percentage of incident light that is reflected from the surface – Table 4.2); and because irradiance is greatest at low latitudes, it is the low-latitude albedo patchwork that is particularly important. The most efficient absorption of solar radiation (and hence the maximum heating effect) would be expected to take place when extensive water surfaces (with low albedo) occur in low latitudes. Moreover, the patchwork of atmospheric components (cloud, water vapour and carbon dioxide) determines the

Table 4.2 Typical albedo values of global surfaces [1,2,42]

	Reflectance (%)
Continental snow and ice	65–95
Desert	25–44
Grassland and agricultural land	8–20
Forest	7–17
Water	2–10

extent to which escaping long-wave radiation is lost to outer space: a system of clear skies and desert vegetation in low latitudes repels far more solar radiation than does a cloud-covered ocean in the same latitudes. The general trend in land/sea configurations over the last 150 Ma has been for the total land area in the northern hemisphere to increase and that in the southern hemisphere to decrease; these changes were most marked at low latitudes in the northern hemisphere and at mid-latitudes in both hemispheres (Table 4.3). It might be expected that the gradual displacement of land masses towards lower latitudes generally would have reduced the efficiency of solar energy utilization in the region where insolation was highest – and thus have led inevitably to declining global temperatures.

Table 4.3 Total land area (million km^2) in three latitudinal zones at different times in the Cenozoic Era [42]

	Ma ago			
	60	40	20	0
High-latitude (60–90°)	33	32	33.5	31.5
Mid-latitude (30–60°)	57.5	54	53	52
Low-latitude (0–30°)	52.5	49	58.5	65.5

This effect was demonstrated mathematically by Barron *et al.* [42]. They calculated global energy receipts at different times in the Cretaceous and Tertiary Periods, using a simple climatic model incorporating known continental positions, inferred vegetation distributions, and a simplified albedo patchwork. Total energy receipts were found to decline by about 2% over the last 100 Ma, but this cooling trend was manifested only in the northern hemisphere, and predominantly at lower latitudes. Barron *et al.* [42] concluded that the increasing land area, particularly of deserts, in the subtropics during the Tertiary Period was a significant factor in the progressive climatic deterioration.

More complex models of global energy circulation have also been used to test the magnitude of the various continental-displacement effects [13,43,44]. The effects of continental position, albedo, global sea level, land topography, and the presence of permanent ice caps have all been studied, whilst the influence of heat transport to high latitudes by ocean currents has been crudely simulated by setting a minimum sea-surface temperature of 10°C for the polar oceans. The models predict a gradual decline in global temperatures similar to that known to have occurred in the later part of the Tertiary Period, but fail to match up the observed high temperatures of the Early Tertiary and preceding Cretaceous Period. All the climatic models tested so far give markedly lower temperatures at high latitudes in these Periods than were known to exist [44,45]. The discrepancy is most marked in continental interiors, where no model system has yet been devised that does not give low temperatures in the interiors of Antarctica, North America and northern Asia in the Cretaceous Period. It is possible that the temperature discrepancies could have been compensated for in reality by more efficient ocean current systems, but many workers do not accept this explanation, and have looked instead to other parameters to 'explain' the ice-free Cretaceous and Early Tertiary climate: an increased input of solar energy, a much reduced tilt to the earth's axis of rotation, higher atmospheric levels of carbon dioxide, or different positions of the North and South Poles [46,47,48,49]. As yet these are suggestions largely unsupported by concrete evidence, and the elucidation of the mechanisms that produced an ice-free planet remains a major challenge to palaeoclimatologists.

REFERENCES

1. Frakes, L. A. (1979) *Climates Through Geological Time*, Elsevier, Amsterdam.
2. Bradley, R. S. (1985) *Quaternary Paleoclimatology. Methods of Paleoclimatic Reconstruction*, Allen and Unwin, Winchester, MA.
3. Hallam, A. (1975) Alfred Wegener and the hypothesis of continental drift. *Scient. Am.*, **232**, 88–97.
4. Heirtzler, J. R. (1968) Sea-floor spreading. *Scient. Am.*, **219**, 60–70.
5. Hurley, P. M. (1968) The confirmation of continental drift. *Scient. Am.*, **218**, 52–64.
6. Smith, A. G. and Bridèn, J. C. (1977) *Mesozoic and Cenozoic Paleocontinental Maps*, Cambridge University Press, Cambridge.
7. Dietz, R. S. and Holden, J. C. (1970) The breakup of Pangaea, *Scient. Am.*, **223**, 30–41.
8. Sahni, A. and Mitra, H. C. (1980) Neogene palaeobiogeography of the Indian subcontinent with special reference to fossil vertebrates. *Palaeogeogr. Palaeoclimatol. Palaeoecol.*, **31**, 39–62.
9. Leopold, E. B. and MacGinitie, H. D. (1972) Development and affinities of Tertiary floras in the Rocky Mountains, in *Floristics and Palaeofloristics of Asia and Eastern North America* (ed. A. Graham), Elsevier, Amsterdam, pp. 147–200.

10. Frakes, L. A. and Kemp, E. M. (1972) The influence of continental positions on Early Tertiary climates. *Nature*, **240**, 97–100.
11. Kemp, E. M. (1978) Tertiary climate evolution and vegetation history in the Southeast Indian Ocean region. *Palaeogeogr. Palaeoclimatol. Palaeoecol.*, **24**, 169–208.
12. McKenna, M. (1980) Eocene paleolatitude, climate, and mammals of Ellesmere Island. *Palaeogeogr. Palaeoclimatol. Palaeoecol.*, **30**, 349–62.
13. Barron, E. J. and Washington, W. M. (1982) Cretaceous climate: a comparison of atmospheric simulations with the geologic problems. *Palaeogeogr. Palaeoclimatol. Palaeoecol.*, **40**, 103–33.
14. Savin, S. M., Douglas, R. G. and Stehli, F. G. (1975) Tertiary marine paleotemperatures. *Geol. Soc. Am. Bull.*, **86**, 1499–1510.
15. Shackleton, N. J. (1986) Paleogene stable isotope events. *Palaeogeogr. Palaeoclimatol. Palaeoecol.*, **57**, 91–102.
16. Axelrod, D. I. and Bailey, H. P. (1969) Paleotemperature analysis of Tertiary floras. *Palaeogeogr. Palaeoclimatol. Palaeoecol.*, **6**, 163–95.
17. Keller, G. and Barron, J. A. (1983) Paleoceanographic implications of Miocene deep-sea hiatuses. *Geol. Soc. Am. Bull.*, **94**, 590–613.
18. Shackleton, N. J. and Opdyke, N. D. (1977) Oxygen isotope and palaeomagnetic evidence for early Northern Hemisphere glaciation. *Nature*, **270**, 216–9.
19. Emiliani, C. (1955) Pleistocene temperatures. *J. Geol.*, **63**, 538–78.
20. Ruddiman, W. F. (1985) Climate studies in ocean cores, in *Paleoclimate Analysis and Modeling* (ed. A. D. Hecht), Wiley, New York, pp. 197–257.
21. Ruddiman, W. F. and Raymo, M. E. (1988) Northern hemisphere climatic regimes during the past 3 Ma: possible tectonic connections. *Phil. Trans. R. Soc. Lond. B*, **318**, 411–30.
22. Denton, G. H. and Armstrong, R. L. (1969) Miocene–Pliocene glaciations in southern Alaska. *Am. J. Sci.*, **267**, 1121–42.
23. Mercer, J. H. (1984) Late Cainozoic glacial variations in South America south of the equator, in *Late Cainozoic Palaeoclimates of the Southern Hemisphere* (ed. J. C. Vogel), Balkema, Rotterdam, pp. 45–58.
24. Ruddiman, W. F. and Wright, H. E. Jr (1987) *North America and Adjacent Oceans during the last Deglaciation*, Geological Society of America, Boulder, CO.
25. Wolfe, J. A. (1971) Tertiary climatic fluctuations and methods of analysis of Tertiary floras. *Palaeogeogr. Palaeoclimatol. Palaeoecol.*, **9**, 27–57.
26. Wolfe, J. A. (1979) Temperature parameters of humid to mesic forests of eastern Asia and relation to forests of other regions of the Northern Hemisphere and Australasia. *US Geol. Survey Prof. Pap.*, **1106**, 1–37.
27. Wolfe, J. A. and Hopkins, D. M. (1967) Climatic changes recorded by Tertiary land floras in northwestern North America, in *Tertiary Correlation and Climatic Changes in the Pacific* (ed. K. Hatai), *Pacific Sci. Congress Symp.*, **25**, pp. 67–76, Sasaki, Sendai, Japan.
28. Schwarzbach, M. (1968) Das Klima des rheinischen Tertiärs. *Z. dt. Geol.*, **118**, 33–88.
29. Tanai, T. (1972) Tertiary history of vegetation in Japan, in *Floristics and Palaeofloristics of Asia and Eastern North America* (ed. A. Graham), Elsevier, Amsterdam, pp. 235–55.

30. Brain, C. K. (1984) The Terminal Miocene Event: a critical environmental and evolutionary episode? in *Late Cainozoic Palaeoclimates of the Southern Hemisphere* (ed. J. C. Vogel), Balkema, Rotterdam, pp. 491–8.
31. Kennett, J. P. (1980) Paleoceanographic and biogeographic evolution of the Southern Ocean during the Cenozoic, and Cenozoic microfossil datums. *Palaeogeogr. Palaeoclimatol. Palaeoecol.*, **31**, 123–52.
32. Gradstein, F. M. and Srivastava, S. P. (1980) Aspects of Cenozoic stratigraphy and paleoceanography of the Labrador Sea and Baffin Bay. *Palaeogeogr. Palaeoclimatol. Palaeoecol.*, **30**, 261–95.
33. Crowley, T. J., Short, D. A., Mengel, J. G. and North, G. R. (1986) Role of seasonality in the evolution of climate through the last 100 million years. *Science*, **231**, 579–84.
34. Flohn, H. (1984) Climate evolution in the Southern Hemisphere and the equatorial region during the Late Cenozoic, in *Late Cainozoic Palaeoclimates of the Southern Hemisphere* (ed. J. C. Vogel), Balkema, Rotterdam, pp. 5–20.
35. Rea, D. K., Leinen, M. and Janacek, T. R. (1985) Geologic approach to the long-term history of atmospheric circulation. *Science*, **227**, 721–5.
36. Parrish, J. T., Ziegler, A. M. and Scotese, C. R. (1982) Rainfall patterns and the distribution of coals and evaporites in the Mesozoic and Cenozoic. *Palaeogeogr. Palaeoclimatol. Palaeoecol.*, **40**, 67–101.
37. Summerhayes, C. P. (1986) Sealevel curves based on seismic stratigraphy: their chronostratigraphic significance. *Palaeogeogr. Palaeoclimatol. Palaeoecol.*, **57**, 27–42.
38. Haq, B. U., Hardenbol, J. and Vail, P. R. (1987) Chronology of fluctuating sea levels since the Triassic. *Science*, **235**, 1156–67.
39. Colquhoun, D. J. and Johnson, H. S. Jr (1968) Tertiary sea-level fluctuation in South Carolina. *Palaeogeogr. Palaeoclimatol. Palaeoecol.*, **5**, 105–26.
40. Fairbridge, R. W. (1960) The changing level of the sea. *Scient. Am.*, **202**, 70–9.
41. Hallam, A. (1984) Pre-Quaternary sea-level changes. *Ann. Rev. Earth Planet. Sci.*, **12**, 205–43.
42. Barron, E. J., Sloan, J. L. II and Harrison, C. G. A. (1980) Potential significance of land–sea distribution and surface albedo variations as a climatic forcing factor; 180 m.y. to the present. *Palaeogeogr. Palaeoclimatol. Palaeoecol.*, **30**, 17–40.
43. Barron, E. J. (1985) Explorations of the Tertiary global cooling trend. *Palaeogeogr. Palaeoclimatol. Palaeoecol.*, **50**, 45–61.
44. Barron, E. J. (1985) Climate models: applications for the pre-Pleistocene, in *Paleoclimate Analysis and Modeling* (ed. A. D. Hecht), Wiley, New York, pp. 397–421.
45. Barron, E. J., Thompson, S. L. and Schneider, S. H. (1981) An ice-free Cretaceous? Results from climatic model simulations. *Science*, **212**, 501–8.
46. Creber, G. T. and Chaloner, W. G. (1985) Tree growth in the Mesozoic and Early Tertiary and the reconstruction of paleoclimates. *Palaeogeogr. Palaeoclimatol. Palaeoecol.*, **52**, 35–60.
47. Donn, W. L. (1982) The enigma of high-latitude paleoclimate. *Palaeogeogr. Palaeoclimatol. Palaeoecol.*, **40**, 199–212.
48. Douglas, J. G. and Williams, G. E. (1982) Southern polar forests: the Early

Cretaceous floras of Victoria and their paleoclimatic significance. *Palaeogeogr. Palaeoclimatol. Palaeoecol.*, **39**, 171–85.

49. Fischer, A. G. (1981) Climatic oscillations in the biosphere, in *Biotic Crises in Ecological and Evolutionary Time* (ed. M. H. Nitecki), Academic Press, New York, pp. 103–31.

CHAPTER FIVE

Environmental change – the Pleistocene Ice Age

5.1 INTRODUCTION

The Pliocene Epoch saw the final icing-up of the polar regions that led to the 'Pleistocene Ice Age'. Even before the start of the Pliocene, however, as high mountain chains were uplifted, so alpine glaciers developed and spread. At higher latitudes in both hemispheres, alpine glaciation was probably present by 7 Ma ago, and in southern South America these glaciers pulsed out on to the lowlands at least three times before the close of the Tertiary Period [1]. Global conditions worsened markedly between 3.2 and 2.4 Ma ago, and as the North Polar ice cap developed and sea-ice formed in the North Atlantic [2], so glaciers formed on lower-latitude mountains as well [3]. Yet most workers would place these events in the Tertiary Period, and seek a later date for the start of the Quaternary Period. Definition of that boundary is not easy [4]. Biological contrasts are evident between Late Tertiary and Early Quaternary floras and faunas, but timing of the changes is variable, and there are no globally synchronous, climatically-induced events such as those which characterize earlier boundaries. Moreover, the defined changes are largely European. Some of the definitions of the start of the Quaternary Period are accordingly vague: the earliest evidence of Man and his culture; the onset of glaciation; or the attainment of specifically modern floras and faunas. Others are more specific: the appearance of taxa adapted to cold conditions in either the Mediterranean Sea (the foram *Hyalinea balthica* and the mollusc *Arctica islandica*) or northern Europe (*Elephas, Bos* and *Equus*); or the extinction in the Netherlands of certain Tertiary 'exotics' (the trees *Sequoia, Taxodium, Nyssa, Liquidambar* and *Aesculus*). Most precise is the recommended adoption of the Italian marine sequence at Le Castella, Calabria, as the international standard. This places the Pliocene/Pleistocene boundary at the base of the Calabrian Stage, and relates it to a clearly-defined biostratigraphic horizon that appears to occur just above the base of the Olduvai Geomagnetic Event. On this basis the close of the Tertiary Period can be dated to about 1.6 Ma ago, and located accurately in the deep-ocean oxygen-isotope sequences [5,6]. Nevertheless, it might be argued that a starting date of 1.6 Ma ago for the Pleistocene delimits a rather arbitrary boundary that bears little relationship to the pattern of icing-up of the polar regions (which was probably almost complete by that time), or to the development of the continental ice sheets (which attained their maximum extent much later).

5.2 THE WORLD AT GLACIAL MAXIMUM

5.2.1 'Planet Ice'

We examine first the nature of the global environment at a time of maximum ice cover ('Planet Ice'). The scenario is for 18–20 ka ago, at the time of the most recent glacial maximum.

Land and sea ice then affected large, and roughly equal, areas of both hemispheres (Figure 5.1). Land-based ice in the northern hemisphere extended south to latitude 52°N in Europe and 39°N in North America, and was linked across the North Pole by sea ice over the Arctic Ocean and the Norwegian, Greenland and Labrador Seas. In the southern hemisphere, on the other hand, sea ice completely encircled the Antarctic ice sheet, and extended north to at least latitude 50°S. The total amount of land-based ice covered about 30% of the earth's land surface, as compared with 10% at the present day (Table 5.1).

Concomitantly, mountain glaciers expanded, so that high-altitude snow and ice spread downwards to lower altitudes. Although there is considerable imprecision in many of the estimates of the position of the snowline at the height of glaciation [13], there is also general agreement that in many parts of the world the snowline was lowered by 900–1000 m [14]. Smaller depressions of the snowline may have been characteristic of some tropical mountains, though this effect was less marked in arid regions [15]. Temperatures 6–8°C lower than at present at middle latitudes 18 ka ago have been estimated from these snowline depressions and from oxygen-isotope analyses of ice cores from Antarctica [13,16,17].

In an extensive area to the south of the northern-hemisphere ice sheet was a zone equivalent to the 'permafrost zone' of Arctic regions today, and characterized by scanty snow cover, intense winter cold, a permanently frozen subsoil, and a range of frost-induced (cryogenic) features such as stone polygons, ice wedges and pingos. For these features to have formed at mid-latitudes in Eurasia and North America during glacial maximum would have required a temperature depression relative to the present day of 10–20°C, concentrated particularly into the winter period [18,19],

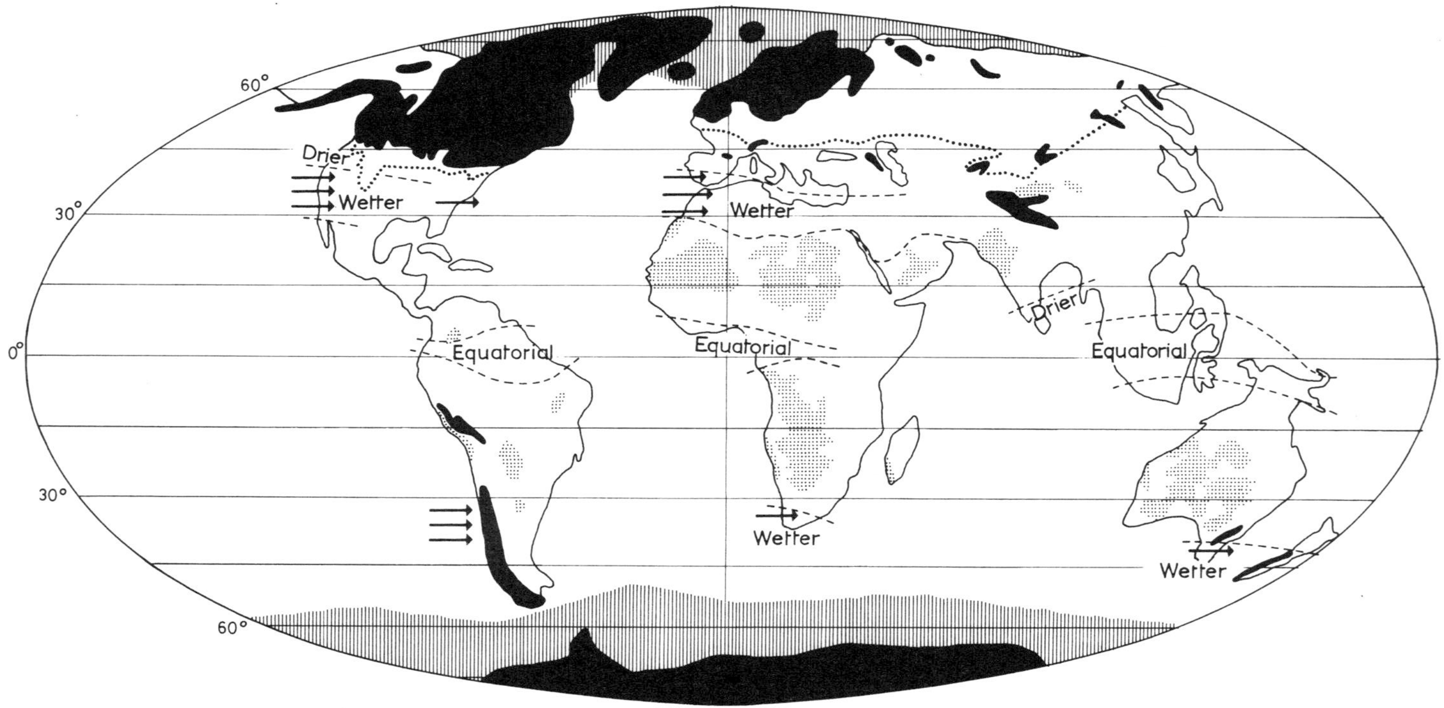

Figure 5.1 The world at glacial maximum. ■, permanent land-based ice; ||||||||||, sea ice; ..., southern limit of permafrost; ░░, fossil dune systems; →, track of Westerlies; ----, postulated climatic zone boundary. Extent of dune systems based on Sarnthein [7], Lancaster [8], Thomas and Goudie [9] and Wasson [10].

Table 5.1 Approximate extent of ice sheets (km^2) at the present day and at maximum glaciation [11,12]

	Present day		Full glacial	
Europe				
Spitsbergen, Faeroes	58 016		311 662	
Iceland	12 550		142 000	
Scandinavia	6 185		6 666 708	
Alps, Pyrenees	3 640		39 000	
Caucasus	1 975		16 000	
	82 366	**(0.06%)***	7 175 370	**(4.82%)***
North America, Greenland				
Greenland	1 630 000		2 295 300	
Canada and Arctic islands	132 660		13 386 964	
USA (mountain glaciers)	520		2 830 127	
	1 763 180	**(1.18%)***	18 512 391	**(12.44%)***
Asia				
Arctic islands	32 850		48 775	
Mountain glaciers	110 700		3 935 000	
	143 550	**(0.10%)***	3 983 775	**(2.68%)***
South America	24 700		870 000	
Africa	20		1 900	
Australasia	1 015		30 000	
Antarctica	12 875 135	**(8.65%)***	13 810 000	**(9.28%)***
Total	**14 889 966 (10.01%)**		**44 383 436 (29.83%)**	

* Percentage values relate to present-day area of global land surface (148 800 000 km^2)

rather than the 6–8°C postulated from the snowline evidence. At low latitudes in tropical regions, however, the temperature changes were more muted: 2–5°C lower only [15,18,20,21].

Similar patterns are evident in the oceans, with the greatest cooling (in excess of 6°C) in subpolar regions and the least (<2°C) in lower latitudes; pronounced temperature gradients were developed between latitudes 35° and 45° (Figure 3.4). The waters of the North Atlantic Ocean, surrounded on three sides by continental and sea ice, experienced the most extreme effects, with summer surface temperatures 10°C or more lower than present ones [22].

A wide body of evidence shows that large areas of the unglaciated land surface were substantially drier 18–20 ka ago. Some of this evidence comes from aerial surveys: the recognition in areas now covered by savanna or even forest of landscapes typical of desert regions today [23,24]. Sarnthein

[7] showed that about half of the total land surface between latitudes 30°N and 30°S 18 ka ago was covered by mobile sand deserts (ergs), as compared with about one-tenth today. In Africa at glacial maximum, dune systems advanced up to 1000 km in both hemispheres to within 5 or 10° latitude of the equator, over areas now covered by rainforest (Figure 5.1).

Mobile dunes occur at the present day in areas receiving less than 150–200 mm rainfall per year [25]. Old fixed dunes of full-glacial age, on the other hand, occur in areas currently receiving up to 1400 mm of rain per year [9], so substantially more arid conditions must have pertained 18–20 ka ago. As moisture declined then, so previously vegetated surfaces would have been bared and freshwater lakes would have dried up. Widespread erosion by wind of the exposed soil and lake sediments must subsequently have taken place, as many low-latitude offshore marine deposits have been found to contain increased concentrations of unweathered soil minerals, freshwater diatom frustules and opal 'phytoliths' (the silica bodies in grass leaves) in the sediments formed 18 ka ago [26–28]. These 'deflation' episodes in the marine sediment column make it clear that lake levels at low latitudes were characteristically low at glacial maximum [13,15,29].

A similar finding comes from modelled simulations of global climate at glacial maximum, which predict a decrease in precipitation at almost all latitudes, varying from 3 to 9% [30]. Several reasons can be adduced for this glacial aridity. Thus less water would have been evaporated from the oceans of the world, because of their lowered surface temperatures, whilst over much of the land, on the other hand, evaporation would have been enhanced as a result of the strengthened wind systems produced by a heightened thermal gradient between equator and poles. Moreover, lowered sea levels 18–20 ka ago would undoubtedly have led to more extensive land surfaces and increased likelihood of mid-continental aridity. The volume of ice and snow then was about three times that at the present day, and sea levels must have been about 130 m lower [12]. Such a fall in sea level would have increased the area of dry land surface by more than 10%, exposing a peripheral strip of land to seaward of present coastlines that was 400–500 km wide in places [13]. Substantial 'land bridges' were thus formed, of which the most important were: Alaska and Siberia across the Bering Straits; the Japanese archipelago and the Asian mainland; Malaysia and the East Indies; Australia, New Guinea and Tasmania; the North and South Islands of New Zealand; Sri Lanka and India; and the British Isles and western Europe across the English Channel and the southern North Sea (Figure 5.1).

5.2.2 Climatic belts

Aridity cannot have been ubiquitous over unglaciated land at glacial maximum, however, as a feature of all global circulation patterns, both now and presumably also in the past, is the increased precipitation associated with the Intertropical Convergence Zone and the polar-frontal zone (Figure 3.2). Climatic modelling suggests that the mid-latitude (polar-frontal zone) precipitation maximum in the northern hemisphere was displaced southwards by 6–8° of latitude 18 ka ago [30,31], whilst a smaller displacement (about 5° of latitude) is deduced by Street-Perrott and Harrison [32] from lake-level evidence. Figure 5.1 shows a possible reconstruction of precipitation patterns at glacial maximum. The reconstruction is incomplete, because for certain geographical regions there are insufficient data to justify any synthesis; it is based on direct biological and non-biological evidence, supplemented by modelling predictions [33]. In the northern hemisphere all land surfaces above about 40°N latitude were characterized by permanent snow and ice or by arid polar desert, and the high-pressure systems developed over the land in these latitudes would have forced the mid-latitude Westerlies to follow a more southerly course than that followed at the present day – across the central and southern United States, North Africa, and perhaps southern Spain [34]. As a result, certain regions that are arid at the present day were appreciably wetter 18 ka ago: the northern edge of the Saharan and Arabian Deserts, and the deserts of the American Southwest, for example (discussed in Chapter 13). In the southern hemisphere, westerly winds would have been displaced northwards to about latitudes 30–40°S [32]*, and brought snow to higher altitudes in the southern Andes, south-east Australia, Tasmania and the South Island of New Zealand. At low latitudes (20°N to 20°S), arid conditions prevailed, and the wet equatorial zone was almost certainly constricted at glacial maximum, particularly in Africa and South America.

Changes in precipitation patterns at glacial maximum were apparent not only as shifts in the positions of the major climatic belts, but also as variations in the intensity and distribution of strongly seasonal rainfall [36]. There is some evidence, based largely on climatic modelling, that monsoon rainfall was less pronounced 18 ka ago [22], and that mediterranean climates would have been somewhat different in character and extent, but little is as yet known about the detailed character of these changes.

* For an alternative view see Markgraf [35].

5.2.3 The timing of glacial maximum

Although 18,000 yr bp* is a convenient date at which to examine global conditions characteristic of the extreme cold of the Pleistocene Ice Age, maximum glaciation was almost certainly not coincident in different parts of the world. It probably occurred rather earlier in the southern hemisphere than in the northern hemisphere: 22–26 ka ago in New Zealand, for example [37], as compared with 17–21 ka ago in North America [38]. Harrison *et al.* [39] use a range of palaeotemperature data to reconstruct temperature trends over the last 25 ka between latitudes 30° and 60° in both hemispheres, and conclude that temperatures were lowest about 16 ka ago in the northern hemisphere and about 20–22 ka ago in the southern hemisphere. In certain parts of the world, however, maximum cold or maximum glaciation may have occurred much earlier in the cycle: 50–60 ka ago in Taiwan [40] and about 73 ka ago in parts of South Africa [1], for example.

5.3 CLIMATIC CHANGE THROUGH THE PLEISTOCENE

The primary characterization of environmental change through the Pleistocene is provided by the record of changes in ice volume in deep-ocean sediments and on formerly glaciated land surfaces. Two continuous sequences of $\delta^{18}O$ values in apparently undisturbed deep-ocean sediments spanning the full duration of the Pleistocene are considered here: one from the mid-equatorial region of the Atlantic Ocean (core V16-205 [41]), and one from the equatorial Pacific (core V28-239 [42]). Both sequences show recurring oscillations between high and low global ice volumes (presumed cold and warm conditions): peaks to the right and left, respectively, in Figure 5.2. The oscillations ('oxygen-isotope stages') are conventionally numbered successively from above downwards, so that cold stages (relatively rich in ^{18}O) are designated by even numerals, and warm stages (relatively deficient in ^{18}O) by odd numerals. (Oxygen-isotope stage 1 is the Holocene.) These stage numbers are the circled numerals in Figure 5.2. A basic timescale is provided by the columns at right and left in Figure 5.2 – which show the palaeomagnetic characteristics of the sediments in the two

* bp = before present, on a radiocarbon timescale. A date in 'yr bp' is more precise than a date in 'yr ago', as it relates to a fixed baseline (1950 AD).

Figure 5.2 Oxygen-isotope analyses of two deep-ocean cores for the Quaternary Period. For explanation, see text. The vertical stripes denote probable glacial stages. Re-drawn from Shackleton and Opdyke [42] and Van Donk [41].

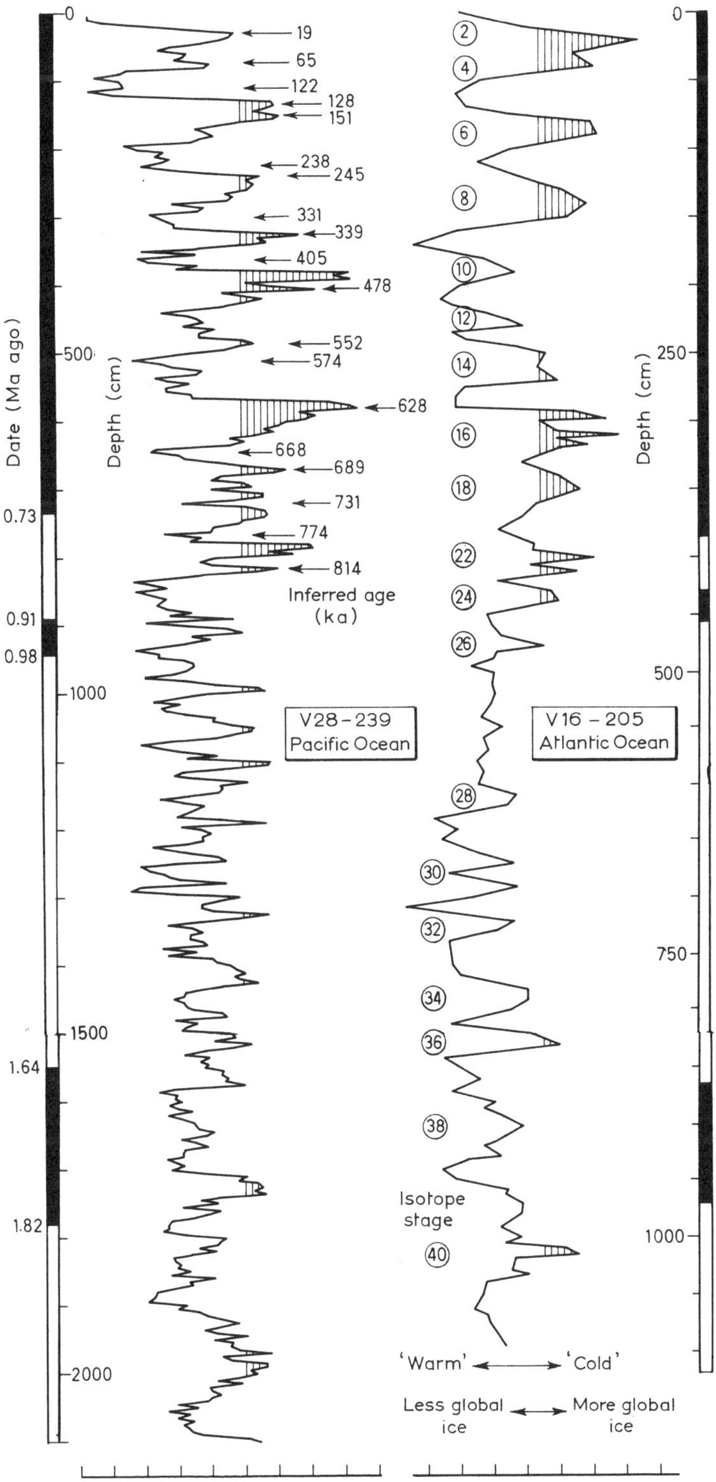
Date (Ma ago)
Depth (cm)
0
500
1000
1500
2000
0.73
0.91
0.98
1.64
1.82
19
65
122
128
151
238
245
331
339
405
478
552
574
628
668
689
731
774
814
Inferred age (ka)
V28–239 Pacific Ocean
V16 – 205 Atlantic Ocean
2
4
6
8
10
12
14
16
18
22
24
26
28
30
32
34
36
38
40
Isotope stage
250
750
'Warm'
'Cold'
Less global ice
More global ice

Table 5.2 Cold and warm stages of the Pleistocene in three regions of Europe†‡. Adapted from Lowe and Walker [44], De Jong [45] and Zagwijn [46]

Subdivision of Pleistocene	Approx. date (ka bp)	Netherlands	Britain	European Alps
——	10			
Upper		WEICHSELIAN*	DEVENSIAN*	WÜRM*
——	130	Eemian	Ipswichian	Riss–Würm
		SAALIAN*	WOLSTONIAN*	RISS*
Middle		Holsteinian	Hoxnian	Mindel–Riss
		ELSTERIAN*	ANGLIAN*	MINDEL*
——	730	*Cromerian*	Cromerian	Günz–Mindel
		Bavelian	BEESTONIAN	GÜNZ*
Early	1000	MENAPIAN		
		Waalian	Pastonian	
——	1600	EBURONIAN	PRE-PASTONIAN	
		Tiglian	Bramertonian	
	2400	PRAETIGLIAN	BAVENTIAN	
			Antian	
(Pliocene)			THURNIAN	
			Ludhamian	

† There is no certain correlation between earlier stages from different regions that are placed on the same horizontal level in the Table. The timescale applies only to the Netherlands sequence.

‡ Upper case = cold stages (* = glacial); lower case = warm stages; italics = complexes of two or more cold/warm cycles.

cores (black = periods of normal polarity; white = periods of reversed polarity), and the estimated dates of the polarity reversals recorded there (see Bradley [13] for details of the method). The more precise inferred ages of the 'cold' and 'warm' peaks in core V28-239 in Figure 5.2 are based on those calculated by Imbrie *et al.* [43] for a neighbouring ocean core.

The patterns of change are very similar in the two cores, to the extent that equivalent cold and warm stages can be recognized in each. Van Donk [41] identified eighteen cold stages during the last 1.64 Ma; Shackleton and Opdyke [42] identified eleven cold stages during the last 800 ka, but declined to give a precise number for the whole of the Pleistocene, as the fluctuations are less clear-cut before 800 ka ago. The date for the most recent ice maximum in these sequences (19 ka ago) is the same as that determined for the last glacial maximum on land in Europe and North America, so that the high-amplitude 'cold' peaks of the last 800 ka must presumably represent major expansions of the continental ice sheets; the fluctuations of lower amplitude in the earlier Pleistocene must also record increases in global ice volume, but of quite unknown extent. The 'warm' peaks, broadly equivalent in amplitude to the Holocene peak (stage 1), must similarly correspond to interglacial periods recognized on land.

Although at least eleven high-amplitude 'cold' stages are recognizable in the deep-ocean record, only three or four major glacial stages on land are recognized in glaciated regions of central and northern Europe (Table 5.2). Four glacial stages are also recognized in North America and in New Zealand, but we cannot be sure that these stages occupy the same time-intervals in the different regions. Indeed, there is evidence that lowland glaciation occurred much earlier in the Pleistocene in North America than in Europe [47], so that the earliest glacial stage there (the Nebraskan) may considerably pre-date the oldest European glacial stage (the Günz [2]).

Continuous Pleistocene sequences on land are most likely to be found in regions far removed from the ice-sheet limits. The best European sequences so far investigated come from sites in Czechoslovakia and Hungary. Thus from a 950-m borehole at Jászladány in the Hungarian Plain east of Budapest, Ronai [48] constructed a pollen diagram extending back without interruption into the Middle Pliocene, and exhibiting twenty or twenty-one cycles of change within the Pleistocene. The pollen changes in each cycle suggested that throughout the Pleistocene there had been regular oscillations of either warm and cold conditions or humid and dry ones. At a number of sites in Czechoslovakia where thick deposits of loess had accumulated during colder stages of the Pleistocene, Kukla [49,50] found regular banding within the loess that he interpreted as recurring cycles of soil development in response to warm/cold oscillations. Seventeen such climatic cycles were identified [51]. Although Kukla thus recorded slightly fewer cycles of change than did Ronai, the difference may not be a real one, but merely a result of uncertainties in the positioning of the Pliocene/Pleistocene boundary in the two sequences. Similar uncertainty pertains to Hooghiemstra's record [52] of twenty-seven alternating cold and warm pollen stages from a deep borehole through Pleistocene lake sediments in the high Andes of Colombia, South America.

Other workers have attempted a cruder subdivision of the Pleistocene, into Lower, Middle and Upper, but even these terms have not been used consistently. In many Pleistocene sequences marked changes are certainly apparent part-way through the period, manifested in temperate latitudes either as the occurrence of lowland glaciation (the change from 'pre-glacial' to 'glacial' Pleistocene [4]), the appearance of 'arctic' mammals (*Coelodonta, Ovibos, Rangifer, Mammuthus*), or the final demise of Tertiary 'exotics' (the trees *Carya, Eucommia, Pterocarya* and *Tsuga*). In deep-ocean cores, as we have seen, changing patterns are evident about 800 ka ago (Figure 5.2). If a formal division into Lower and Middle Pleistocene is to be made, Butzer and Isaacs' suggestion [53] that the boundary be placed at the base of the Brunhes Geomagnetic Epoch (730 ka ago) is an eminently sensible pragmatic one. The Upper Pleistocene comprises the last cold/warm cycle – extending from about 130 ka ago to 10 ka ago (Table 5.2).

5.4 CLIMATIC CHANGES DURING A GLACIAL/INTERGLACIAL CYCLE

Deep-ocean sediments spanning the Upper and Middle Pleistocene (the last 730 ka) consistently show eight or nine regularly-spaced cycles in oxygen-isotope values that record the recurrent waxing and waning of the terrestrial ice sheets. The mean length of each cycle must thus have been 80–90 ka. In many ocean cores, particularly those from lower latitudes, each cycle is typically asymmetric, with a gradual but fluctuating decline to minimum $\delta^{18}O$ values, followed by a sudden increase to a rather short-lived maximum. This 'saw-tooth' pattern is interpreted as representing a protracted build-up of ice to a glacial maximum, followed by rapid deglaciation and an early but brief temperature optimum in the ensuing interglacial. The sudden increase in oxygen-isotope values at the close of each cycle (deglaciation) is referred to by palaeoceanographers as a **termination** [54]. A similar phenomenon can be seen in the Czechoslovakian loess sequences studied by Kukla [49], where each cycle is delimited above by a pronounced 'markline', resulting from an abrupt transition from loess deposited under polar desert conditions to overlying temperate forest soil – and hence a rapid change from glacial to interglacial conditions. The subsequent cooling phase, represented by successive steppe and tundra soils, is more drawn-out.

A consequence of this temperature pattern is that optimum interglacial conditions, rather similar to those occurring globally today, can have occupied only a small part of the timespan of each cycle [55]. Thus Kellogg [56] found in ocean cores from the Norwegian Sea, at latitude 65°N, that the planktonic fauna for most of the last 150 ka was dominated by foraminifers found today in cold polar seas much further north; the current foraminiferal assemblage in the Norwegian Sea has been present only since about 12 ka ago, and occurred otherwise only as a brief pulse between 127 and 110 ka ago. Bé and Duplessy [57] similarly found that off the east coast of Africa there were only four short episodes recorded in the ocean cores over the last 540 ka where temperatures were reckoned to have been as high as at the present day. In the earlier part of the Quaternary (the Lower Pleistocene – before 730 ka ago), where inferred climatic fluctuations were of smaller amplitude (Figure 5.2), such optimal conditions might have been even less frequent. It is likely, therefore, that less than one-fifth, and probably as little as one-twentieth, of the Quaternary Period enjoyed climatic conditions as favourable as those today.

The oxygen-isotope curves of sediments with a faster-than-average rate of accumulation through the last glacial/interglacial cycle show additional smaller fluctuations within the cycle (Figure 5.3). The fluctuations portray

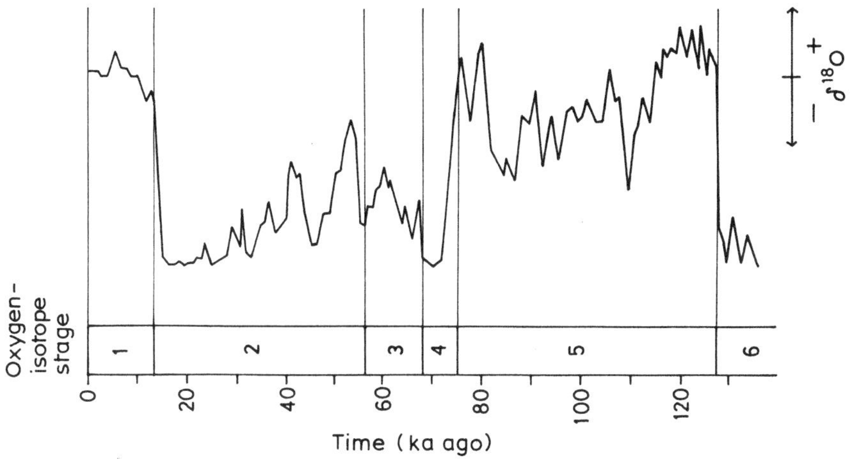

Figure 5.3 Oxygen-isotope analyses of ocean sediments spanning the last 130 ka. Re-drawn from Bradley [13].

a complex series of changes in ice volume (and thus presumably in climate) over the last 128 ka. The picture is a global one, however, and cannot easily be translated into precise temperature terms, as the conditions governing the expansion and contraction of the world's ice sheets are complex; particularly during times of rapid ice change, the $\delta^{18}O$ values may bear no close relationship to sea-surface temperature [58,59]. Hence we would not expect any necessary connection between this global marine record and the terrestrial record – which gives a strictly regional picture of changes during the last glacial/interglacial cycle. Nevertheless, similarities can be detected. Thus in the Netherlands up to four thin peaty bands are included within aeolian 'coversands' deposited to the south of the Scandinavian Ice Sheet during the last glaciation. On the basis of pollen analyses and the character of the sediments, van der Hammen *et al.* [60] proposed a subdivision of the last glacial into a cool **Early Glacial**, with two mild interstadials* (earlier than 60 ka ago), and a succeeding colder **Pleniglacial**, with two cool interstadials within it. The most severe conditions occurred in the Upper Pleniglacial, and were associated with the maximum expansion of the Scandinavian Ice Sheet. These and comparable records from other regions [61–63] all indicate that conditions of extreme cold (and perhaps also of aridity) during a glacial/interglacial cycle were as restricted in duration as peak interglacial conditions. The detailed character

* An interstadial is a period of milder climate within a glacial stage that fails to develop the full interglacial vegetation characteristic of a particular latitude – either because interglacial temperatures are not achieved or because the interstadial is of insufficient duration.

of climatic changes within each cycle revealed by the records suggests that plants at higher latitudes would have been called on to respond not to gradual pendulum swings between extreme cold and warmth spread over perhaps 50 ka, but to a complex series of shorter-duration oscillations, included within which were some periods of extremely rapid change – when temperatures could have changed by 6–10°C in 1 ka or less. This aspect is considered further in Chapter 8.

5.5 THE MILANKOVITCH MECHANISM

Climatic cycles of varying duration and amplitude have been identified from the Cenozoic era. The most striking are the regular cold/warm oscillations ('glacial cycles') of the Middle and Upper Pleistocene, with a wavelength of about 100 ka and an amplitude (on land) of at least 6–8°C. Within the most recent glacial cycle, and almost certainly within earlier ones as well, are further small oscillations (stadials and interstadials); at least four have been described in both oceanic and terrestrial sequences. Where long records of oscillatory change are available, as in ocean cores, the technique of *spectral analysis* can be used to characterize statistically the mean wavelengths of the fluctuations. Various such analyses have now been performed on ocean cores, using oxygen-isotope measurements, calcium carbonate concentrations, frequencies of particular 'indicator' protistan species, and the relative proportions of distinctive foraminiferan or radiolarian assemblages, as estimators of ocean temperatures over periods of several hundred thousand years. All the analyses identify cycle lengths of about 100 ka, 41–42 ka and 19–23 ka [64]; some also give a muted signal at 400 ka [65].

The possibility that regular variations in the geometry of the earth's orbit around the sun, such as are known to occur, could lead to cyclical variations in climate, was demonstrated theoretically by the Yugoslav astronomer Milutin Milankovitch in the 1920s. The maxima and minima of the cycles he calculated probably correspond to present-day and ice-age conditions, respectively. These geometrical variations include changes in the precise shape of the earth's elliptical orbit and in the degree of tilt of the earth's axis of rotation (for details, see Imbrie and Imbrie [66] and Ruddiman and Wright [59]), and generate changes in the seasonal and geographical distribution of solar radiation over the earth's surface. Milankovitch showed that the periodicity of these phenomena led to solar-insolation cycles of 21–25 ka ('precession of the equinoxes'), 41 ka ('obliquity of the ecliptic') and 90–100 ka ('eccentricity of the orbit'). At the time Milankovitch propounded his theory, geological data were not adequate for it to be critically evaluated, and the theory was not widely accepted. Only later, when undisturbed sediment cores from the deep oceans became available for study was it realized that Milankovitch's predictions agreed closely with the ice-volume (= climatic) oscillations recorded in those cores (with wavelengths of 19–23 ka, 41–42 ka and about 100 ka), and hence that some of the climatic changes during the Quaternary Period could be explained on the basis of changes in the way in which the earth orbited the sun [67]. For the earlier part of the Quaternary Period the predominant rhythm was one of 41 ka; over the last 700–900 ka it has been the 100-ka cycle, with the shorter-length and weaker cycles superimposed [59]. Moreover, the rhythms recorded in the ocean cores were different at high and low latitudes: at higher latitudes the 100-ka and 41-ka cycles were the strongest signal, whereas at lower latitudes the 23-ka cycle predominated. This too is what Milankovitch theory would predict [22,30].

The Milankovitch mechanism does not involve changes in the total solar radiation input to the earth's atmosphere, only changes in the geographical and seasonal distribution of that radiation. The predicted seasonal pattern of input at latitudes 45° and 65°N over the last 130 ka shows periods of strongly seasonal climate (i.e., large differences between January and July temperatures) at the start of the last interglacial stage and the Holocene, whilst glacial maximum at about 20 ka ago is associated with much reduced seasonality [68]. The initiation of all interglacial periods is probably associated with a summer maximum and a winter minimum in radiation input [69].

There is growing acceptance among palaeoecologists now that the Milankovitch mechanism underlies many of the climatic features and episodes of the Quaternary Period. However, seasonal variation in solar radiation input, by itself, does not appear to be sufficient to explain the most prominent feature of the oxygen-isotope record of the later Pleistocene, the sudden deglaciations that occurred approximately every 100 ka. Additional feedback mechanisms, such as changes in atmospheric CO_2 concentrations, moisture starvation to the waning ice sheets, and iceberg calving from their margins, are currently being considered [59]. Moreover, the Milankovitch mechanism probably never allows the development of a completely ice-free planet, such as is known to have existed in the Early Tertiary Period. Different mechanisms, as yet unclear, must be sought for the climatic changes then.

REFERENCES

1. Mercer, J. H. (1984) Late Cainozoic glacial variations in South America south of the equator, in *Late Cainozoic Palaeoclimates of the Southern Hemisphere* (ed. J. C. Vogel), Balkema, Rotterdam, pp. 45–58.

2. Ruddiman, W. F. and Raymo, M. E. (1988) Northern hemisphere climatic regimes during the past 3 Ma: possible tectonic connections. *Phil. Trans. R. Soc. Lond. B*, **318**, 411–30.
3. Curry, R. R. (1966) Glaciation about 3,000,000 years ago in the Sierra Nevada. *Science*, **154**, 770–1.
4. West, R. G. (1977) *Pleistocene Geology and Biology*, 2nd edn, Longman, London.
5. Haq, B. U., Hardenbol, J. and Vail, P. R. (1987) Chronology of fluctuating sea levels since the Triassic. *Science*, **235**, 1156–7.
6. Tauxe, L., Opdyke, N. D., Pasini, G. and Elmi, C. (1983) Age of the Plio – Pleistocene boundary in the Vinca section, southern Italy. *Nature*, **304**, 125–9.
7. Sarnthein, M. (1978) Sand deserts during glacial maximum and climatic optimum. *Nature*, **272**, 43–6.
8. Lancaster, N. (1984) Aridity in southern Africa: age, origins and expression in landforms and sediments, in *Late Cainozoic Palaeoclimates of the Southern Hemisphere* (ed. J. C. Vogel), Balkema, Rotterdam, pp. 433–44.
9. Thomas, D. S. G. and Goudie, A. S. (1984) Ancient ergs of the Southern Hemisphere, in *Late Cainozoic Palaeoclimates of the Southern Hemisphere* (ed. J. C. Vogel), Balkema, Rotterdam, pp. 407–18.
10. Wasson, R. J. (1984) Late Quaternary palaeoenvironments in the desert dunefields of Australia, in *Late Cainozoic Palaeoclimates of the Southern Hemisphere* (ed. J. C. Vogel), Balkema, Rotterdam, pp. 419–32.
11. Derbyshire, E. (1979) Glaciers and environment, in *The Winters of the World* (ed. B. John), David and Charles, Newton Abbot, Devon, pp. 58–106.
12. Flint, R. F. (1971) *Glacial and Quaternary Geology*, Wiley, New York.
13. Bradley, R. S. (1985) *Quaternary Paleoclimatology. Methods of Paleoclimatic Reconstruction*, Allen and Unwin, Winchester, MA.
14. Rind, D. and Peteet, D. (1985) Terrestrial conditions at the last glacial maximum and CLIMAP sea-surface temperature estimates: are they consistent? *Quat. Res.*, **24**, 1–22.
15. Hamilton, A. C. (1982) *Environmental History of East Africa. A Study of the Quaternary*, Academic Press, London.
16. Lorius, C., Merlivat, L., Jouzel, J. and Pourchet, M. (1979) A 30,000-yr isotope climatic record from Antarctic ice. *Nature*, **280**, 644–8.
17. Robin, G. de Q. (1977) Ice cores and climatic change. *Phil. Trans. R. Soc. Lond. B*, **280**, 143–68.
18. Peterson, G. M., Webb, T., Kutzbach, J. E. *et al.* (1979) The continental record of environmental conditions at 18,000 yr bp: an initial evaluation. *Quat. Res.*, **12**, 47–82.
19. Washburn, A. L. (1979) Permafrost features as evidence of climatic change. *Earth Sci. Rev.*, **15**, 327–402.
20. Newsome, J. and Flenley, J. R. (1988) Late Quaternary vegetational history of the Central Highlands of Sumatra. II. Palaeopalynology and vegetational history. *J. Biogeogr.*, **15**, 555–78.
21. Whitmore, T. C. (ed.) (1981) *Wallace's Line and Plate Tectonics*, Clarendon Press, Oxford.
22. Ruddiman, W. F. (1985) Climate studies in ocean cores, in *Paleoclimate Analysis and Modeling* (ed. A. D. Hecht), Wiley, New York, pp. 197–257.

23. Colinvaux, P. (1979) The Ice-age Amazon. *Nature*, **278**, 399–400.
24. Damuth, J. E. and Fairbridge, R. W. (1970) Equatorial Atlantic deep-sea arkosic sands and Ice-Age aridity in tropical South America. *Geol. Soc. Am. Bull.*, **81**, 189–206.
25. Lancaster, N. (1981) Paleoenvironmental implications of fixed dune systems in southern Africa. *Palaeogeogr. Palaeoclimatol. Palaeoecol.*, **33**, 327–46.
26. Bonatti, E. and Gartner, S. Jr (1973) Caribbean climate during Pleistocene ice ages. *Nature*, **244**, 563–5.
27. Parmenter, C. and Folger, D. W. (1974) Eolian biogenic detritus in deep sea sediments: a possible index of equatorial Ice Age aridity. *Science*, **185**, 695–8.
28. Pokras, E. M. and Mix, A. C. (1985) Eolian evidence for spatial variability of Late Quaternary climates in tropical Africa. *Quat. Res.*, **24**, 137–49.
29. Street, F. A. and Grove, A. T. (1979) Global maps of lake-level fluctuations since 30,000 yr bp. *Quat. Res.*, **12**, 83–118.
30. Hecht, A. D. (1985) Paleoclimatology: a retrospective of the last 20 years, in *Paleoclimate Analysis and Modeling* (ed. A. D. Hecht), Wiley, New York, pp. 1–25.
31. Spaulding, W. G. and Graumlich, L. J. (1986) The last pluvial climatic episodes in the deserts of southwestern North America. *Nature*, **320**, 441–4.
32. Street-Perrott, F. A. and Harrison, S. P. (1985) Lake levels and climate reconstruction, in *Paleoclimate Analysis and Modeling* (ed. A. D. Hecht), Wiley, New York, pp. 291–341.
33. Kutzbach, J. E. and Guetter, P. J. (1986) The influence of changing orbital parameters and surface boundary conditions on climate simulations for the past 18,000 years. *J. Atmos. Sci.*, **43**, 1726–59.
34. Van Campo, M. (1984) Relations entre la végétation de l'Europe et les températures de surface océaniques après le dernier maximum glaciaire. *Pollen Spores*, **26**, 497–518.
35. Markgraf, V. (1989) Reply to C. J. Heusser's 'Southern Westerlies during the Last Glacial Maximum'. *Quat. Res.*, **31**, 426–32.
36. Van Zinderen Bakker, E. M. (1982) African palaeoenvironments 18,000 yrs bp. *Palaeoecol. Africa*, **15**, 77–99.
37. Salinger, M. J. (1984) New Zealand climate: the last 5 million years, in *Late Cainozoic Palaeoclimates of the Southern Hemisphere* (ed. J. C. Vogel), Balkema, Rotterdam, pp. 131–50.
38. Michelson, D. M., Clayton, L., Fullerton, D. S. and Borns, H. W., Jr (1983) The Late Wisconsin glacial record of the Laurentide Ice Sheet in the United States, in *Late-Quaternary Environments of the United States*, Vol. 1, *The Late Pleistocene* (ed. S. C. Porter), Longman, London, pp. 3–37.
39. Harrison, S. P., Metcalfe, S. E., Street-Perrott, F. A. *et al.* (1984) A climatic model of the last Glacial/Interglacial transition based on palaeotemperature and palaeohydrological evidence, in *Late Cainozoic Palaeoclimates of the Southern Hemisphere* (ed. J. C. Vogel), Balkema, Rotterdam, pp. 21–34.
40. Tsukada, M. (1967) Vegetation in subtropical Formosa during the Pleistocene glaciations and the Holocene. *Palaeogeogr. Palaeoclimatol. Palaeoecol.*, **3**, 49–64.
41. Van Donk, J. (1976) O^{18} record of the Atlantic Ocean for the entire Pleistocene Epoch, in *Investigation of Late Quaternary Paleoceanography and Paleoclimatology*

(eds R. M. Cline and J. D. Hays), Geological Society of America, Boulder, CO, pp. 147–63.

42. Shackleton, N. J. and Opdyke, N. D. (1976) Oxygen-isotope and paleomagnetic stratigraphy of Pacific core V28-239 Late Pliocene to latest Pleistocene, in *Investigation of Late Quaternary Paleoceanography and Paleoclimatology* (eds R. M. Cline and J. D. Hays), Geological Society of America, Boulder, CO, pp. 449–64.
43. Imbrie, J., Hays, J. D., Martinson, D. G. *et al.* (1984) The orbital theory of Pleistocene climate: support from a revised chronology of the marine $\delta^{18}O$ record, in *Milankovitch and Climate,* Part I (eds A. L. Berger and others), Reidel, Dordrecht, pp. 269–305.
44. Lowe, J. J. and Walker, M. J. C. (1984) *Reconstructing Quaternary Environments,* Longman, London.
45. De Jong, J. (1988) Climatic variability during the past three million years, as indicated by vegetational evolution in northwest Europe and with emphasis on data from the Netherlands. *Phil. Trans. R. Soc. Lond. B,* **318**, 603–17.
46. Zagwijn, W. H. (1985) An outline of the Quaternary stratigraphy of the Netherlands. *Geol. Mijn.*, **64**, 17–24.
47. Cooke, H. B. S. (1973) Pleistocene chronology: long or short? *Quat. Res.,* **3** 206–20.
48. Ronai, A. (1970) Lower and Middle Pleistocene flora in the Carpathian Basin. *Palaeogeogr. Palaeoclimatol. Palaeoecol.*, **8**, 265–85.
49. Kukla, G. J. (1975) Loess stratigraphy of central Europe, in *After the Australopithecines: Stratigraphy, Ecology and Culture Change in the Middle Pleistocene* (eds K. W. Butzer and G. L. Isaacs), Mouton, The Hague, pp. 99–188.
50. Kukla, G. J. (1977) Pleistocene land–sea correlations. *Earth Sci. Rev.,* **13** 307–74.
51. Fink, J. and Kukla, G. J. (1977) Pleistocene climates in central Europe: at least 17 interglacials after the Olduvai Event. *Quat. Res.*, **7**, 363–71.
52. Hooghiemstra, A. (1984) A palynological registration of climatic change of the last 3.5 million years, in *Milankovitch and Climate* (eds A. L. Berger and others), Reidel, Dordrecht, pp. 371–7.
53. Butzer, K. W. and Isaacs, G. L. (1975) *After the Australopithecines: Stratigraphy, Ecology and Culture Change in the Middle Pleistocene,* Mouton, The Hague.
54. Broecker, W. S. and van Donk, J. (1970) Insolation changes, ice volumes, and the O^{18} record in deep-sea cores. *Rev. Geophys. Space Physics,* **8**, 169–98.
55. Emiliani, C. (1955) Pleistocene temperatures. *J. Geol.*, **63**, 538–78.
56. Kellogg, T. B. (1976) Late Quaternary climatic changes: evidence from deep-sea cores of Norwegian and Greenland Seas. *Mem. geol. Soc. Am.*, **145**, 77–110.
57. Bé, A. W. H. and Duplessy, J. C. (1976) Subtropical convergence fluctuations and Quaternary climates in the middle latitudes of the Indian Ocean. *Science,* **194**, 419–22.
58. Ruddiman, W. F. and McIntyre, A. (1979) Warmth of the subpolar North Atlantic Ocean during Northern Hemisphere ice-sheet growth. *Science,* **204**, 173–5.
59. Ruddiman, W. F. and Wright, H. E., Jr (1987) Introduction, in *North America and Adjacent Oceans during the Last Deglaciation* (eds W. F. Ruddiman and H. E. Wright, Jr), Geological Society of America, Boulder, CO, pp. 1–12.

60. Van der Hammen, T., Maarleveld, G. C., Vogel, J. C. and Zagwijn, W. H. (1967) Stratigraphy, climatic succession and radiocarbon dating of the Last Glacial in the Netherlands. *Geol. Mijn.*, **46**, 79–95.
61. Benson, L. V. (1978) Fluctuation in the level of pluvial Lake Lahontan during the last 40,000 years. *Quat. Res.*, **9**, 300–18.
62. Farrand, W. R. (1981) Pluvial climates and frost action during the last glacial cycle in the Eastern Mediterranean – evidence from archaeological sites, in *Quaternary Paleoclimate* (ed. W. C. Mahaney), Geo Abstracts, Norwich, pp. 393–410.
63. Van der Hammen, T., Barelds, J., de Jong, H. and de Veer, A. A. (1981) Glacial sequence and environmental history in the Sierra Nevada del Coquy (Colombia). *Palaeogeogr. Palaeoclimatol. Palaeoecol.*, **32**, 247–340.
64. Hays, J. D., Lozano, J. A., Shackleton, N. and Irving, G. (1976) Reconstruction of the Atlantic and western Indian Ocean sectors of the 18,000 B.P. Antarctic Ocean, in *Investigation of Late Quaternary Paleoceanography and Paleoclimatology* (eds R. M. Cline and J. D. Hays), Geological Society of America, Boulder, CO, pp. 337–72.
65. Moore, T. C., Pisias, N. G. and Dunn, D. A. (1982) Carbonate time series of the Quaternary and Late Miocene sediments in the Pacific Ocean: a spectral comparison. *Marine Geol.*, **46**, 217–33.
66. Imbrie, J. and Imbrie, K. P. (1979) *Ice Ages. Solving the Mystery*, Macmillan, London.
67. Denton, G. H. and Hughes, T. J. (1983) Milankovitch theory of ice ages: hypothesis of ice-sheet linkage between regional insolation and global climate. *Quat. Res.*, **20**, 125–44.
68. Bonnichsen, R., Stanford, D. and Fastook, J. L. (1987) Environmental change and developmental history of human adaptive patterns; the Paleoindian case, in *North America and Adjacent Oceans During the Last Deglaciation* (eds W. F. Ruddiman and H. E. Wright, Jr), Geological Society of America, Boulder, CO, pp. 403–24.
69. Watts, W. A. (1988) Europe, in *Vegetation History* (eds B. Huntley and T. Webb III), Kluwer Academic Publishers, Dordrecht, pp. 155–92.

CHAPTER SIX

Environmental change – the Holocene

6.1 INTRODUCTION

For several thousand years following maximum glaciation, temperatures oscillated rather irregularly, and the major ice sheets contracted and expanded in a complex fashion several times. The last pronounced cold pulse in many regions occurred 14–15 ka ago [1,2]. Thereafter, deglaciation proceeded rapidly. The southern-hemisphere ice sheets had shrunk to their present extent by about 11 ka ago [1], and the Scandinavian Ice Sheet had almost vanished by 9 ka ago [3]. The Laurentide Ice Sheet in North America persisted rather longer, but it too had largely vanished by 6.5 ka ago [4]. The oxygen-isotope curves in Figure 6.1 show this pattern very clearly, with the major increase in $\delta^{18}O$ values accompanying deglaciation occupying only 2–3 ka, between about 14 and 11.5 ka ago [5,6].

In several parts of the world temperatures plummetted again about 11 ka ago, and did not recover for another 1 ka. This short cold period was originally recorded from north-west Europe as the 'Younger *Dryas* period' or 'Loch Lomond Stadial', and was widely regarded as unique to the region [7]. Convincing arguments have been put forward as to why this might have been so [8]. However, comparable climatic reversals for the period 9.5–11.5 ka ago are now known also from the northern United States [9], Canada [10], Peru [1] and Chile [11,12], whilst in both ocean cores and ice cores there are claims of a temporary check in the pattern of rapidly-changing $\delta^{18}O$ values between 11.5 and 10.7 ka ago [5,6,13]. The preceding period of warmer climate from about 14 to 11 ka ago has been called the 'Lateglacial Interstadial' [3], and was itself of complex character in parts of north-west Europe, where it included two intervals of warmer climate (the Bölling and Allerød Oscillations) and an intervening cold period (the Older *Dryas*). These variations epitomize the non-uniform nature of the deglaciation process.

Somewhere along the rapidly-rising temperature curve accompanying deglaciation occurs the start of the Holocene Epoch. This is pragmatically

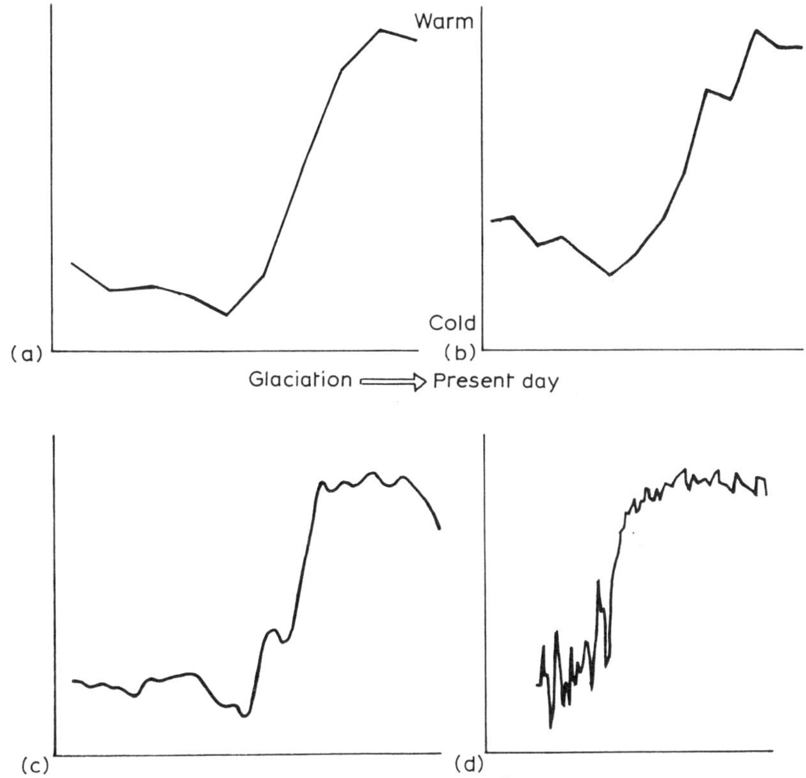

Figure 6.1 Oxygen-isotope analyses of sediments spanning the last 25 ka, from (a) Caribbean core V12-122 (re-drawn from Imbrie *et al.* [14]); (b) equatorial Pacific core V28-238 (re-drawn from Shackleton and Opdyke [15]); (c) Dome C, Antarctica (re-drawn from Lorius *et al.* [16]); and (d) Camp Century, Greenland ice-sheet (re-drawn from Dansgaard *et al.* [17]). The vertical axis is equivalent to a temperature scale and the horizontal axis to a time scale, but the four curves are not calibrated to a common time and temperature scale.

placed at 10 000 yr bp. To some extent such a separation creates an artificial termination to the process of deglaciation, though in many parts of the world major climatic change is indeed recognizable at this time – as recovery from the cold 'Loch Lomond stadial' or from glacial-stage aridity, for example. Nevertheless, further short cold snaps, marked by limited re-advance of mountain glaciers, occurred in some regions 9–10 ka ago and around 8 ka ago, and it was not until after 8000 yr bp that the world's residual ice sheets shrank to a position of approximate stability that was then maintained for nearly 3 ka.

The deep-ocean oxygen-isotope record shows unequivocally that changes over the last 18 ka have been similar to those occurring in the earlier stages of previous glacial/interglacial cycles (sections 5.3 and 5.4). Peak interglacial conditions, achieved early in the Holocene after the rapid initial rise in temperature accompanying deglaciation, have now already been passed, and the predicted course of change should be a slow progression towards a new ice age (Figure 5.3). That prediction is unlikely to be correct, as explained in Chapter 15. Nevertheless, global trends prior to the present century conform broadly to the pattern stated.

The unique character of the Holocene, that differentiates it from all earlier interglacials of the Quaternary, is the increasing prevalence of human interference with the environment. That aspect is considered in more detail in Chapters 12, 13 and 15.

6.2 CLIMATIC CHANGE

6.2.1 Glacial pulses

The process of deglaciation in most countries effectively terminated about 8 ka ago, after a minor glacial advance that is recorded from the Swiss and New Zealand Alps; residues of the Laurentide Ice Sheet in north-eastern North America, however, lingered on until about 6.5 ka ago [4]. Over the next 3 ka only one further glacial re-advance is known to have occurred – the so-called Misox or Frosnitz Oscillation, in the Swiss and Austrian Alps, about 6–7 ka ago [18]. Otherwise, the world's glaciers remained nearly quiescent, and reduced in extent to values similar to the minima of earlier interglacials. The first major episode of renewed glacial activity, heralding the onset of what has been called the **Neoglacial** period, occurred from about 5300 to 4000 yr bp, and was followed by further pulses between about 2800 and 2200 yr bp, 1900 and 1200 yr bp, and 650 and 200 yr bp* [1,9,18,20–22]. In the European Alps these glacial phases have been designated the Piora, Göschenen I and Göschenen II Oscillations, and the Little Ice Age, respectively.

6.2.2 The oxygen-isotope record

Examples of fine-resolution oxygen-isotope series from land sediments spanning the full duration of the Holocene are shown in Figure 6.1, for ice cores from Antarctica and from Greenland. The general patterns of change in the two cores are very similar: an initial rise in $\delta^{18}O$ values beginning

* These are radiocarbon dates; calibrated to 'calendar years' AD or BC [19], they are approximately 4200–2500 BC, 100–200 BC, 50–750 AD and 1300–1750 AD.

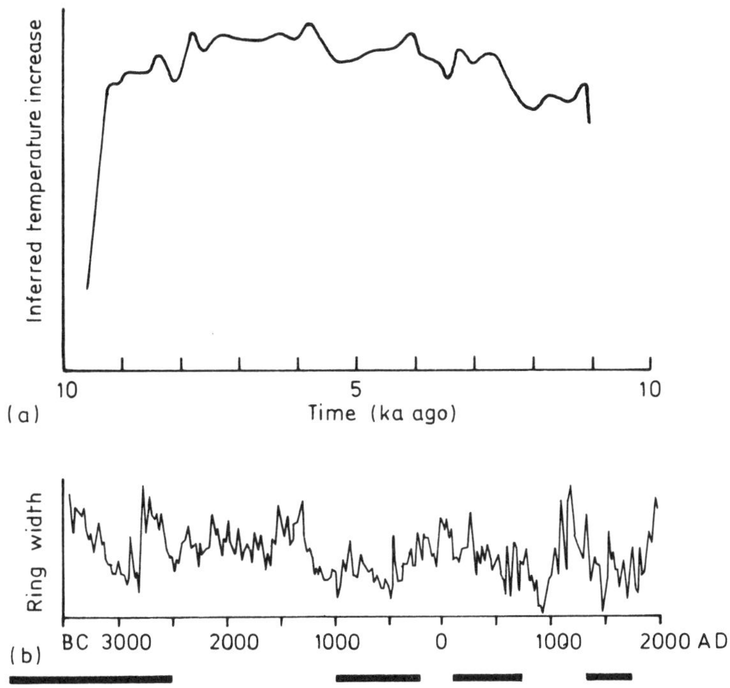

Figure 6.2 Proxy data of climatic change in the Holocene. (a) Oxygen-isotope analyses of calcareous sediments in a southern Swedish lake (re-drawn from Mörner and Wallin [25]). (b) Tree-ring widths (20-year means) in bristlecone pine, California (re-drawn from Lamb [29]). The horizontal bars below the tree-ring record are the approximate dates of glacial pulses, calibrated to the same timescale (footnote to section 6.2.1.).

about 14 ka ago [23], a subsequent period of maximum values, and a final period of slow decline. Despite the problems associated with the accurate dating of ice cores [19], it is nevertheless clear that the date of the (inferred) temperature maximum is different at the two sites: 8–11 ka ago at the southern-hemisphere site [16], and 4.6–8.0 ka ago at the northern-hemisphere site [24]. These differences presumably reflect differences in the overall patterns of change in ice volume in the two hemispheres. By contrast, the curve in Figure 6.2(a) – of $\delta^{18}O$ values for lake sediments in southern Sweden [25] – gives a strictly regional picture. At this site the $\delta^{18}O$ values were largely unaffected by local ice-volume changes after 9.3 ka bp, and are consequently a closer record of temperature changes through the Holocene – though probably reflecting mean summer temperatures rather than mean annual temperatures. The general pattern of change is similar to that in the two ice cores, but the period of maximum values extends from 5.8 to 7.8 ka bp (when mean summer temperatures may have been about 2°C higher than at present).

On theoretical considerations, a peak in solar radiation input in the summer months in the northern hemisphere would be expected between about 12 and 9 ka ago [26–28], and in regions distant from the waning ice sheets there is indeed some evidence for elevated temperatures at this time. A delayed date of the temperature optimum is recorded, however, for many northern-hemisphere sites, and may be the result of proximity to the long-persistent residues of the continental ice sheets [4], or, in the case of high-altitude sites, of the slow migration rates of certain of the subalpine tree taxa forming the tree-line. For the southern hemisphere the situation is less clear: warming of land surfaces here may have been delayed generally until after 9 ka bp, and even then the effect was probably of smaller magnitude than that in the northern hemisphere [27].

6.2.3 The tree-ring record

An even higher-resolution proxy record of climatic change over long timespans is provided by the growth-ring record of certain long-lived trees [30]. That of bristlecone pine (*Pinus longaeva*) growing in the White Mountains of California is shown in Figure 6.2. The striking fluctuations in annual ring-width of this slow-growing subalpine tree are thought to be primarily a response to variations in warm-season temperature [31]. The tree-ring record shows clearly the abrupt nature of some of the temperature changes in the last 5 ka, albeit in a climatically sensitive site: for example, the marked decreases in temperature that set in about 3400, 2600 and 1300 BC, and 300 and 1200 AD; and the intervening recoveries, particularly at about 2800 BC and 950 AD (solar calendar, not radiocarbon calendar [19]).

Although the primary picture of the bristlecone pine record relates to the American Southwest, similarities are apparent in Figure 6.2 between the timing of the temperature changes and those of the neoglacial ice pulses (provided the radiocarbon dates of the glacier events have been calibrated to bring them into line with the growth-ring calendar). The most recent major decline in the bristlecone pine record, beginning about 1200 AD, is also visible in the temperature record for the last millennium for southern England compiled from documentary evidence by Lamb [32] – where the decline led ultimately to the onset of the so-called 'Little Ice Age' (Figure 8.2). There is thus some reason to hope that the bristlecone pine record may give a picture of temperature changes that is broadly applicable over quite a large geographical area.

6.2.4 The lake-level record

Associated with these temperature changes during the Holocene Epoch were equally pronounced variations in precipitation. In part, the variations can be explained by the increased amounts of water in global circulation following deglaciation. As the ice sheets melted, so sea levels rose, and so too did lake levels in many parts of the world, particularly in endoreic basins (interior landlocked basins where water loss is almost entirely due to evaporation). However, lake levels in other regions fell at the same time, as Figure 6.3 shows. This diagram shows lake levels in a series of lake basins in tropical Africa and in the western United States over a timespan encompassing the Holocene, the preceding deglaciation (when lake levels in Africa rose dramatically, but those in the United States fell), and the last glacial maximum. The records for the two regions are almost mirror-image

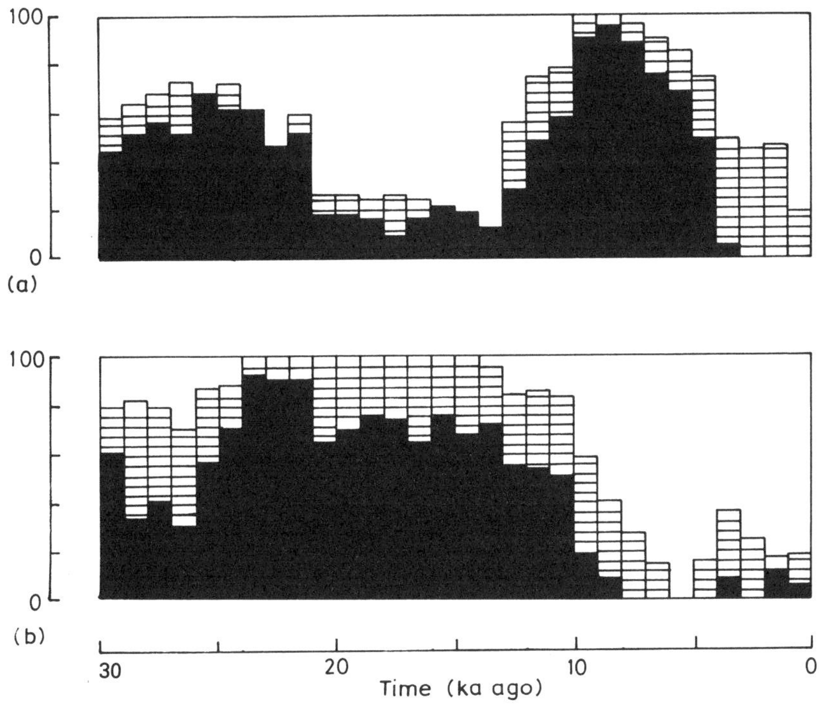

Figure 6.3 Lake levels over the past 30 ka in (a) tropical Africa, (b) the western United States. In each histogram is shown the proportion of lakes in a 1-ka radiocarbon-dated time-interval with high (black), intermediate (shaded) and low (white) lake levels. Re-drawn from Street and Grove [33] and Smith and Street-Perrott [34].

versions of each other, with alternating periods of high and low lake levels in both regions. If the time-interval occupied by deglaciation is considered as well as the 10-ka timespan of the Holocene, then a tripartite division of the last 14 ka can be clearly recognized in Figure 6.3: a period of rapidly-changing lake levels from 14 to 9 ka ago; a period of rather stable lake levels from 9 to 5 ka ago; and a period of more irregular change over the last 5 ka.

At the present day the world's lakes (on an areal basis) are concentrated into three latitudinal belts, centred on 40°N, 38°S and 6°N. These correspond to the wet climatic belts influenced by the northern and southern hemisphere Westerlies, and the equatorial trough, respectively. Few lakes are found in the subtropical high-pressure belts, between about latitudes 20 and 35° in both hemispheres [35]. The patterns of change in lake levels over time shown in Figure 6.3 might thus be seen as the result in part of shifts in the position and extent of these climatic belts. Thus the formation of a wide and strongly-developed belt of enlarged lakes in the tropics in the Early Holocene, which reached its northernmost position centred on latitude 14°N about 6 ka ago, was the result of a progressive withdrawal northwards of the subtropical high-pressure belt in the northern hemisphere, coupled with an increased monsoonal precipitation in the lake belt. Over the same time-period, the westerly-wind belt in the northern hemisphere was displaced polewards from a mean position at latitude 35°N at 18 ka ago to about 55°N at 6 ka ago [35], and as a result the western United States became progressively more arid. These displacements can be related to variations in the thermal gradient between equator and poles through the Holocene, consequent upon the overall patterns of temperature change revealed by the ice-core records, as discussed below.

6.2.5 Thermal gradients

At the present day the thermal gradient (ΔT) between equator and pole is of greater magnitude for the southern hemisphere than for the northern hemisphere [19]. As a result, the mean annual positions of the major air-pressure belts are displaced towards the northern hemisphere. The difference in ΔT between the two hemispheres is a consequence of the greater volume of ice in south polar regions. At glacial maximum 18–20 ka ago, when ice volumes were similar over both poles (Table 5.1), the difference in ΔT was minimized, and the air-pressure belts were almost symmetrically disposed between the two hemispheres – though compressed meridionally towards the equator because of the presence of continental ice sheets at higher latitudes. When these ice sheets melted the air-pressure belts expanded in width again; but because deglaciation was more extreme in the northern hemisphere, the difference in ΔT between

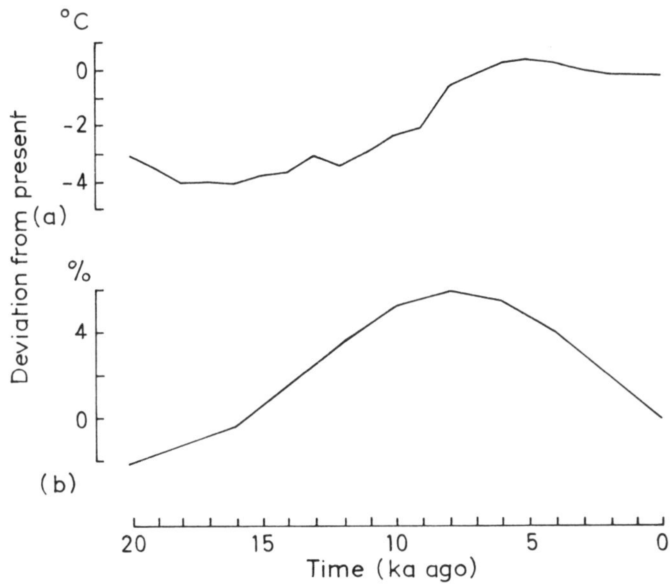

Figure 6.4 Thermal input and thermal gradients over the last 20 ka. (a) Calculated differences between the northern and southern hemisphere thermal gradients, equator/pole; values are expressed as deviations from present-day values. Re-drawn from Harrison *et al.* [36]. (b) Calculated inputs of solar energy for the summer months (June–August), based on Milankovitch theory, for latitude 65°N; values are expressed as deviations from present-day values, and broadly parallel those for seasonality of climate. Data of Newell and Hsiung [28].

the two hemispheres was accentuated, and the pressure belts were displaced asymmetrically towards the northern hemisphere.

Theoretical calculations have been carried out to try and assess the precise character and timing of these changes. Harrison *et al.* [36] used a range of palaeotemperature estimates for land and sea surfaces over the last 20 ka to derive values of ΔT for the northern and southern hemispheres. The temperature estimates were aggregated into 30° latitudinal bands, and averaged using an area-weighting for the relative proportions of land and sea in each band. A single mean value of palaeotemperature for each latitudinal band was thus produced, and from these values ΔT was calculated. The difference in ΔT between the two hemispheres over the last 20 ka is shown in Figure 6.4. Maximum disparity between the two hemispheres was achieved about 5 ka ago – after the Laurentide Ice Sheet had finally melted from North America. The air-pressure belts must correspondingly have been displaced northwards to a Holocene maximum at this time; since then they have tended to shift southwards again.

6.3 SUBDIVISIONS OF THE HOLOCENE

Three major types of evidence have been used to subdivide the 10-ka timespan of the Holocene: climatic, cultural and pollen-analytical. Within a given geographical region, particularly one where human activity has been long-sustained, it is often easier to delimit a sequence of episodes based on cultural (i.e., land-use and settlement) or pollen (i.e., vegetation) patterns than to try and reconstruct climatic patterns. Indeed, the pollen evidence is sometimes used to try and deduce climatic changes within the region [37]. However, climate has a governing influence on both cultural and pollen patterns, and hence it is important first to establish a framework of climatic change within the region. Examples of different regional frameworks are given below.

One widely-used American scheme [38] recognizes an **Early Holocene** period, from about 10 to 8 ka bp, characterized by the lingering effects of the waning ice sheets; an **Altithermal** period (often very dry), showing the maximum expression of the interglacial climate; and a **Neoglacial** period of generally decreasing warmth and increasing wetness (climatic 'deterioration') over the last 5–6 ka. Such a framework corresponds closely to the tripartite division of the Holocene proposed many years ago by Lennart Von Post [39], on the basis of patterns of change in pollen diagrams from several different parts of the world. He recognized an early phase of increasing warmth (characterized by **protocratic** elements among the pollen taxa), a central phase of optimal warmth (characterized by **mediocratic** elements), and a recent phase of diminished warmth (characterized by the resurgence of certain protocratic pollen taxa and the appearance of new **terminocratic** ones). Other more detailed climatic subdivisions of the Holocene can generally be regarded as elaborations of these two schemes. In Europe, the sequence of **climatic periods** originally proposed at the end of the last century by Axel Blytt and Rolf Sernander is widely used, and has been closely linked to pollen-analytical, archaeological and radiometric data (Table 6.1). The **Pre-Boreal** and **Boreal** periods of this sequence are broadly equivalent to the Early Holocene of North American terminology, the **Atlantic** to the Altithermal, and the **Sub-Boreal** and **Sub-Atlantic** to the Neoglacial. The original characterization of these climatic periods by Blytt and Sernander was in a Scandinavian context, though many workers have subsequently assumed that the same characterization would be applicable also to other parts of Europe. Such a viewpoint now seems rather naive in view of the complexities of global climatic patterns. Nevertheless, if global climate does indeed form an integrated system, then we might expect regional patterns such as those described above to be just different small-scale expressions of much

Table 6.1 The climatic periods recognized for the Holocene of north-western Europe, and their relationship to pollen-analytically-determined vegetation changes (Pollen Zones) and archaeological cultures. Modified from Godwin [40]

Date (yr bp)	Climatic period	Type of climate	Pollen Zone	Major pollen taxa	Archaeological culture
10000	Pre-Boreal	Cold and dry	IV	Birch–(pine)	Mesolithic
8000	Boreal	Increasing warmth, dry	V,VI	Hazel–pine–(birch)	
6000	Atlantic	Warm and wet (oceanic)	VIIa	Alder–oak–elm–lime	
4000	Sub-Boreal	Warm, drier (continental)	VIIb	Alder–oak–lime	Neolithic; Bronze Age
2000	Sub-Atlantic	Cooler, wet	VIII	Alder–birch–oak–(beech)	Iron Age; Historic
0					

larger-scale trends of climate affecting hemispheric or even global patterns in temperature, precipitation or air circulation. These patterns are considered in the next section.

6.4 CLIMATIC MODELS FOR THE LAST 18 000 YEARS

In recent years, with general acceptance of the Milankovitch mechanism as the major 'forcing factor' of global climatic change through the Quaternary, various attempts have been made to model global climatic patterns at glacial maximum and for more recent times. The 'General Circulation Models' so constructed have used independently-derived estimates of orbital parameters, global ice volume and location, sea-surface temperature, sea level and surface albedo as input data. A recent detailed study by Kutzbach and Guetter [27] utilizes data for 18, 15, 12, 9, 6, 3 and 0 ka bp, and provides a theoretical basis for possible subdivision of Late Pleistocene and Holocene climates on a global scale.

Kutzbach and Guetter's model demonstrates that the primary response to orbital 'forcing' over the last 18 ka has been a variation in seasonal temperature patterns [41]. This seasonal variation was most pronounced in the northern hemisphere, with the greatest difference between summer and winter temperatures reached about 9 ka ago. The basic pattern is

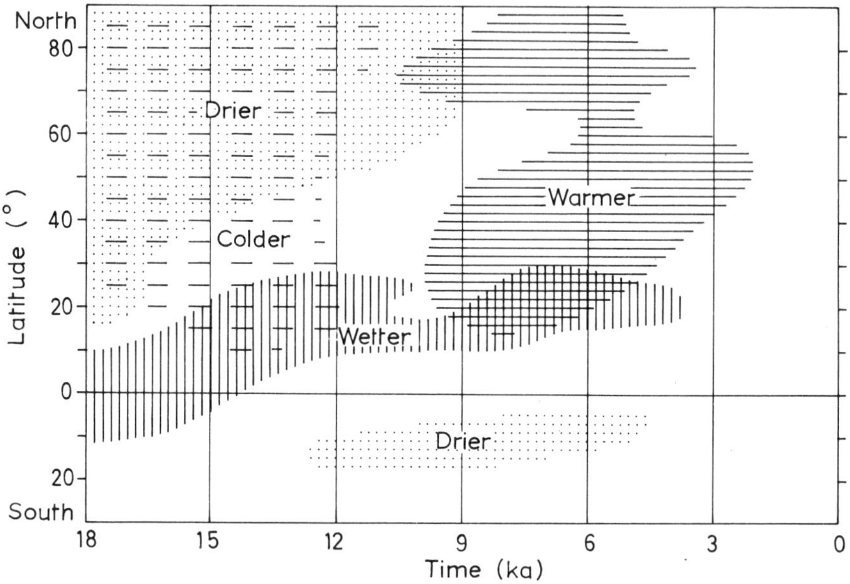

Figure 6.5 Summary of predicted changes in global climate based on modelling. The chart shows regions of the land surfaces between 30°S and 90°N latitude for which zonally-averaged values of temperature and precipitation over the last 18 ka depart appreciably from modern values. — — January land temperatures ⩾4°C lower than present; :::::: July precipitation lower than present; |||||| annual P – E higher than present; ☰ July land temperatures ⩾1°C higher than present. Re-drawn from Kutzbach and Guetter [27].

approximated by the values for summer radiation input shown in Figure 6.4. Seasonal rainfall patterns were also affected, particularly at low latitudes, where the increased solar radiation input during the summer months from 12 to 6 ka bp resulted in substantially increased monsoonal precipitation.

Seasonal patterns of solar radiation input over the last 18 ka can be divided arbitrarily into three phases: an initial phase of increasing seasonal extremes in the northern hemisphere (from about 18 to 10 ka bp); a subsequent more stable phase of strongly seasonal climate (10–6 ka bp); and a final phase of decreasing extremes (6–0 ka bp). The expression of the three phases in different parts of the world has clearly been very variable, as the chart in Figure 6.5 illustrates. The chart is based on the results of modelling with rather coarse temporal (3-ka intervals) and spatial (a 4.5° latitude/longitude grid) resolution. Climatic changes conforming most closely to the predicted three-phase pattern are registered in regions remote from the continental ice sheets – and particularly at low latitudes, where the middle phase is broadly coincident with Kutzbach and Street-

Perrott's 'lacustral phase' of increased monsoonal precipitation and high lake levels [42]. At mid and high latitudes in the northern hemisphere the clarity of the tripartite subdivision is more blurred, and the onset of the middle phase is delayed (Figure 6.5). The blurring stems in part from the poor time-resolution of the model, and in part from the delayed effects on climate of ice-volume changes at mid and high latitudes in the northern hemisphere.

Nevertheless, the general agreement of these simulations with the values for observed variables from 18 to 0 ka bp that have been discussed earlier supports the belief that the model reproduces some of the major processes that have governed climatic change over this period.

A quite different approach to understanding palaeoclimates, advocated by Professor H. H. Lamb, is to seek analogies for prevalent atmospheric circulation patterns of the past in specific climatic episodes (short-term weather patterns) of the present day. Knox [43] used this approach to study Holocene climates in North America. He suggested that parallels to the windflow patterns over North America at 8 ka and 5–6 ka ago could be found in the present-day circulation patterns for March and June in normal years, respectively (Figure 6.6). Major climatic patterns in North America are determined by the interplay of three air masses: the Pacific (warm and moist along the west coast, but dry eastwards beyond the Western Cordillera), the Canadian (cold and dry), and the Gulf of Mexican (warm and moist). With extensive snow-covered and ice-covered surfaces in north-central North America (as occur only in the late winter today, but occurred throughout the year 8 ka ago), the steep temperature gradient to the south of that area generates a prevailing latitudinal (west–east) airflow which draws in the Pacific air mass over much of the United States. In summer today, and more generally 5–6 ka ago, Knox suggested that a more meridional (north–south) airflow develops, and that the interplay between the Canadian and Gulf air masses gives rise to intense slow-moving cyclones and periodic heavy rainfall in middle latitudes.

A similar approach has been used by Magny [44] to explain the different character of the Atlantic and Sub-Boreal periods (section 6.3) in various parts of Europe. The Sub-Boreal period was originally envisaged by Blytt and Sernander for Scandinavia as drier and slightly less warm than the preceding Atlantic period. Evidence of increased climatic dryness in the Sub-Boreal for the British Isles is not convincing, however, and the major change is seen as a shift from oceanic to more continental conditions [40]. In the Jura Mountains of the French–Swiss borderland, the trend was in the opposite direction, to a damper cooler climate in the Sub-Boreal [44]. These regional differences can be seen as a consequence of the changing relationships of the principal air masses over Europe through the Holocene. Magny's reconstruction of the atmospheric circulation over

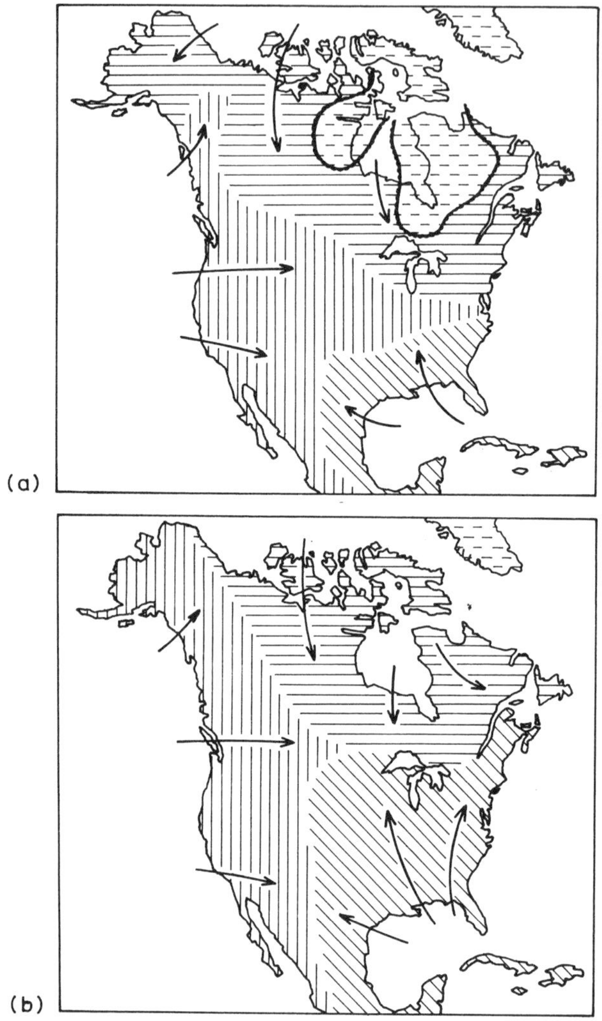

Figure 6.6 Regions of North America influenced by three principal air masses (a) 8 ka ago, (b) 5–6 ka ago; reconstructions based on present-day March and June wind circulation patterns, respectively. Pacific Air Mass, |||||; Canadian Air Mass, ☰; Gulf Air Mass, \\\\\\; Laurentide Ice Sheet, -/. Re-drawn from Knox [43].

Europe in the Atlantic and Sub-Boreal periods (Figure 6.7) shows a prevailing maritime influence over Britain in both periods, but a replacement of maritime (wet) by continental (drier) polar air over Scandinavia in the Sub-Boreal, and a concomitant southward shift of the polar frontal zone (giving a more disturbed climate) to affect much of west-central Europe.

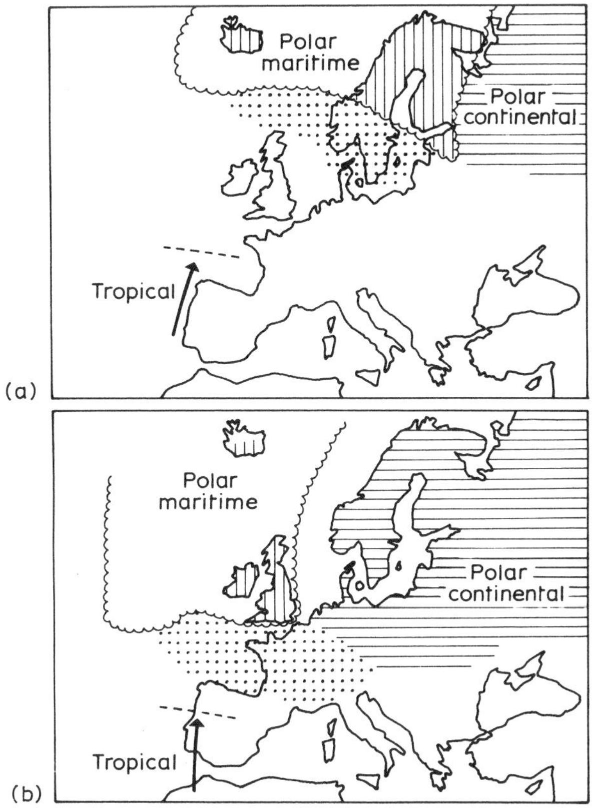

Figure 6.7 Reconstructed atmospheric circulation over Europe in (a) the Atlantic period, and (b) the Sub-Boreal period. Continental air, ═══; polar maritime air, ⅢⅢ; polar frontal zone, ⁝⁝⁝⁝; tropical air, ‾↑‾. Re-drawn from Magny [44].

The danger of this approach is that it assumes that the full range of Holocene climate is encapsulated in present-day 'weather'. While this assumption may be true for climate, it certainly is not so for vegetation – as we shall see in subsequent chapters.

REFERENCES

1. Mercer, J. H. (1984) Late Cainozoic glacial variations in South America south of the equator, in *Late Cainozoic Palaeoclimates of the Southern Hemisphere* (ed. J. C. Vogel), Balkema, Rotterdam, pp. 45–58.
2. Michelson, D. M., Clayton, L., Fullerton, D. S. and Borns, H. W., Jr (1983) The Late Wisconsin glacial record of the Laurentide Ice Sheet in the United States,

in *Late-Quaternary Environments of the United States*, Vol. 1, *The Late Pleistocene* (ed. S. C. Porter), Longman, London, pp. 3–37.

3. Lowe, J. J. and Walker, M. J. C. (1984) *Reconstructing Quaternary Environments*, Longman, London.
4. Gray, J. T. and Lauriol, B. (1985) Dynamics of the Late Wisconsin ice sheet in the Ungava Peninsula interpreted from geomorphological evidence. *Arctic Alpine Res.*, **17**, 289–310.
5. Mix, A. C. (1987) The oxygen-isotope record of deglaciation, in *North America and Adjacent Oceans during the Last Deglaciation* (eds W. F. Ruddiman and H. E. Wright Jr), Geological Society of America, Boulder, CO, pp. 111–35.
6. Paterson, W. S. B. and Hammer, C. U. (1987) Ice core and other glaciological data, in *North America and Adjacent Oceans during the Last Deglaciation* (eds W. F. Ruddiman and H. E. Wright Jr) Geological Society of America, Boulder, CO, pp. 91–109.
7. Watts, W. A. (1980) Regional variation in the response of vegetation to Lateglacial climatic events in Europe, in *Studies in the Lateglacial of North-West Europe* (eds J. J. Lowe, J. M. Gray and J. E. Robinson), Pergamon, Oxford, pp. 1–21.
8. Ruddiman, W. F. and McIntyre, A. (1976) Northeast Atlantic paleoclimatic changes over the past 600,000 years. *Mem. Geol. Soc. Am.*, **145**, 111–46.
9. Burrows, C. J. (1979) A chronology for cool-climate episodes in the Southern Hemisphere 12,000–1000 yr B.P. *Palaeogeogr. Palaeoclimatol. Palaeoecol.*, **27**, 287–347.
10. Mott, R. J., Grant, D. R., Stea, R. and Occhietti, S. (1986) Late-glacial climatic oscillation in Atlantic Canada equivalent to the Allerød/Younger Dryas event. *Nature*, **323**, 247–50.
11. Heusser, C. J. (1984) Late glacial–Holocene climate of the Lake District of Chile. *Quat. Res.*, **22**, 77–90.
12. Heusser, C. J. and Rabassa, J. (1987) Cold climatic episode of Younger Dryas age in Tierra del Fuego. *Nature*, **328**, 609–11.
13. Ruddiman, W. F. and Duplessy, J.-C. (1985) Conference on the last deglaciation: timing and mechanism. *Quat. Res.*, **23**, 1–17.
14. Imbrie, J., Van Donk, J. and Kipp, N. G. (1973) Paleoclimatic investigation of a Late Pleistocene Caribbean deep-sea core: comparison of isotopic and faunal methods. *Quat. Res.*, **3**, 10–38.
15. Shackleton, N. J. and Opdyke, N. D. (1973) Oxygen isotope and palaeomagnetic stratigraphy of Equatorial Pacific core V28-238: oxygen isotope temperatures and ice volumes on a 10^5 and 10^6 year scale. *Quat. Res.*, **3**, 39–55.
16. Lorius, C., Merlivat, L., Jouzel, J. and Pourchet, M. (1979) A 30,000-yr isotope climatic record from Antarctic ice. *Nature*, **280**, 644–8.
17. Dansgaard, W., Johnsen, S. J., Møller, J. and Langway, C. C., Jr (1969) One thousand centuries of climatic record from Camp Century on the Greenland ice sheet. *Science*, **166**, 377–81.
18. Beug, H.-J. (1982) Vegetation history and climatic changes in central and southern Europe, in *Climatic Change in Later Prehistory* (ed. A. F. Harding), Edinburgh University Press, Edinburgh, pp. 85–102.

19. Bradley, R. S. (1985) *Quaternary Paleoclimatology. Methods of Paleoclimatic Reconstruction*, Allen and Unwin, Winchester, MA.
20. Denton, G. H. and Hughes, T. J. (1986) Global ice-sheet system interlocked by sea-level. *Quat. Res.*, **26**, 3–26.
21. Denton, G. H. and Karlén, W. (1973) Holocene climatic variations: their pattern and possible cause. *Quat. Res.*, **3**, 155–205.
22. Heuberger, H. (1974) Alpine Quaternary glaciation, in *Arctic and Alpine Environments* (eds J. D. Ives and R. G. Barry), Methuen, London, pp. 319–38.
23. Ruddiman, W. F. (1985) Climate studies in ocean cores, in *Paleoclimate Analysis and Modeling* (ed. A. D. Hecht), Wiley, New York, pp. 197–257.
24. Johnsen, S. J., Dansgaard, W., Clausen, H. B. and Langway, C. C., Jr (1972) Oxygen isotope profiles through the Antarctic and Greenland ice sheets. *Nature*, **235**, 429–34.
25. Mörner, N.-A. and Wallin, B. (1977) A 10,000-year temperature record from Gotland, Sweden. *Palaeogeogr. Palaeoclimatol. Palaeoecol.*, **21**, 113–38.
26. Berger, A. (1977) Power and limitation of an energy-balance climate model as applied to the astronomical theory of palaeoclimates. *Palaeogeogr. Palaeoclimatol. Palaeoecol.*, **21**, 227–35.
27. Kutzbach, J. E. and Guetter, P. J. (1986) The influence of changing orbital parameters and surface boundary conditions on climatic simulations for the past 18,000 years. *J. Atmos. Sci.*, **43**, 1726–59.
28. Newell, R. E. and Hsiung, J. (1984) Sea surface temperature, atmospheric CO_2 and the global energy budget: some comparisons between the past and present, in *Climatic Changes on a Yearly to Millennial Basis* (eds N.-A. Mörner and W. Karlén), Reidel, Dordrecht, pp. 533–61.
29. Lamb, H. H. (1982) Reconstruction of the course of postglacial climate over the world, in *Climatic Change in Later Prehistory* (ed. A. F. Harding), Edinburgh University Press, Edinburgh, pp. 11–32.
30. Stockton, C. W., Boggess, W. R. and Meko, D. M. (1985) Climate and tree rings, in *Paleoclimate Analysis and Modeling* (ed. A. D. Hecht), Wiley, New York, pp. 71–161.
31. Lamarche, V. C. (1974) Paleoclimatic inferences from long tree-ring records. *Science*, **183**, 1043–8.
32. Lamb, H. H. (1985) Climate and landscape in the British Isles, in *The English Landscape: Past, Present and Future* (ed. S. R. J. Woodell), Oxford University Press, Oxford, pp. 148–67.
33. Street, F. A. and Grove A. T. (1979) Global maps of lake-level fluctuations since 30,000 yr B.P. *Quat. Res.*, **12**, 83–118.
34. Smith, G. I. and Street-Perrott, A. (1983) Pluvial lakes of the western United States, in *Late-Quarternary Environments of the United States*, Vol. 1, *The Late Pleistocene* (ed. S. C. Porter), Longman, London, pp. 190–212.
35. Street-Perrott, F. A. and Harrison, S. P. (1985) Lake levels and climate reconstruction, in *Paleoclimate Analysis and Modeling* (ed. A. D. Hecht), Wiley, New York, pp. 291–341.
36. Harrison, S. P., Metcalfe, S. E., Street-Perrott, F. A. *et al.* (1984). A climatic model of the Last Glacial/Interglacial transition based on palaeotemperature

and palaeohydrological evidence, in *Late Cainozoic Palaeoclimates of the Southern Hemisphere* (ed. J. C. Vogel), Balkema, Rotterdam, pp. 21–34.

37. Webb, T. III (1985) Holocene palynology and climate, in *Paleoclimate Analysis and Modeling* (ed. A. D. Hecht), Wiley, New York, pp. 163–95.
38. Burke, R. M. and Birkeland, P. W. (1984) Holocene glaciation in the mountain ranges of the western United States, in *Late-Quaternary Environments of the United States*, Vol. 2, *The Holocene* (ed. H. E. Wright Jr), Longman, London, pp. 3–11.
39. Von Post, L. (1946) The prospect for pollen analysis in the study of the earth's climatic history. *New Phytol.*, **45**, 193–217.
40. Godwin, H. (1975) *History of the British Flora*, 2nd edn, Cambridge University Press, Cambridge.
41. Bonnichsen, R., Stanford, D. and Fastook, J. L. (1987) Environmental change and developmental history of human adaptive patterns; the Paleoindian case, in *North America and Adjacent Oceans during the Last Deglaciation* (eds W. F. Ruddiman and H. E. Wright Jr), Geological Society of America, Boulder, CO, pp. 403–24.
42. Kutzbach, J. E. and Street-Perrott, F. A. (1985) Milankovitch forcing of fluctuations in the level of tropical lakes from 18 to 0 kyr BP. *Nature*, **317**, 130–4.
43. Knox, J. C. (1984) Responses of river systems to Holocene climates, in *Late-Quaternary Environments of the United States*, Vol. 2, *The Holocene* (ed. H. E. Wright Jr), Longman, London, pp. 26–41.
44. Magny, M. (1982) Atlantic and Sub-boreal: dampness and dryness? in *Climatic Change in Later Prehistory* (ed. A. F. Harding), Edinburgh University Press, Edinburgh, pp. 33–43.

PART THREE

Plant Responses to Environmental Change

CHAPTER SEVEN

The development of diversity

7.1 INTRODUCTION

The floristic and vegetational diversity of the world's plant cover, as we saw in Chapters 1 and 2, is a product of the available species-pools in particular regions and the long-term action on them of evolutionary and distributional change. Accordingly, in this chapter and the next we examine some of the ways in which global diversity has arisen. The longer-term aspects, extending back to the initial appearance of the angiosperms more than 100 Ma ago, are considered in this chapter; distributional and evolutionary changes over the last 20 ka are considered in Chapter 8.

7.2 THE EARLIEST ANGIOSPERMS

Part of the floristic heterogeneity visible today on a global scale is of very considerable antiquity, dating back 50–100 Ma to when angiosperms were establishing themselves as the dominant components of much of the world's vegetation. Most palaeobotanists believe that the first flowering plants evolved rather rapidly in the Early Cretaceous, over a period of about 10 Ma around 120 Ma ago [1,2]. Claims of angiospermous fossils from earlier geological time are not generally accepted. The Early Cretaceous was a period of extremely benign climate [3], with land temperatures at high latitudes up to 30°C higher than at present [4]; as a result, permanent polar ice was absent. Sea levels were high (the highest at any time over the last 200 Ma), and overall surface relief was low. Nevertheless, rainfall was modest over extensive areas [4], and seasonal drought conditions were probably widespread in lower-middle latitudes (30–50°N and S). Tropical non-seasonal, subtropical seasonal, and temperate climatic zones have accordingly been recognized [5–7].

The Early Cretaceous landscape was dominated by ferns, cycads and conifers. At least four major vegetation types have been identified: evergreen conifer forests (at lower latitudes); deciduous ginkgophyte forests (at higher latitudes); cycadophyte shrublands; and fern marshes. These four

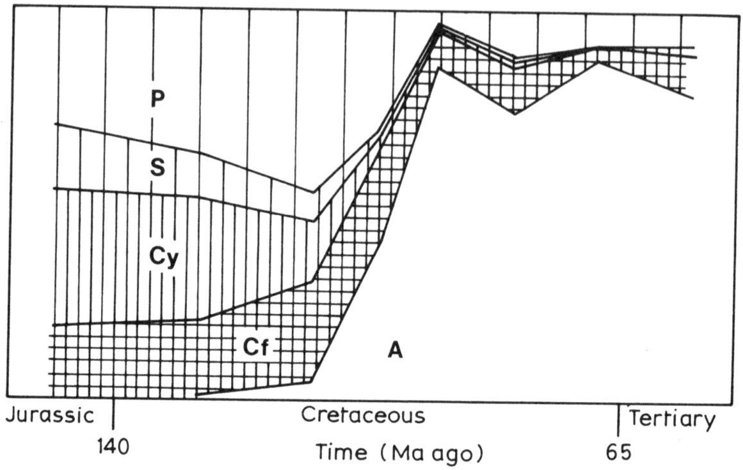

Figure 7.1 The rise of the angiosperms. Changes in mean percentage contribution of major plant groups to Late Jurassic, Cretaceous and Early Tertiary floras. A = angiosperms, Cf = conifers, Cy = cycadophytes, S = other seed plants, P = pterophytes. Re-drawn from Crane [9].

types were broadly equivalent to the modern araucarian forests of Queensland, the swamp forests of Florida, the redwood forests and the chaparral of California, and the *Dicksonia* swamps of Queensland [8].

The first flowering plants were probably small-leaved, small-seeded, evergreen shrubs with 'weedy' tendencies, growing in streamside and aquatic habitats, in early successional thickets, and as forest understorey species [1,9,10]. Browsing and disturbance by the dominant dinosaurs (particularly of the cycadophyte shrublands and fern marshes [8]) could have been instrumental in allowing the initial ingress of angiosperms [11]. As larger more competitive angiosperms evolved in the Middle and Late Cretaceous, rapid replacement of the cycadophytes and pterophytes occurred (Figure 7.1), and the whole aspect of the global plant cover was dramatically altered. The reasons for this replacement are unclear, but a crucial factor might have been the superior ability of the early angiosperms to exploit the disturbed environments created, for example, by mountain uplift, the ebb and flow of shallow epicontinental seas, and dinosaur browsing, by reason of their superior seed dispersal and targeted pollination (by birds and insects, respectively) [7,12]. Nevertheless, even in the Late Cretaceous the predominant angiosperms probably occurred mainly as densely-packed small trees and shrubs (with evergreen laurel-like leaves), overtopped by occasional emergent conifers [9]; deciduous taxa were present only at high latitudes, and complex multistratal forests of rainforest-type were extremely restricted in distribution [13]. Not until the Early Tertiary did recognizably modern forest-types begin to develop.

The Cretaceous world consisted of more tightly-knit land surfaces than the modern one, with only incipient fragmentation of the northern and southern supercontinents of Laurasia and Gondwanaland. The gradual separation and fragmentation of Laurasia and Gondwanaland through the Cretaceous led to an early differentiation of discrete northern and southern hemisphere floras. Modern distribution patterns referable to this event are considered in section 7.5. Nevertheless, in the initial stages of separation the potential was still there for plants to build up continuous populations over enormous areas. Despite the conceptual difficulties involved in envisaging how such distributions could have been achieved in a forested landscape, especially by the ancestors of modern genera with large poorly-dispersed fruits, the existence today of families and genera of angiosperms that range widely in one or both hemispheres demands such an achievement. The family Fagaceae is a typical example. Its modern members (forest trees with heavy seeds and fruits) occur worldwide today, in Europe, Asia, North and South America, and Australasia; species of *Fagus* itself (beech) occur disjunctly in Europe, eastern Asia, eastern North America and Mexico. A common ancestry for the various genera of Fagaceae and the various species of *Fagus* requires highly successful dispersal and establishment at some stage in the past; and as fossil remains attributable to both *Fagus* and *Nothofagus* have been recorded from the Late Cretaceous [7], this wide ancestral range must have been achieved very early in angiosperm evolution.

7.3 THE RISE OF THE ANGIOSPERMS

The takeover of the world's land surfaces by flowering plants was strengthened by the sudden changes recorded in the fossil record at the Cretaceous/Tertiary boundary some 65 Ma ago. In what must have been a global catastrophe of monumental proportions, up to half the genera of certain animals and plants living at the time may have perished, within a time-period variously estimated at between 0.5 Ma and a few decades [14]. Extinction was particularly severe among animals of large size in terrestrial communities (notably the dinosaurs), planktonic organisms with calcareous skeletons, and tropical-reef invertebrates [15]. Though land plants were less affected overall, nevertheless between latitudes 30° and 60° in the northern hemisphere 50–90% of the gymnosperm and evergreen angiosperm taxa may have perished [16]. Two explanations have been proposed to account for this global catastrophe. One possibility that has attracted much attention recently [17] is that a massive asteroid up to 10 km in diameter struck the earth and formed an impact crater from which vast quantities of dust were injected into the stratosphere. This dust was spread around the globe, and effectively prevented sunlight from reaching the surface for a period of several years until the dust settled to the earth. Loss of sunlight suppressed photosynthesis, and as a result most food chains collapsed and mass extinction of plants and animals resulted [18–20]. The asteroid impact may also have generated worldwide wildfires which burned off much of the earth's vegetation, and accentuated the temporary opacity of the earth's atmosphere to sunlight [21]. An alternative, more orthodox, hypothesis is of intense volcanic activity at the Cretaceous/ Tertiary boundary, which, by rather similar mechanisms, led to a global fall in temperature, the production of acid rain, a decrease in pH values at the ocean surface, and increased ultra-violet irradiation; a major sea-level regression at the same time could have resulted in markedly more continental climates [15,22]. Both explanations involve major disruptions of the world's biomes, and the creation, in some regions at least, of a partial ecological desert, which the fossil record indicates was later colonized initially by plant assemblages dominated by ferns [20]. In the forest vegetation that subsequently returned, woody angiosperms were the major component; and in the northern hemisphere at latitudes above 55°N (where extinction had been most severe) there was a predominance of deciduous taxa [23]. At lower latitudes, true rainforest developed, with a greater potential for diversification among its component species than any of the Cretaceous forests possessed [13].

The dramatic replacement of large dinosaurs by small mammals at the Cretaceous/Tertiary boundary, moreover, led to a fundamental change in plant/animal interactions that accompanied and intensified these changes: specialized herbivory, often of fruits and seeds, supplanted generalized herbivory (browsing), and far-reaching changes in the dispersal potential of angiosperm taxa probably ensued [11]. The angiosperm rise to dominance was thus further aided.

7.4 EVOLUTIONARY CHANGE

Recognizably modern species and genera of angiosperms occur as fossils for the first time in any numbers in the Eocene forests of about 50 Ma ago. Typically the range of variation is small, and considerable diversification must have taken place subsequently, as a much wider range of morphologically distinct fossils is present by the Late Tertiary. This increase in morphological diversity was the result of **speciation**, and can be illustrated by the data in Table 7.1, which show the numbers of species identified from Tertiary deposits of different ages in western North America

Table 7.1 Number of species in three families of modern temperate-zone trees that have been recorded in Tertiary floras from western North America (data of Wolfe [24])

	Aceraceae	Betulaceae	Salicaceae
Palaeocene	2	2	1
Eocene:			
tropical	2	4	3
temperate	3	4	3
Early Oligocene:			
tropical	0	3	0
upland	5	2	8
Late Oligocene	6	15	10
Early/Middle Miocene	11	12	25
Late Miocene	6	12	35

for three families of trees (comprising four main genera: *Acer*, maple; *Alnus*, alder; *Betula*, birch; and *Salix*, willow). Numbers increased from five species at the start of the Tertiary to fifty-three by the Late Miocene (about 10 Ma ago).

Speciation in many genera involved both an increase in the number of species and an increase in the range of habitats occupied. Thus many tree genera (including the four listed above) that are recorded as fossils from tropical and subtropical environments in the Early Tertiary of the western United States [24] subsequently diversified in the Middle and Late Tertiary Period to produce additional species that are now characteristic of modern broadleaved forests of temperate regions. Several genera spread into other habitats also. Thus modern species of *Quercus* (oak), *Alnus*, *Betula* and *Salix* are now found additionally in North America in subarctic scrub, boreal forest and xerophytic woodland.

In other genera, however (and particularly in gymnosperms), the range of present-day morphological variation is apparently less than that found in fossil material. Certain lineages have become totally extinct over the last 65 Ma. At least twelve species of redwood are recorded from fossil material of Early and Middle Tertiary age, whereas only three (one in each of the genera *Metasequoia*, *Sequoia* and *Sequoiadendron*) occur today.

Various attempts have been made to utilize the fossil record to calculate rates of evolution. All the calculations assume that where a genus is undergoing specific diversification, the pattern of increase in species-number can be likened to exponential or logistic growth in living plant and animal populations; if the age of the genus is known, then the rate of increase, *r*, can be estimated [26]. Most studies have been on animals, where mean rates of increase varying from about 0.05 to 0.35 new species per million years have been calculated [26]. Levin and Wilson [25] present a similar analysis for higher plants, based on fossil records of 201 angiosperm genera and 37 gymnosperm genera. Their results, summarized in Table 7.2, show that speciation has apparently proceeded at a faster rate in herbaceous angiosperms than in woody ones, and at a faster rate in angiosperms generally than in gymnosperms. The results indicate that speciation is a slow process, occupying millions rather than thousands of years, and appear to contradict other data (from animal populations evolving in isolation over a known time-period), where new species have evolved in a few thousand years [26]. Both data-sets should be viewed with some scepticism. The fossil evidence necessarily uses morphological criteria to distinguish species, and morphological differentiation may either precede, accompany, or postdate the separation of populations on breeding behaviour (the normal criterion for species-differentiation). The fossil evidence also requires information on the time of first appearance of a genus in the fossil record – which may be known only very approximately. Moreover, the fossil evidence yields only mean values for rates of speciation, averaged over genera and through time. The fossil record shows clearly that evolution has not proceeded at a steady rate through time, but has exhibited periodic bursts of **adaptive radiation**, when speciation rates have been considerably enhanced. Characterization of speciation rates by mean values obscures this potential of plant populations for rapid differentiation under certain circumstances. Levin and Wilson's values include rates for individual angiosperm taxa that range from 0.06 to 2.41 new species per million years; perhaps this latter value reflects more truly the rate of speciation that can be achieved by angiosperms under favourable conditions. Even if calculated rates of speciation cannot be regarded too seriously, however, they should nevertheless encourage caution against accepting too uncritically hypotheses (such as the 'refuge hypothesis' – section 8.5) that require the differentiation of large numbers of

Table 7.2 Mean speciation rates (number of new species per million years) in 238 genera of higher plants (data of Levin and Wilson [25])*†

Angiosperms:	herbaceous	1.15
	shrubs	0.28
	trees	0.12
Gymnosperms:	conifers	0.04
	cycads	0.03

* Rates have been corrected to allow for extinction.

† Levin and Wilson use a formula different from that used by Stanley [26] to calculate the rate of increase, r; their values should be divided by 3.3 to make them comparable with Stanley's values.

related species-populations from a single ancestral one in the space of , thousand years.

Rates of evolutionary change are known to be influenced by partic features of the breeding systems of higher plant populations – short versus long generation time, inbreeding versus outbreeding, etc.; and different types of breeding system have themselves undoubtedly evolved as adaptations to different habitats – transient versus stable, uniform versus non-uniform, etc. Accordingly, different rates of evolutionary change may be characteristic of different types of habitat. Levin and Wilson [25] conclude that 'a system of small, semi-isolated populations in transient habitats or ecotonal regions experiencing marked fluctuations in habitat hospitality is most conducive for the fixation of chromosomal and morphological novelties and rapid speciation. Conversely, systems of large continuous populations where climatic and biotic pressures are stable through time are likely to be much more conservative.' Suitable small populations probably existed in the earliest stages of angiosperm evolution, when the fossil evidence indeed indicates rapid diversification.

The results of rapid evolutionary change within small isolated populations are sometimes referred to as **quantum speciation**: 'speciation in which most evolution is concentrated within an initial interval of time that is very brief with respect to the total longevity of the new lineage that is produced' [26]. We can expect small populations to be associated particularly with plant taxa with widely-dispersed seeds, as such taxa are most likely to establish new colonies well outside their main range. The overall genetic constitution of such a new colony must be determined by that of its founder member(s), and may well differ from that of the parent population, as colonization of new **ecospace** (section 9.4) may either confer a selective advantage on genetically-aberrant individuals, or remove them altogether from competition with better-adapted individuals. If a new colony is sufficiently distant from the parent population for gene exchange with it to be excluded, then the new colony may quickly diverge genetically and/or chromosomally from it (the **founder principle**).

Empirically we can surmise that quantum speciation might be triggered by the formation of new ecospace, the emptying of existing ecospace by the extinction of certain lineages, or the sudden achievement of an adaptive breakthrough [26]. What is known about environmental changes over the last 65 Ma (Chapters 4–6) suggests that these conditions would have been met repeatedly during the Cenozoic Era, as a result of orogenesis, fluctuating sea-levels, and increasing cold and aridity. Moreover, these processes, coupled with continental displacements, must also have been conducive to the fragmentation of plant ranges, with increasing numbers of geographical barriers producing the segregated populations postulated as necessary for quantum speciation. The probable scenario for plant evolution within the Tertiary and Quaternary Periods has accordingly been summarized as follows [27]: 'large-scale dispersals to produce widespread ancestral biotas and later allopatric speciation events which . . . fragmented the ancestral biotas into their present highly subdivided states'.

7.5 FRAGMENTATION AND FUSION OF FLORAS

The major fragmentation of the Cretaceous flora was achieved by the progressive break-up of the supercontinents of Laurasia and Gondwanaland. Any panglobal continuity was early sundered by the formation of the Tethys Sea linking the proto-Atlantic Ocean to the Indo-Pacific Ocean along the line of the future Mediterranean Sea. This Tethys Sea must have been in existence, at least during times of high sea-level, long before the angiosperms evolved, and its efficiency as a barrier can be seen in the persistence of segregated northern-hemisphere and southern-hemisphere gymnosperm floras through to the present-day (Table 7.3)*. Segregated angiosperm floras quickly arose also, and even many supposedly primitive angiosperm families show clear northern-hemisphere or southern-hemisphere distributions at the present-day (Table 7.3). It is easy to appreciate in a general way that the time of separation of a particular land mass from the main Laurasian or Gondwanic bloc as it began to break up, relative to the time of origin of a particular family or genus, might have determined whether that taxon was represented in the flora of the land mass or not; and in certain instances this process can be seen to have operated. Thus both India and Africa separated early from Gondwanaland (about 90 Ma and 140 Ma ago, respectively), and as a result contain few of the families that characterized the southern supercontinent in the Late Cretaceous and Early Tertiary Periods. Of the sixteen southern-hemisphere families listed in Table 7.3, only four (Araucariaceae, Podocarpaceae, Proteaceae and Restionaceae) are recorded as fossils from Africa [28], and only six (Araucariaceae, Casuarinaceae, Myrtaceae, Podocarpaceae, Proteaceae and Restionaceae) from India [29–31]. There is no evidence that the important southern-hemisphere genus *Nothofagus* was ever present in either India or Africa.

The somewhat later separation of South America and Australia–New Zealand from Antarctica resulted in a much greater representation of these old austral families in their floras. It has been suggested that the modern native flora of the New Zealand lowlands, in particular, has preserved much of the character of the Late Cretaceous forest floras of Gondwana-

* Records of southern gymnosperms as Tertiary fossils from the northern hemisphere are mostly erroneous.

Table 7.3 Examples of modern families of higher plants with ranges confined (with rare exceptions) to either the northern or the southern hemisphere

	Northern hemisphere	Southern hemisphere
Gymnosperm families [32]	Abietaceae Taxaceae Taxodiaceae	Araucariaceae Podocarpaceae
Families of primitive woody angiosperms [31]*	Calycanthaceae Cercidiphyllaceae Eucommiaceae Eupteleaceae Illiciaceae Magnoliaceae Schisandraceae Tetracentraceae Trochodendraceae	Amborellaceae Austrobaileyaceae Degeneriaceae Epacridaceae Eupomatiaceae Himantandraceae Monimiaceae Proteaceae Trimeniaceae Winteraceae
Families of primitive herbaceous angiosperms [31]*	Saururaceae	Gunneraceae Restionaceae
Other notable angiosperm families		Casuarinaceae Myrtaceae

* This list relates primarily to South-east Asia and Australasia.

land. Rather constant and highly equable climatic conditions were maintained in New Zealand throughout its 80-Ma period of isolation, with a gradual northward movement of its land surfaces into lower latitudes effectively counteracting the worldwide temperature decline during the Tertiary. The only notable extinctions in the New Zealand flora during this period were the gymnosperms *Araucaria* and *Athrotaxis*, certain species-lineages of *Nothofagus* (the 'brassii' group), and *Casuarina* [28]. The smaller islands of the South Pacific, and especially New Caledonia, that were thrust up along the eastern plate margin as the Australian Plate moved northwards, have retained even closer links with the Cretaceous flora of Gondwanaland.

In Australia, the temperate and subtropical rainforest floras that were widespread in the Eocene became progressively restricted in extent as drier conditions spread in the later Tertiary. Nevertheless, the modern rainforest flora of Australia still retains much of the character of its Early Tertiary predecessors, and has been claimed to represent 'an essentially relict flora surrounded by a sea of evolutionary novelty, the *Eucalyptus*-dominated forests and woodlands' [33].

By contrast, there is much greater uniformity in the floras of the various land masses of the northern hemisphere, as a result of the delayed break-up of Laurasia compared with Gondwanaland. Whereas ancestral links in the floras of the southern continents are now visible mainly at the family level, the northern land masses share many common genera and even species. Similarities are most marked in the boreal zone [34], but the persistence of an intermittent land bridge across the Bering Straits throughout the Cenozoic Era led to continuous floristic interchange between North America and Asia. Table 7.4 lists thirty-two woody genera recorded from Early Miocene deposits of Japan [35] that have also been recorded from Tertiary deposits in North America, and are still present in either eastern Asia, western North America, or both regions. These genera form part of the very rich flora of gymnosperms and angiosperms comprising the forest vegetation at higher latitudes in the northern hemisphere for much of the Tertiary Period (the so-called 'Arctotertiary Geoflora'). The formerly continuous distributions of these genera have now been fragmented by continental separations and the development of mid-continental aridity (discussed in the next section), and they are now represented by different species in eastern Asia, western North America and eastern North America.

Fragmentation of range is not the only process that has operated through the Cenozoic Era to mould present-day plant distribution patterns. At least three instances are known where previously isolated land masses 'drifted' into contact with each other, and intermingling of their floras and faunas resulted. In one case – the union of the North American and South American subcontinents across the Panama Isthmus about 6 Ma ago – the floristic consequences appear to have been rather small [36], and interest has centred on the faunas. The 'Great American Interchange' of two distinct faunas led to competition between animal taxa formerly occupying separate but similar ecological niches in the two subcontinents, and resulted in widespread extinctions, major expansions of range by the successful taxa, and subsequent adaptive radiation [37–40]. Interchange in the other two cases of secondary continental contact (India with south-central Asia 40–50 Ma ago, and Australia–New Guinea with south-east Asia 5–15 Ma ago) did not apparently result in comparable levels of extinction, though considerable range expansion and adaptive radiation of plants subsequently ensued.

In both cases a mobile land mass contacted a nearly stationary one, bringing Gondwanic taxa into proximity with Laurasian taxa. Each mobile land mass behaved as a sort of Noah's Ark, rafting a whole flora northwards into warmer climates, at a rate prior to contact that was sufficiently slow to allow most taxa to survive and to adapt to the changing conditions [31]. Along the zone of impact between the two land masses massive tectonic

Table 7.4 Fragmentation of range. Modern ranges of tree taxa widespread in Miocene deposits of Japan, and also recorded in Tertiary deposits of North America [35]

	Japan, China	Western North America	Eastern North America
Picea	+	+	+
Pinus	+	+	+
Tsuga	+	+	+
Acer	+	+	+
Alnus	+	+	+
Betula	+	+	+
Corylus	+	+	+
Fraxinus	+	+	+
Quercus	+	+	+
Rhus	+	+	+
Salix	+	+	+
Sorbus	+	+	+
Carpinus	+		+
Carya	+		+
Castanea	+		+
Diospyros	+		+
Fagus	+		+
Ilex	+		+
Sophora	+		+
Tilia	+		+
Ulmus	+		+
Pseudotsuga	+	+	
Glyptostrobus	+		
Metasequoia	+		
Alangium	+		
Cercidiphyllum	+		
Cinnamomum	+		
Machilus	+		
Paliurus	+		
Zelkova	+		
Comptonia			+
Liquidambar		+	+

disturbance occurred. Extensive new ecospace – the mountains of the Himalaya, New Guinea, Indonesia and Borneo – was thus created, which permitted certain elements from both the Laurasian and Gondwanic floras to expand their ranges. The new mountain chains also provided high-level corridors along which temperate plant taxa could penetrate deeply into subtropical and tropical regions – where, freed from their normal competitors, some underwent adaptive radiation to produce a gamut of new species [41].

The consequences of secondary contact have been worked out in detail only for South-east Asia [28,31,42,43]. As the Australian Plate 'drifted' equatorwards through the Tertiary, it carried on it a land mass of prevailing low relief and composed largely of hard old rocks. The flora that developed was accordingly a lowland one, adapted to acidic base-poor soils; and because of the long passage of the Australian Plate through the arid subtropics, the component plants became increasingly xeromorphic. The lush evergreen Gondwanic forests were progressively reduced in extent [44,45], and although rainforest communities persisted in north-eastern Australia they were floristically poor. The leading edge of the Australian Plate – subsequently to become New Guinea – remained submerged for much of the Tertiary. The Australian flora was thus one of temperate origin that became secondarily adapted to warmer (and drier) climates. The Malesian (South-east Asian) flora, by contrast, developed in an essentially stable moist tropical climate throughout the Tertiary, to produce a luxuriant and diversified assemblage of lowland rainforest plants. At contact of the Asian and Australian land masses, intermingling of floras occurred, principally in the islands lying between the Asian and Australian mainlands. The northern Australian rainforest flora was probably little affected, however [45]. Contact produced new land surfaces (notably New Guinea) and new ecospace (the highlands of New Guinea, Borneo and Indonesia). In the lowlands the Malesian flora had an enormous competitive advantage, and became predominant throughout; however, at higher altitudes components of the Gondwanic flora were able to take advantage of the newly-created ecospace and spread northwards in the montane forest zone (Table 7.5).

The New Guinea flora, in particular, is a blend of Laurasian and Gondwanic elements, segregated to some extent by altitude. The flora is characterized by a high percentage of endemic genera and species, presumably the result of adaptive radiation in new habitats. Its predominantly recent origin is betrayed by the low density of both genera and species compared with the floras of the older land surfaces to the west (Table 7.6).

The region of contact between the Asian and Australian Plates corresponds to 'Wallace's Line', a deep-water barrier of great faunistic significance [46], but one which has functioned far less effectively as a barrier to plant migration [42]. Nevertheless, many plants with poor dispersal mechanisms have been unable to cross Wallace's Line. Most notable are certain supposedly primitive angiosperm families. Thus of the ten Gondwanic families listed in Table 7.3, only one (Winteraceae: one species of *Drimys*) has crossed the line; likewise, members of only one Laurasian

Table 7.5 Genera of predominantly southern distribution that occur at higher altitudes in Indo-Malesia [41]

Araucariaceae	*Agathis*
Podocarpaceae	*Dacrydium*
	Podocarpus
Cunoniaceae	*Weinmannia*
Ericaceae	*Gaultheria*
Fagaceae	*Nothofagus*
Gunneraceae	*Gunnera*
Myrtaceae	*Leptospermum*
	Metrosideros
Rubiaceae	*Coprosma*
Winteraceae	*Drimys*

family of those listed (Magnoliaceae: three species) have spread southwards [31].

The present-day floristic richness of the Asian tropics has thus arisen by a variety of processes, involving both separation and mingling of plant populations. The scenario here has been on a continental scale, featuring continental displacements. Other processes have operated at smaller geographical scales throughout the Cenozoic Era, and in particular, other processes besides continental displacement have acted to subdivide wide-ranging plant populations. Some of these are illustrated schematically in Figure 7.2 in relation to the so-called Arctotertiary Geoflora. As climate worsened through the Tertiary, so the component plants of this geoflora were forced southwards, increasingly channelled along defined migration pathways by newly-uplifted mountain ranges, sea barriers (both epicontinental and oceanic) and developing drought areas (Figure 7.2(a)). Plant ranges thus became discontinuous, and many tree taxa died out in particular regions. Frenzel [47] maps the fossil records for nine of these taxa (*Carya, Fagus, Ilex, Juglans, Liquidambar, Liriodendron, Nyssa, Pterocarya* and

Table 7.6 Some comparisons of the higher-plant floras of New Guinea, Java and the Malayan peninsula [43]

	New Guinea	Java	Malayan peninsula
Total number of genera	1465	1370	1500
species	>9000	4100	7900
Number of endemic genera	98	4	41
endemic species	c. 4500	few	c. 1820
Genera per 10^5 km^2	288	1054	1144
Species per 10^5 km^2	c. 1400	3155	6075

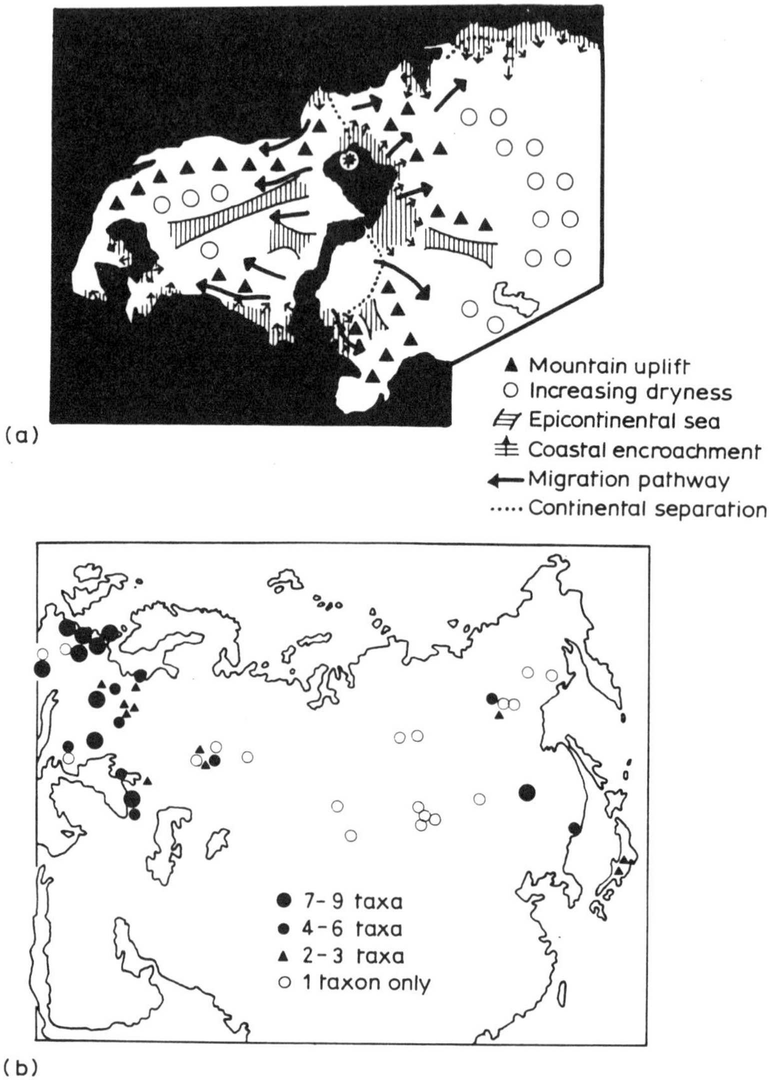

Figure 7.2 Fragmentation of range in the Tertiary Period. (a) The northern hemisphere land mass 60 Ma ago (re-drawn from Smith and Bridén [48]), and the subsequent migration of the Arctotertiary Geoflora in relation to developing geographical barriers. NA = North America; G = Greenland; A = Asia; * = North Pole. (b) Fossil records of nine thermophilic tree taxa belonging to the Arctotertiary Geoflora at the end of the Tertiary Period (modified from Frenzel [47]).

Tsuga) in Eurasia at the close of the Tertiary Period (Figure 7.2(b)). The effects of mid-continental aridity are clearly seen then in the restriction of species-rich sites to two main areas: Europe–Colchis (the Black Sea region) and western Asia. How far the clustering of intervening sites in the Soviet Union is an artefact of incomplete fossil data and how far it represents surviving isolated pockets of forest is uncertain. What is clear is that at the least favourable sites (those with only one tree taxon surviving) no one taxon had a monopoly. Although *Tsuga* was the commonest tree at these sites, six of the other eight taxa also occurred at individual sites. The forest 'residues' at the close of the Tertiary Period must thus have been of very varied composition, as a result of differential extinctions during the long migration southwards.

7.6 MODERNIZATION OF FLORAS

The various processes of evolutionary and environmental change that we have been considering acted in concert to produce a progressively more heterogeneous global plant cover through the Cenozoic Era. Plant populations were subdivided by land and water barriers, and forced to migrate long distances by changing climates. At the level of resolution provided by the fossil record, many taxa appear to have responded in unison, so that plant assemblages were formed which maintained some semblance of unity over long periods of time and through great displacements of range ('geofloras' [49]). As spatially-segregated assemblages developed, however, so differential evolutionary and extinction patterns between assemblages generated progressively divergent floristic compositions, and by the end of the Tertiary Period different geographical regions had markedly different floras. As new habitats were created, so additional entirely novel assemblages came into being, their members bonded together by common responses and common ecospace. By the close of the Tertiary the disposition and topography of the earth's land masses were not very different from what they are today. On these land masses were recognizably modern floras, awaiting their final shaping by the events of the Pleistocene Epoch.

'Modernization' of the flora of a particular region through the Tertiary Period involved gradual progress towards representation of the more important present-day taxa by closely similar fossil forms, and towards loss of taxa no longer found in the region today (plant species known or suspected to have been introduced by man are excluded from consideration). Modernization was achieved, in other words, by the dual processes of elimination of exotic taxa from the region and recruitment, by evolution and immigration, of now-native taxa, in response to climatic and tectonic changes. Cooling global climate through the Tertiary Period shifted plant ranges equatorwards in both hemispheres, thus eliminating some taxa from a region, leaving others behind to adapt, and drawing in yet others from higher latitudes. Elimination of ill-adapted taxa and recruitment of new better-adapted ones were almost certainly conjoint ongoing processes throughout the Cenozoic Era. Nevertheless, there is some evidence that one or other process was accentuated at certain times. This evidence is based on floristic comparisons of fossil assemblages of different age, and is largely independent of conclusions about environmental change derived from other data. It is encouraging, therefore, that accelerated elimination has been recognized in some floras at times when the ocean-temperature record also indicates accelerated change – at the end of the Eocene Epoch, the end of the Miocene Epoch, and the end of the Pliocene Epoch (Figure 4.3). Changes at the end of the Eocene, about 38 Ma ago, led to the elimination of tropical elements from many higher-latitude floras, particularly in regions where seasonally dry climates developed for the first time [50]. Final elimination of tropical and subtropical elements from most middle-latitude floras, together with some warm-temperate elements, occurred in the later Miocene, about 10 Ma ago (Table 7.7). However, several tropical/subtropical taxa persisted locally in middle-latitude floras until the Late Pliocene – when some warm-temperature elements were also lost from more northerly floras. Nevertheless, many warm-temperate taxa lingered on in Europe into the Pleistocene, succumbing eventually to one of the documented cold periods then (Table 7.7). Post-Miocene extinctions were notably muted in Japan and New Zealand (and probably other similar island floras as well), where the investing oceans prevented extremes of cold and dryness from developing; the

Table 7.7 Examples of differential plant extinctions through the Cenozoic Era: at the end of the Miocene (M) or Pliocene (P) Epochs, or in the Early (E) or Late (L) Pleistocene. Data of Suchlik [51], Wolfe [52], Straka [53], Shatilova [54]

	Alaska	Netherlands	Poland	Caucasus	Modern range
Subtropical taxa					
Alangium	M		M	P	Africa, Asia
Engelhardtia			M	P	Asia, America
Symplocos			M	P	Asia, America
Warm-temperate taxa					
Cunninghamia			M	P	China
Glyptostrobus			M	P	China
Pterocarya	M	E			Asia
Sciadopitys		E		L	Japan
Carya		E		L	E Asia, E North America

modern forest floras of Japan and New Zealand have accordingly retained more of the character of their Tertiary plant cover than have the larger land masses [35,44].

These dual processes of elimination and recruitment were most marked, and probably most effective, at middle latitudes. High-latitude land areas, by contrast, although continually depleted of plants by emigration, recruited new taxa only slowly, by gradual evolution of cold-resistant forms. Low-latitude land areas, on the other hand, functioned as gathering grounds for plant taxa throughout the Cenozoic Era, yet at the same time probably lost relatively few taxa by extinction. The richness of low-latitude floras might thus be seen as a consequence of continuous recruitment accompanied by protracted evolutionary differentiation under uninterruptedly favourable conditions. However, the deep-ocean oxygen-isotope record suggests that land temperatures at lower latitudes might have fallen below present-day levels by the Miocene, and then subsequently risen (Figure 4.3), so that tropical floras and biomes could have been more restricted in range in the mid-Tertiary than at present. It is possible that the floristic poverty of the African tropics, as compared with those of America and Asia, arose at this time as a result of markedly cooler drier climates developing in the Miocene in equatorial Africa [55]. The presence of 'temperate elements' (e.g., *Myrica, Podocarpus*) in the rainforests of Central America [56] might also be explained on this basis: taxa drawn into the rainforests as they contracted under unfavourable climates, and subsequently persisting as relicts. Many of the differences between the present-day tropical floras of America, Africa and Asia, however, date back almost certainly to the early (pre-Tertiary) separation of these land masses – though it must not be forgotten that secondary contacts subsequently occurred in all these regions also. Moreover, the possibility of rapid speciation in tropical floras relatively recently in response to repeated changes of range during the Quaternary Period (discussed in the next chapter) has gained favour in recent years [57], and has highlighted the complexity of speciation processes in low latitudes. Thus although the basic framework for low-latitude floras may have been laid down in the Early Tertiary, much change has probably occurred subsequently.

Modern high-latitude floras have clearly been assembled only recently, by immigration of plants into regions formerly covered by ice. Moreover, most cold-adapted plants are unlikely to have evolved at high latitudes in the first place, as these regions were forested until late in the Tertiary. More probably, the uplift of high-mountain areas in middle latitudes from the Miocene onwards provided ecospace above the forests for alpine floras to develop, and these subsequently spread into polar regions (refer also to Chapter 14).

Tectonic instability, registered most dramatically in periodic mountain uplift, undoubtedly played as important a role in modernization of floras as did deterioration of global climate. Extensive new ecospace, both at high altitudes and in rainshadow areas experiencing seasonal drought, and initially free of competition from an established plant cover, was created by orogenesis, and these harsher environments encouraged the emergence and spread of new genotypes. Adaptive radiation thus probably accompanied orogenesis. Particularly in the Pliocene Epoch we find an extended range of new plant taxa, many of them herbaceous, associated with the developing sclerophyll woodlands, savannas and grasslands – in, for example, California [28], the central United States [58], New Zealand [59] and South Africa [55]. These drier-climate biomes formed in juxtaposition to more mesic forest communities, and during the subsequent Pleistocene Epoch repeatedly expanded and contracted their ranges, in response to the recurrent climatic fluctuations. The mingling of adjacent floras thus brought about could have led to increased hybridization and further speciation [55].

Where mountain uplift occurred early on in the Tertiary Period, however, it may have hastened modernization by accentuating regional climatic deterioration and increasing the rate of elimination of tropical and subtropical taxa. Thus the montane coniferous forests of the central Rocky Mountains, occupying areas already partially elevated in the Early Tertiary, had achieved a largely modern aspect by the Miocene [50], 10–15 Ma earlier than the west-coast forests where mountain uplift was delayed until the Pliocene.

What is clear from the fossil record is that modernization has proceeded at different rates in different geographical regions, dependent upon the nature and timing of the climatic and tectonic changes in those regions. Modernization was notably delayed in regions with a strong herbaceous component to their present-day floras, and in land masses that 'drifted' into secondary contact late in the Cenozoic. Modernization, conversely, was probably achieved relatively early in tectonically more stable regions with predominantly woody angiosperm floras.

REFERENCES

1. Hickey, L. J. and Doyle, J. A. (1977) Early Cretaceous fossil evidence for angiosperm evolution. *Bot. Rev.*, **43**, 3–104.
2. Hughes, N. F. (1977) Palaeo-succession of earliest angiosperm evolution. *Bot. Rev.*, **43**, 105–27.
3. Kerr, R. A. (1984) How to make a warm Cretaceous climate. *Science*, **223**, 677–8.
4. Parrish, J. T. (1987) Global palaeogeography and palaeoclimate of the Late Cretaceous and Early Tertiary, in *The Origins of Angiosperms and their Biological*

Consequences (eds M. E. Friis, W. G. Chaloner and P. R. Crane), Cambridge University Press, Cambridge, pp. 51–73.

5. Axelrod, D. I. (1966) Origin of deciduous and evergreen habits in temperate forests. *Evolution*, **20**, 1–15.
6. Creber, G. T. and Chaloner, W. G. (1985) Tree growth in the Mesozoic and Early Tertiary and the reconstruction of paleoclimates. *Palaeogeogr. Palaeoclimatol. Palaeoecol.*, **52**, 35–60.
7. Takhtajan, A. (1969) *Flowering Plants. Origin and Dispersal* (transl. by C. Jeffrey), Oliver and Boyd, Edinburgh.
8. Krasilov, V. A. (1981) Changes of Mesozoic vegetation and the extinction of dinosaurs. *Palaeogeogr, Palaeoclimatol. Palaeoecol.*, **34**, 207–24.
9. Crane, P. R. (1987) Vegetational consequences of the angiosperm diversification, in *The Origins of Angiosperms and their Biological Consequences* (eds M. E. Friis, W. G. Chaloner and P. R. Crane), Cambridge University Press, Cambridge, pp. 107–44.
10. Stebbins, G. L. (1974) *Flowering Plants: Evolution above the Species Level*, Arnold, London.
11. Wing, S. L. and Tiffney, B. H. (1987) The reciprocal interaction of angiosperm evolution and tetrapod herbivory. *Rev. Palaeobot. Palynol.*, **50** 179–210.
12. Regal, P. J. (1977) Ecology and evolution of flowering plant dominance. *Science*, **196**, 622–9.
13. Upchurch, G. R. and Wolfe, J. A. (1987) Mid-Cretaceous to Early Tertiary vegetation and climate: evidence from fossil leaves and woods, in *The Origins of Angiosperms and their Biological Consequences* (eds M. E. Friis, W. G. Chaloner and P. R. Crane), Cambridge University Press, Cambridge, pp. 75–105.
14. Russell, D. A. (1979) The enigma of the extinction of the dinosaurs. *Ann. Rev. Earth Planet. Sci.*, **7**, 163–82.
15. Officer, C. B., Hallam, A., Drake, C. L. and Devine, J. D. (1987) Late Cretaceous and paroxysmal Cretaceous/Tertiary extinctions. *Nature*, **326**, 143–9.
16. Hickey, L. J. (1981) Land plant evidence compatible with gradual, not catastrophic, change at the end of the Cretaceous. *Nature*, **292**, 529–31.
17. Kerr, R. A. (1987) Asteroid impact gets more support. *Science*, **236**, 666–8.
18. Alvarez, L. W., Alvarez, W., Asaro, F. and Michel, H. V. (1980) Extraterrestrial cause for the Cretaceous–Tertiary extinction. *Science*, **208**, 1095–1108.
19. Alvarez, W., Alvarez, L. W., Asaro, F. and Michel, H. V. (1984) The end of the Cretaceous: sharp boundary or gradual transition. *Science*, **223**, 1183–6.
20. Tschudy, R. H., Pillmore, C. L., Orth, C. J. *et al.* (1984) Description of the terrestrial plant ecosystem at the Cretaceous–Tertiary boundary, Western Interior. *Science*, **225**, 1030–2.
21. Wolbach, W., Lewis, R. S. and Anders, E. (1985) Cretaceous extinctions: evidence for wildfires and search for meteoritic material. *Science*, **230**, 167–70.
22. Duncan, R. A. and Pyle, D. G. (1988) Rapid eruption of the Deccan flood basalts at the Cretaceous/Tertiary boundary. *Nature*, **333**, 841–3.
23. Wolfe, J. A. and Upchurch, G. R., Jr (1986) Vegetation, climatic and floral changes at the Cretaceous–Tertiary boundary. *Nature*, **324**, 148–52.
24. Wolfe, J. A. (1977) Palaeogene floras from the Gulf of Alaska region. *US Geol. Survey Prof. Pap.*, **997**, 1–108.

25. Levin, D. A. and Wilson, A. C. (1976) Rates of evolution in seed plants: net increase in diversity of chromosome numbers and species numbers through time. *Proc. Nat. Acad. Sci. USA*, **73**, 2086–90.
26. Stanley, S. M. (1979) *Macroevolution. Pattern and Process*, Freeman, San Francisco.
27. Rosen, D. E. (1978) Vicariant patterns and historical explanation in biogeography. *Syst. Zool.*, **27**, 159–88.
28. Raven, P. H. and Axelrod, D. I. (1972) Plate tectonics and Australasian paleobiogeography. *Science*, **176**, 1379–86.
29. Bande, M. B. and Prakash, U. (1986) The Tertiary flora of Southeast Asia with remarks on its palaeoenvironment and phytogeography of the Indo-Malayan region. *Rev. Palaeobot. Palynol.*, **49**, 203–33.
30. Lakhanpal, R. N. (1970) Tertiary floras of India and their bearing on the historical geology of the region. *Taxon*, **19**, 675–94.
31. Schuster, R. M. (1974) Plate tectonics and its bearing on the geographical origin and dispersal of angiosperms, in *Origin and Early Evolution of Angiosperms* (ed. C. B. Beck), Columbia University Press, New York, pp. 48–138.
32. Li, H.-L. (1953) Present distribution and habitats of the conifers and taxads. *Evolution*, **7**, 245–61.
33. Christophel, D. C. and Greenwood, D. R. (1989) Changes in climate and vegetation in Australia during the Tertiary. *Rev. Palaeobot. Palynol.*, **58**, 95–109.
34. Kornás, J. (1972) Corresponding taxa and their ecological background in the forests of temperate Eurasia and North America, in *Taxonomy, Phytogeography and Evolution* (ed. D. H. Valentine), Academic Press, London, pp. 37–59.
35. Tanai, T. (1972) Tertiary history of vegetation in Japan, in *Floristics and Palaeofloristics of Asia and Eastern North America* (ed. A. Graham), Elsevier, Amsterdam, pp. 235–55.
36. Gentry, A. H. (1982) Phytogeographic patterns as evidence for a Chocó Refuge, in *Biological Diversification in the Tropics* (ed. G. T. Prance), Columbia University Press, New York, pp. 112–36.
37. Marshall, L. G., Webb, S. D., Seposki, J. J. and Raup, D. M. (1982) Mammalian evolution and the Great American Interchange. *Science*, **215**, 1351–7.
38. Webb, S. D. (1976) Mammalian faunal dynamics of the Great American Interchange. *Paleobiol.*, **2**, 220–34.
39. Webb, S. D. (1977) A history of savanna vertebrates in the New World. Part I: North America. *Ann. Rev. Ecol. Syst.*, **8**, 355–80.
40. Webb, S. D. (1978) A history of savanna vertebrates in the New World. Part II: South America and the Great Interchange. *Ann. Rev. Ecol. Syst.*, **9**, 393–426.
41. Troll, C. (1960) The relationship between the climates, ecology and plant geography of the southern cold temperate zone and of the tropical high mountains. *Proc. Roy. Soc.*, Series B, **152**, 529–32.
42. Van Steenis, C. G. G. J. (1979) Plant-geography of east Malesia. *Bot. J. Linn. Soc.*, **79**, 97–178.
43. Walker, D. (1982) Speculations on the origin and evolution of Sunda–Sahul rain forests, in *Biological Diversification in the Tropics* (ed. G. T. Prance), Columbia University Press, New York, pp. 554–75.
44. Couper, R. A. (1960) Southern hemisphere Mesozoic and Tertiary Podocar-

paceae and Fagaceae and their palaeogeographic significance. *Proc. Roy Soc.*, Series B, **152**, 491–500.

45. Kershaw, A. P. (1988) Australasia, in *Vegetation History* (eds B. Huntley and T. Webb III), Kluwer Academic Publishers, Dordrecht, pp. 237–306.
46. Whitmore, T. C. (ed.) (1981) *Wallace's Line and Plate Tectonics*, Clarendon Press, Oxford.
47. Frenzel, B. (1973) *Climatic Fluctuations of the Ice Age* (transl. by A. E. M. Nairn), Case Western Reserve University Press, Cleveland.
48. Smith, A. G. and Bridén, J. C. (1977) *Mesozoic and Cenozoic Paleocontinental Maps*. Cambridge University Press, Cambridge.
49. Chaney, R. W. (1940) Tertiary forests and continental history. *Geol. Soc. Am. Bull.*, **51**, 469–88.
50. Leopold, E. B. and MacGinitie, H. D. (1972) Development and affinities of Tertiary floras in the Rocky Mountains, in *Floristics and Palaeofloristics of Asia and Eastern North America* (ed. A. Graham), Elsevier, Amsterdam, pp. 147–200.
51. Suchlik, L. (1964) Pollen analyses of the Miocene deposits at Rypin. *Acta Palaeobot.*, **5**, 1–111.
52. Wolfe, J. A. (1972) An interpretation of Alaskan Tertiary floras, in *Floristics and Palaeofloristics of Asia and Eastern North America* (ed. A. Graham), Elsevier, Amsterdam, pp. 201–33.
53. Straka, H. (1975) *Pollen- und Sporenkunde. Eine Einführung in die Palynologie*, Fischer, Stuttgart.
54. Shatilova, I. I. (1986) The palynological base of stratigraphical subdivision of Late Cainozoic deposits of the Western Transcaucasus. *Rev. Palaeobot. Palynol.*, **48**, 409–14.
55. Axelrod, D. I. and Raven, P. H. (1978) Late Cretaceous and Tertiary vegetation history of Africa. *Monogr. Biol.*, **31**, 77–130.
56. Graham, A. (1975) Late Cenozoic evolution of tropical lowland vegetation in Veracruz, Mexico. *Evolution*, **29**, 723–35.
57. Prance, G. T. (1982) Forest refuges: evidence from woody angiosperms, in *Biological Diversification in the Tropics* (ed. G. T. Prance), Columbia University Press, New York, pp. 137–57.
58. Axelrod, D. I. (1985) Rise of the grassland biome, central North America. *Bot. Rev.*, **51**, 163–201.
59. Mildenhall, D. C. and Pocknall, D. I. (1984) Palaeobotanical evidence for changes in Miocene and Pliocene climates in New Zealand, in *Late Cainozoic Palaeoclimates of the Southern Hemisphere* (ed. J. C. Vogel), Balkema, Rotterdam, pp. 159–71.

CHAPTER EIGHT

Plant migration and survival

8.1 INTRODUCTION

Because land plants are attached to their substrate, they cannot move to new locations if environmental conditions change. As a result, worsening conditions must lead ultimately to the death of an individual plant; and it is the progeny of that plant that adapt, migrate, or perish. At the risk of oversimplification we can say that the individual plants in any population can be of three types: reproductive individuals, non-reproductive individuals, and seedlings. Seedlings are plants that have recently entered the population, and may not survive to maturity; reproductive individuals are the norm in a favourable environment, whilst individuals that fail to flower and fruit (the non-reproductive individuals) may be immature, senile, or mature but growing in locally unfavourable conditions. The life-span of an individual plant normally encompasses successive seedling, non-reproductive, reproductive and non-reproductive stages, and may extend over several months, a few years, or many decades. Within a circumscribed plant population (i.e., one occupying a defined geographical area), moreover, environmental conditions at the periphery of the range might be expected to be less favourable than those in the central part, and accordingly we might predict that reproductive individuals would be relatively more frequent centrally in the population and non-reproductive individuals more frequent peripherally. Seedlings are characteristically restricted within the population to particular localized situations – of favourable light intensity, substrate, etc. Seedlings also occur beyond the range boundary; however, in a stable environment (and hence with a static population) these seedlings fail to become established. If environmental conditions change, then parts of the plant range (at, for example, the northern limits) will normally be subjected to worsening conditions, and other parts (at e.g., the southern limits) to ameliorating conditions. Under worsening conditions more individual plants become non-reproductive, whilst mortality, both of established plants and of seedlings within the range boundary, increases. When established plants die they are not replaced by seedlings: consequently a **retreating range margin** develops.

Under ameliorating conditions, at other parts of the range boundary, the local reproductive capacity of the population may improve, as more individuals become fully reproductive, but, more importantly, seedlings establish themselves beyond the range boundary ('extra-territorial seedlings'). An **advancing range margin** develops, and the plant population expands into new territory. Both advancing and retreating range margins constitute **plant migration**; as a result of migration the spatial location of the range changes. If the retreating range margin moves more rapidly than the advancing margin, **extinction** of the population ultimately occurs. If as a result of the changing environmental conditions certain seedlings in the population with aberrant genetic constitution become preferentially established, either within the existing range of the population or beyond its boundaries, then the genetic composition of the population may gradually alter, and evolutionary change occurs as new **biotypes** develop.

The major changes in environmental conditions that are likely to induce plant migration are those of climate. In the following pages, accordingly, we examine the role of climatic change in modifying plant distribution patterns, and discuss some of the consequences arising from these modifications.

8.2 THE ENVIRONMENTAL CONTROL OF PLANT MIGRATION

8.2.1 The speed and magnitude of climatic change

The relative speeds of the retreating and advancing range margins of a plant population in a changing environment determine whether survival or extinction of that population occurs. These rates in turn are determined by the relative speeds of environmental change and plant response. The speed of plant response depends largely on the behaviour of the extra-territorial seedling population (i.e., those seedlings that establish beyond the current range boundary) – which is affected by the 'seed inoculum' (i.e., the numbers of seeds dispersed to different distances beyond the current range margin), the 'environmental resistance' to seedling establishment (i.e., competition from other plants already present, unfavourable soil conditions, etc.), and the 'generation time' of the establishing population (i.e., the time elapsing between germination and first seed production). The speed of plant response is also affected by the balance between reproductive and non-reproductive individuals in the population. Rather little information exists about the possible climatic control of this balance, though Iversen's classic work [1] on the distributions of ivy (*Hedera helix*), holly (*Ilex aquifolium*) and mistletoe (*Viscum album*) in Europe sheds some light on the problem. Iversen examined the performance of populations of these three species in locations throughout Europe adjacent to meteorological stations, characterizing each population as fully reproductive, reproductive only in favourable microhabitats, largely non-reproductive, or totally non-reproductive. Iversen expressed his observations graphically in relation to the mean temperature of the warmest and coldest months at each location. For ivy (Figure 8.1) he found that the degree of winter cold was crucial, with a difference of only 1°C in the mean temperature of the coldest month separating non-reproductive and fully reproductive populations in certain regions. A change of this magnitude thus represents the difference between success and failure for ivy, in one sense, at the limits of its range.

Climatic changes of 1°C magnitude are known to have occurred in historical times within the course of a few centuries. Thus Professor H. H. Lamb's estimates of temperature changes in central England since 800 AD (Figure 8.2) show that mean summer temperatures declined by about 0.5°C between 1230 and 1330 AD, and by a further 0.4°C by 1480 AD. A rate of change of 0.5°C per century, or less, is indicated. This rate can be translated approximately into geographical displacement (of, for example, plant ranges), on the basis that in mid-latitudes in the northern hemisphere the annual temperature gradient is about 0.7°C per 1° of latitude. A

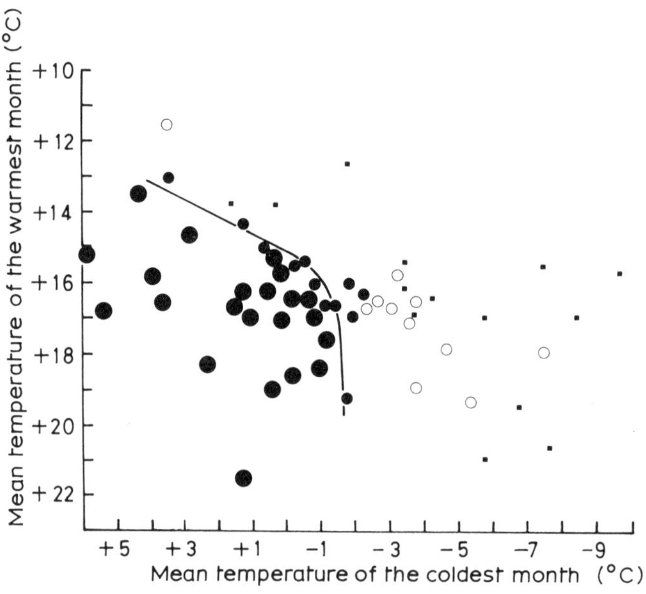

Figure 8.1 Thermal correlation curve for European populations of ivy (*Hedera helix*). Re-drawn from Godwin [2]. ● Regularly fertile; • fertile only in exceptional situations; ○ never bears ripe fruit; ▪ absent within the study area.

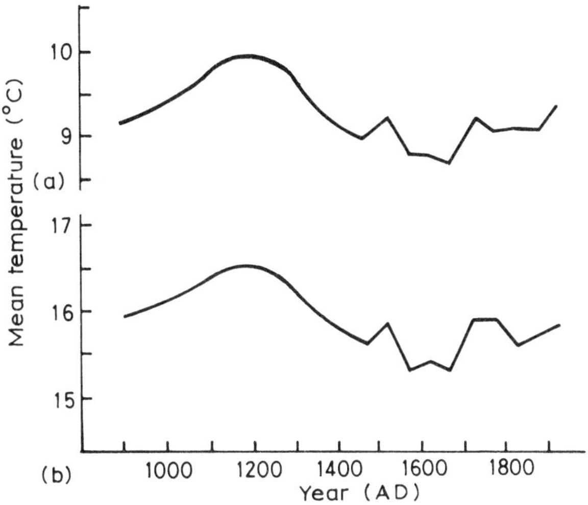

Figure 8.2 Estimates of prevailing temperatures in central England since 800 AD. (a) Averages for the whole year; (b) averages for high summer (July–August). Re-drawn from Lamb [6].

potential latitudinal displacement of about 80 km per century (or a corresponding altitudinal displacement of 70–80 m per century) is thus indicated. An independent check that this estimate is of the right order of magnitude is provided by observations of glacier retreat at Glacier Bay, Alaska. There the position of the ice front was displaced inland by nearly 100 km between the late 1700s and the early 1900s [3].

Although the rate of ice retreat at Glacier Bay is often regarded as rather exceptional, estimates of the rate of retreat of the great Pleistocene ice sheets of North America and northern Europe show that this is not so. Comparable rates (up to 65 km per century) at the height of ice-recession have been calculated [4]. Even faster rates of change for that time-period (the latest Pleistocene) have recently been calculated for Britain, with estimates of temperature change of at least 2.6°C per century at 13 ka ago (and perhaps as high as 7.2°C per century), and 1.7°C per century at 10 ka ago [5]. These rates of change, moreover, were maintained over several (5–10) centuries.

Indirect evidence that rapid temperature change has been a recurring feature of at least the last 5 ka is provided by the long-term proxy record of climatic change contained in the tree-ring widths of the Californian bristlecone pine (Figure 6.2). Major pulses, where increases or decreases in ring width were maintained consistently for time-periods of 100–200 years, are easily discernible at intervals throughout the 5-ka record. Two such pulses can be seen in the last millennium: one from about 900 to 1100

AD (increasing ring width) and one from about 1250 to 1400 AD (decreasing ring width). These pulses match up with the temperature trends in Lamb's record for central England (Figure 8.2 – where the rate of change was about 0.5°C per century), and we may perhaps assume accordingly that earlier 'punctuations' in the bristlecone pine record also were characterized by similar rates of change.

Climatic cycles of longer duration (often in excess of 10^4 yr) result necessarily in overall temperature changes of much greater magnitude than 1°C; usually, however, the rate of change is less [7]. Evidence for the magnitude of these temperature changes has been discussed in preceding chapters. The extent of the resulting geographical displacements can be gauged from the positions of the major climatic belts at specified times in the past – which are discussed at various places in this book. Displacements of climatic belt boundaries through up to 20° of latitude (more than 2000 km) have probably occurred through the Quaternary Period, though over the last 10 ka the displacements have been on a more modest scale – over distances of hundreds rather than thousands of kilometres. During the Tertiary Period the prevailing climatic feature over millions of years at higher latitudes was a trend towards lower temperatures, accompanied by a unidirectional shift of plant ranges away from the poles towards the equator. If the Spitsbergen fossil floras are to be believed (section 1.2), then this shift involved displacements of up to 5000 or 6000 km.

8.2.2 Measuring rates of plant migration

Because plant migration is a long-term process it can rarely be measured by direct observation. Well-documented examples of recent plant migration in response to climatic change, as at Glacier Bay, Alaska [3,8,9], yield at best a short-term estimate of migration rates (50–70 km per century for the pioneer plants at Glacier Bay). The rapid rates of spread of alien plants introduced to far-off countries within the last 100 years (*Senecio squalidus* in Britain, *Opuntia* spp. in Australia, for example), do not represent comparable situations, as climatic change is not normally involved in the spread of such aliens. Long-term rates of plant migration in the past can be assessed, however, from pollen evidence. The immigration of a major pollen-producing species into a region not previously occupied by it is manifested in pollen diagrams by a characteristic rising curve of frequency, which is preceded by a prolonged period of sporadic low values and terminates in a near-steady state of pollen abundance. The pattern shows a striking resemblance to the theoretical and observed curves for **logistic growth** in plant and animal populations, and has been so interpreted [10]. Given suitable radiocarbon-dating control, the pattern can be quantified in various ways: by the date of initial population increase, the date of

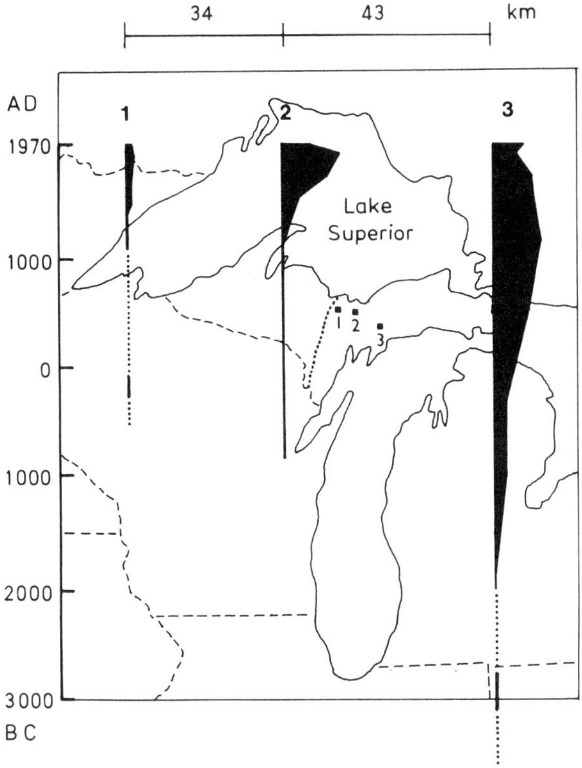

Figure 8.3 The pollen record of plant migration: pollen percentages of beech (*Fagus*) plotted against time at three sites in Upper Michigan, USA. The dotted line just to the west of site 1 is the present western limit of beech. Re-drawn from Davis [15].

maximum increase, the total duration of the growth phase, the rate of population increase (r), or the time for doubling of the population size during the growth phase [11]. The growth phase for temperate trees in the Holocene typically lasted several hundred years, and was thus spread over several generations: 400–500 yr in *Pinus sylvestris* and *Corylus* in eastern England [11], about 900 yr in *Pinus strobus* in Minnesota [12]. Bennett [13] gives a mean doubling time for sixteen temperate tree taxa of 158 yr (with a range of 31 to 462 yr). A very similar value (145 yr, range 63–355 yr) is recorded by Chen [14] for eleven Queensland rainforest trees; the data suggest shorter doubling times for successional taxa than for canopy taxa.

A major problem in determining rates of plant migration from pollen evidence is that a tree species may be present in small numbers in an area, and yet not be visible in the pollen record until its numbers have increased

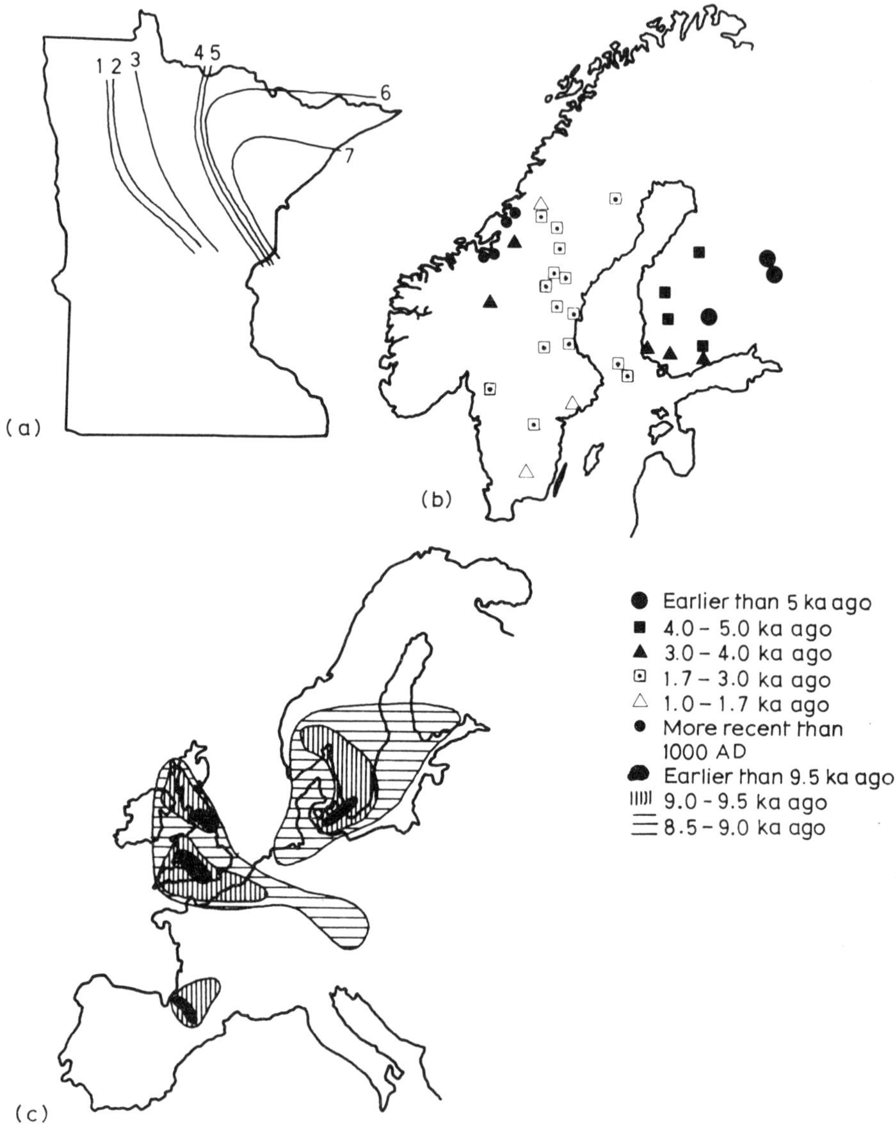

Figure 8.4 Holocene migration rates. (a) *Pinus strobus* in Minnesota (isochrones in ka ago); (b) *Picea abies* in Scandinavia; (c) *Corylus avellana* in western Europe. Re-drawn from Jacobson [16], Tallantire [17] and Deacon [18].

substantially [13]. Thus some doubt exists as to the validity of the customary methods of determining migration rates – which utilize either the date of the beginning of the first sustained increase in pollen frequency, or of the attainment of a specific pollen frequency. Evidence of plant migration from pollen data may indeed relate to the rates of change of a certain (unspecified) level of abundance of the population [13] rather then to overall range changes. Nevertheless, it is normally believed that a comparison of dated pollen curves for a particular pollen taxon from different locations within a single geographical region across which it is migrating enables average rates of migration for that taxon to be calculated. Thus the three curves in Figure 8.3 for beech (*Fagus grandifolia*) in Upper Michigan, dated approximately by radiocarbon, show an extension of range by 77 km in 3500 years [15]. More wide-ranging data for three other taxa are shown in Figure 8.4. For white pine (*Pinus strobus*) spreading through Minnesota [16], and for spruce (*Picea abies*) spreading through Scandinavia [17], a maximum rate of migration of about 50 km per century in the later Holocene was achieved; for hazel (*Corylus*) in northern Europe, rather faster rates were achieved in the Early Holocene (>100 km per century), but rates in southwestern Europe at the same time were considerably less [18].

8.2.3 Factors influencing migration rate

(a) Climatic change

Huntley and Birks [19] list 43 determinations of migration rate for 16 European trees in the Holocene. Values range from 2.5 km per century for ash (*Fraxinus ornus*) to 200 km per century for alder (*Alnus*), with a mean of 40.6 km per century; 56% of the values lie between 20 and 50 km per century. In eastern North America, Holocene migration rates for 20 tree taxa range from 8.7 to 41.6 km per century, with a mean value of 20.5 km per century [20]. The rates thus differ by nearly two orders of magnitude; although these rates are for a very limited and non-representative set of woody taxa (there are almost no data for tropical and subtropical trees, or for growth-forms other than trees), they nevertheless indicate that plants may not always have the capacity to respond conformably with (i.e., to migrate at the same rate as) climatic change. As a result, ranges may develop that are not in equilibrium with the contemporary climate.

Theoretical reasons for such differences between taxa are not difficult to find. Thus in general terms we can appreciate that the efficient seed-dispersal mechanisms and short generation times of many annual plants should enable those species to respond more rapidly to climatic change than can forest trees with heavy seeds and long generation time (10–50 yr), and that light-seeded trees such as birch (*Betula*) should migrate more rapidly than heavy-seeded trees such as oak (*Quercus*) and beech (*Fagus*). However, an underlying climatic control is also apparent in the data. The effects of climatic change on the pattern of migration of a single species are elegantly shown by the progress of white pine (*Pinus strobus*) across Minnesota in the mid-Holocene (Figure 8.4). The westward-moving invasion front of this species was held almost stationary for about 2500 yr after 6.5 ka bp, when markedly drier conditions are known to have prevailed over much of the North American subcontinent (section 13.3); these drier conditions prevented the spread of the pine forest into the adjoining prairie [16]. The importance of environmental factors in governing rates of migration is also indicated by Delcourt and Delcourt's work [20] on eastern North American trees. The calculated rates of migration of four tree taxa (*Carya, Juglans, Populus* and *Quercus*) varied considerably between five different migration pathways (Figure 8.5); two of the trees achieved their fastest rates along the easternmost track and two along the westernmost.

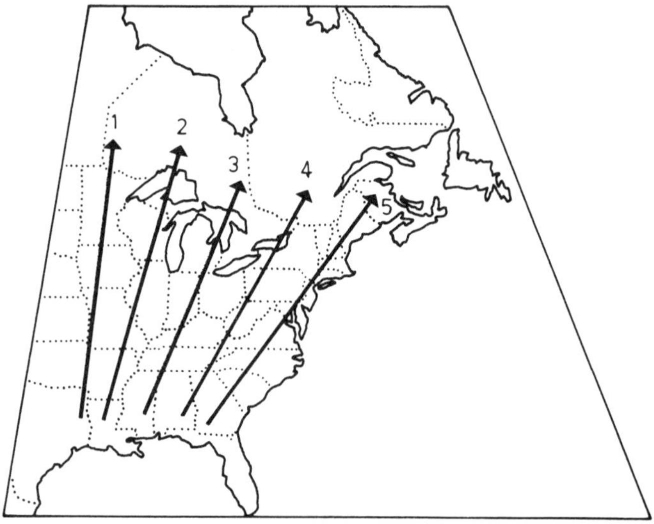

	Migration track 1	2	3	4	5
Carya	19.3	9.6	15.4	7.6	7.4
Juglans	22.3	22.1	7.1	7.0	11.4
Populus	16.7	22.3	27.6	29.7	35.0
Quercus	11.9	11.0	11.7	12.1	16.3

(all values km per century)

Figure 8.5 Migration rates (km per century) of four forest trees along five different tracks in eastern North America in the Holocene. (Data of Delcourt and Delcourt [23].)

At the beginning of this period of migration, a number of trees in eastern North America showed pulses of more rapid migration at approximately the same time – at 13.5, 12.3 and 10.0 ka ago – when major shifts in the patterns of atmospheric circulation were also occurring [21]. Climatic control of migration, over limited time-spans within the total duration of the last warm stage, is again indicated.

(b) Long-distance dispersal

In a changing environment, favourable conditions for seedling establishment are most likely to pertain close to the current range margin (where the greatest seed inoculum is), and consequently the likelihood of seeds dispersed to exceptional distances beyond the range margins becoming established is remote. The logical conclusion is that much of plant migration must take place by gradual advancement of the leading range margin through local seed dispersal. Many foresters believe, in fact, that seed dispersal by a forest tree is almost entirely confined within an area around the parent tree of radius equal to the height of the tree. Dispersal beyond 100 m is thus exceptional. Even for a forest tree with a short generation time (10–15 yr), migration rates in excess of 1 km per century are unlikely then to be achieved by normal free-fall dispersal mechanisms [22]. Such a value is considerably less than any recorded for temperate forest trees in the Holocene [19,20], and suggests that long-distance dispersal must also be involved; indeed, some past distribution patterns can be explained only by assuming long-distance dispersal to have taken place. Thus during the last glacial maximum in the western United States many coniferous tree species were displaced well to the south of the Great Basin. Subsequently they returned, to form montane conifer forest at higher altitudes on the peaks and ranges of the Great Basin; some of them (e.g., lodgepole pine, *Pinus contorta*; pinyon pine, *P. monophylla*; white fir, *Abies concolor*) migrated up to 640 km northwards, covering this distance in 8000 years. Wells [22] has calculated that free-fall dispersal of seeds would have been quite inadequate to achieve this rate of migration, and concludes that other agents were almost certainly involved – birds such as Clark's Nutcracker (*Nucifraga columbiana*) and the Pinyon Jay (*Gymnorhinus cyanocephalus*). If Wells' contentions about unaided seed dispersal are correct, then much tree migration in the Holocene may have involved longer-distance dispersal by animals.

However, animal dispersal may not be necessary to explain the exceptionally rapid spread of white spruce (*Picea glauca*) through western Canada, documented by Ritchie and MacDonald [24]. Spruce initially migrated rather slowly (20–30 km per century) northwestwards through eastern North America in the Late Pleistocene, in the wake of the waning

Laurentide Ice Sheet, but was prevented from spreading northwards into unglaciated Alaska until an ice-free corridor opened up between the Laurentide and Cordilleran Ice Sheets about 12 ka ago [25]. Once access to the north was possible, spruce migrated from southern Manitoba to the Mackenzie Delta on the northern coast of Canada (a distance of nearly 3000 km) in less than 2000 years. Ritchie and MacDonald suggest that the light seeds of spruce were blown long distances over frozen ground in winter by the persistent strong southeasterly winds around the western margin of the melting Laurentide Ice Sheet.

The higher overall rates for European trees than for North American trees noted above may be in part an artefact, a consequence of the rather different methods used in their derivation [20], but may also be a result of the rather different environmental conditions encountered by the northward-migrating trees in the two regions. Thus water-dispersal of seeds or fruits of streamside trees would have been more effective in Europe, with a network of northward-flowing rivers, than in eastern North America, where most rivers drain to the south. Moreover, much of central and northern Europe at the start of the Holocene was covered by tundra-like vegetation offering little resistance to invading trees, whereas in eastern North America a widespread cover of coniferous trees was already present [20].

(c) Competition

Any expansion of the range of a plant population, in fact, can be achieved only by displacing plants of other species growing adjacently – by overcoming competitive resistance. Differing competition from the established vegetation has been invoked, for example [18], to explain the very different migration rates of hazel (*Corylus avellana*) in northern and western Europe in the Early Holocene. In Scandinavia and Britain, where the only woody taxa already present were of rather low competitive ability – birch (*Betula*), pine (*Pinus*), poplar (*Populus*), willow (*Salix*) and juniper (*Juniperus*) – hazel spread at rates of 90–126 km per century. In south-central Europe, facing competition from oak (*Quercus*), hazel spread at only 34 km per century (Figure 8.2). Across the Atlantic, in the American Midwest, hornbeam (*Ostrya/Carpinus*) apparently spread more slowly into areas where ash (*Fraxinus nigra*) was already established. Thus the estimated doubling time for populations of hornbeam at sites with and without existing ash populations was 650 and 145 yr, respectively [20].

In Nova Scotia, a mixed conifer – hardwood forest developed gradually through the Holocene by progressive incorporation of new tree species. The first appearances of several of these species (fir, *Abies*; maple, *Acer*; beech, *Fagus*; ash, *Fraxinus*; hop-hornbeam, *Ostrya*; oak, *Quercus*; and elm,

Ulmus) coincided with, or immediately followed, pronounced peaks of charcoal in the lake sediments, suggesting that individual tree species were able to enter the existing forest only when it had been opened up by fire [26]. Further to the south, in the New England states, an abrupt decline in pollen of hemlock (*Tsuga*) about 4.8 ka ago apparently records a major opening-up of the forest which allowed the ingress of several of the present-day components of the mixed conifer–hardwood forest. Different species-combinations were favoured in different areas; thus, beech and maple increased in abundance in northern New Hampshire following the hemlock decline, but pine and oak spread in the south of the state. A lethal outbreak of a fungal or insect pathogen may have been responsible for the sudden demise of the hemlock [27] (section 12.3.4).

The part played by prehistoric forest clearance in opening up the forests of northern Europe, and changing the proportions of the component tree species, has been stressed by many people. Ash (*Fraxinus excelsior*), for example, was a rare member of many English forests until limited forest clearance by Neolithic and Bronze Age settlers, followed by tree regeneration when the settlements were subsequently abandoned, allowed this opportunistic species to spread [28].

The plants of other species that flank a migrating plant population will almost certainly be different at the northern and southern range boundaries, and may well be growing on different soil types. Consequently, a plant population migrating south in the northern hemisphere, for example, under deteriorating climate will encounter different problems from those faced by the same population migrating north under ameliorating climate. Published rates of plant migration refer almost exclusively to the latter situation. It should not be forgotten that this is only one of a number of possible scenarios, and that a given plant may respond very differently to the various scenarios. For example, trees advancing northwards into tundra vegetation may have little competitive resistance to overcome, whereas the same trees advancing southwards will have to displace established forest or grassland species. In the northward advance, seedlings capable of establishing themselves on mineral soil will be favoured, and certain tree species may be held back for many years until organic matter and nitrogen in the soil have built up above a critical level [3]. On the other hand, tundra plants advancing southwards under deteriorating climate (i.e. a receding tree-line) may have to colonize the mor litter characteristically formed under boreal forest. Because this litter decays so slowly, the advance of the tundra plants may be delayed – perhaps until fire destroys the surface litter, often many years after the trees themselves have died [29].

If long-distance dispersal is indeed a valid agent of plant migration, then we can surmise, finally, that the likelihood of seedling establishment far from the parent plant will be enhanced where new habitats are being extensively created (by orogenesis, for example), as these new habitats are more likely to experience climatic conditions that differ from the 'norm of unfavourableness' for that latitude ('unfavourableness' being expressed relative to the climatic tolerances of the migrating plant population). Thus a northern temperate forest species migrating southwards under conditions of worsening climate will find suitable temperatures for seedling establishment far to the south of its range only at high altitudes in mountain areas. If such a mountain area has been recently uplifted, competition from the native flora of that region may be reduced, and the probability of successful establishment by seedlings of the migrating species correspondingly increased.

8.3 CASE-STUDIES OF PLANT MIGRATION

8.3.1 Movements of the treeline*

The values for rates of plant migration and climatic change discussed above are of similar magnitude, suggesting that plants may be capable of responding very sensitively to climatic change. If climatic change can in theory produce displacements of plant range of up to 80 km latitudinally, or 80 m altitudinally, within a century, then responses of this magnitude should produce detectable changes in plant distributions within a human lifespan. Unfortunately, in most situations where reliable documentation of change exists, the observed changes in plant distribution can also be explained in terms of variations of grazing intensity, frequency of major fires, soil deterioration, etc. Nevertheless, sufficient evidence exists for one particular range-boundary, the treeline (strictly speaking, a vegetation boundary, but where the treeline is formed by a single tree species then it is also a range-boundary), to show that plants have indeed responded sensitively to climatic change within recent centuries; enough is now known about the physiology of tree growth at the treeline [30,31] to be reasonably certain of some of the climatic control mechanisms involved. Moreover, as former

* Considerable confusion exists in the literature over the terms 'treeline' and 'timberline' (discussed also in Chapter 14). It might be better to abandon their usage altogether in favour of the more precise terms 'forest-limit' or 'forest-border' (the altitudinal or latitudinal termination of regions with 50% or more tree cover) and 'tree-limit' (the altitudinal or latitudinal limit of individual tall trees – single-stemmed woody plants at least 2 m tall). In mountain regions these two boundaries are often nearly coincident, and 'timberline' can then be used to denote the narrow zone between the two where conditions for tree growth quickly become limiting. In arctic regions, however, the two boundaries are typically separated by an extensive 'forest–tundra' zone. In this chapter and subsequently, 'treeline' is used as synonymous with 'tree-limit'.

positions of the treeline can also be reconstructed from the presence of preserved plant remains (timber, charcoal, leaves, cones) and from buried soil profiles, and can be dated by radiocarbon or by growth-ring counts, observed long-term displacements of this boundary in the past can be compared with deduced displacements of climatic boundaries over the same time-period.

The last millennium in temperate latitudes of the northern hemisphere has encompassed the major part of a complete climatic cycle (albeit one of rather low amplitude), from warm to cold conditions and back again to warm conditions (Figure 8.2). Maximum warmth at the onset of this cycle occurred between about 1150 and 1300 AD, and maximum cold between about 1550 and 1700 AD. There is clear evidence that in many parts of the northern hemisphere the treeline responded in parallel. Thus in the Scandes Mountains of central Sweden there are many reports from the last century of dead and dying trees of pine (*Pinus sylvestris*) and spruce (*Picea abies*) at higher altitudes. Most of these trees have now vanished, but it seems likely that their initial demise was caused by the cold conditions of the preceding three centuries (the 'Little Ice Age'). Moreover, growth-ring counts of surviving trees at selected localities showed from their ages that no seedlings had become established at higher altitudes between 1680 and 1720 AD. Since 1760 AD there had been progressive colonization of the higher altitudes by seedlings, so that now there is a broad belt of young trees (all less than 230 yr old) within and above the zone where the mature trees had earlier died. The treeline had risen by perhaps 50 m by 1915 AD, and by a further 25 m subsequently [32,33]. In the Khibiny Mountains of northwest Russia there is evidence (less good than that for the Scandes) that trees of pine (*Pinus sibirica*) and birch (*Betula tortuosa*) could be found 700 or 800 years ago up to altitudes of 520–550 m; subsequently the treeline was depressed to 250–300 m altitude during the Little Ice Age, but then started to rise again 200–250 yr ago towards its present-day limits of 360–365 m altitude [34].

A longer perspective is provided by the investigations of Bryson *et al.* [35] and Nichols [36,37] in northern Canada. They reconstructed the changing positions of the arctic treeline there over the last 8 ka, on the basis of soil and peat stratigraphy, pollen and macrofossil analyses, and numerous radiocarbon dates, at a series of sites along the 100°W meridian. In the Early Holocene, forest spread rapidly northwards over ground recently uncovered by the melting Laurentide Ice Sheet, to reach a position 280 km north of the present treeline by 5.5 ka ago. Subsequently there was a marked recession of the treeline about 3.5 ka ago, and again about 600 yr ago; between these two recessions there were other, smaller, fluctuations (Figure 8.6). Rather similar long-term changes in the position of the alpine treeline have been revealed by studies in a number of widely-separated

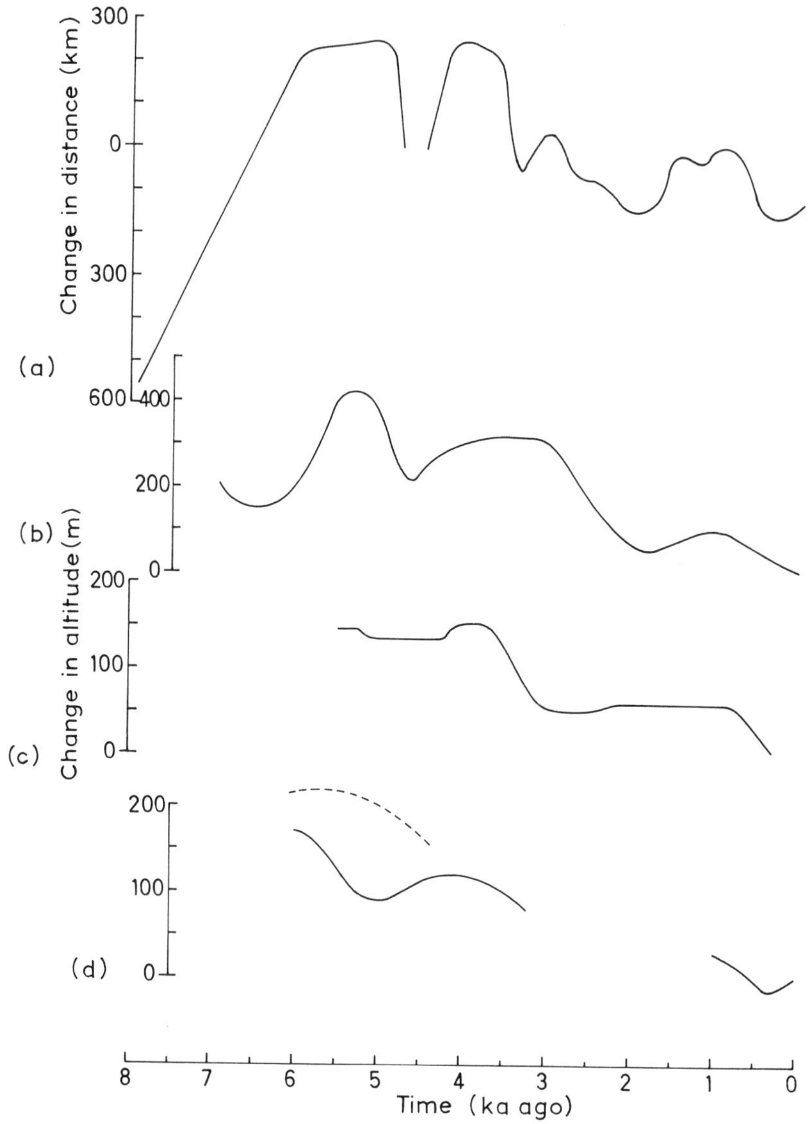

Figure 8.6 Changing positions of the treeline over the last 8 ka. (a) The arctic treeline in northern Canada [38,39]. (b) The alpine treeline in the central Swiss Alps [40]. (c) The alpine treeline in the White Mountains, California [41]. (d) The alpine treeline in the Scandes Mountains, Sweden; ——, *Pinus sylvestris*; – – –, *Betula tortuosa* [42]. The vertical axis shows distance relative to the position of the present-day treeline (the zero value).

mountain regions: the White Mountains of California (bristlecone pine), the Scandes Mountains of Sweden (pine and birch), and the Central Alps of Switzerland (pine, *Pinus cembra*, and larch, *Larix decidua*), for example. Over the last 5 ka the position of the treeline in these regions has varied by 150 m, 180 m and 400 m altitude, respectively (Figure 8.6). These and other observations suggest fluctuations in the height of the treeline by more than 100 m altitude within a few centuries in response to climatic change. Where the climatic change was long-sustained, as at the start of the Holocene Epoch, displacements were considerably greater; even so, the northward migration of the treeline by 700 km shown in Figure 8.6 represents rather less than half of its total displacement resulting from the melting of the North American Ice Sheet.

8.3.2 The Holocene migration of beech in Europe

The complexities of plant migration are well illustrated by the Holocene history of beech (*Fagus sylvatica*) in Europe. The pollen record documents a slow progression northwards and westwards through central Europe in the Mid-Holocene, from glacial-stage refugia in the Balkans and perhaps in Italy (Figure 8.12), at a rate of about 50 km per century (Figure 8.7). As beech is a heavy-fruited tree, that slow migration rate is not unexpected, and Godwin [2] believes that the rate of progress of beech lagged behind the development of climatic conditions favourable for its growth. In

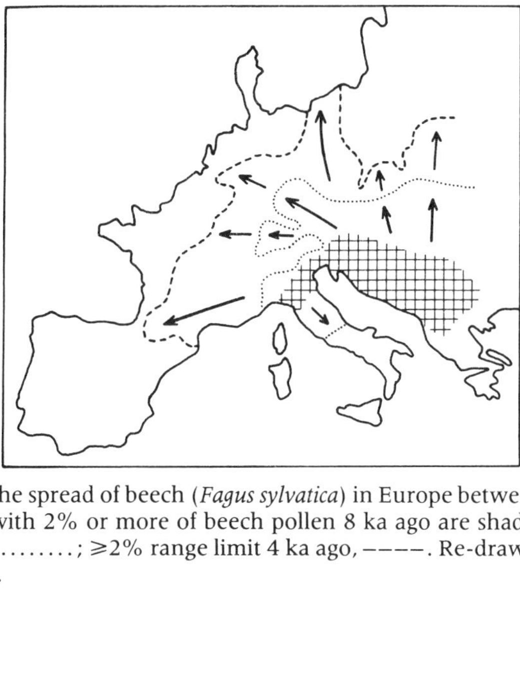

Figure 8.7 The spread of beech (*Fagus sylvatica*) in Europe between 8 ka and 4 ka ago. Regions with 2% or more of beech pollen 8 ka ago are shaded; ⩾2% range limit 6 ka ago, ...; ⩾2% range limit 4 ka ago, ––––. Re-drawn from Huntley and Birks [19].

support of this contention is the clear evidence from the pollen record that outposts of beech were established well to the north of the advancing front in Europe, and these outposts much later became foci for the spread of beech in these regions. Thus in southern England, the major rise of beech in pollen diagrams did not take place until about 2.5 ka ago, but isolated trees or stands must have been present for many centuries before that, as beech timber was used in the construction of wooden trackways across the flooded peatlands of the Somerset Levels around 4.5 ka ago [2]. Long-distance dispersal of the fruits by animals or birds must surely have been involved in this initial spread of beech, with competitive exclusion by previously-established forest trees preventing subsequent expansion of the pioneer beech populations. In north-west Europe the major expansion of beech occurred at a time when the climate might actually have been becoming less favourable (colder and wetter) for its growth. As the detailed pattern of expansion of beech in pollen diagrams from northern Germany differs from site to site [43,44], it is possible that prehistoric forest clearance could have been involved. In the Iron Age in southern England, for example, there was abandonment of cultivation of thin chalk downland soils previously cleared and farmed in the Neolithic and Bronze Ages, as more efficient ploughs enabled other deeper more fertile soils to be cultivated; the change in farming practice probably allowed beech to establish itself in the regenerating woodland on the chalk downs [2]. However, climatic change cannot be ruled out, as Huntley and Webb [45] suggest that the progressive reduction in seasonal extremes of climate over the last 5 ka (Figure 6.4) could have meant that winters in northern Europe were actually becoming milder (i.e., more favourable) at the time of the major beech expansion. Huntley and Webb point to a similar late expansion of the related *Fagus grandifolia* in eastern North America as support for the belief that climate was the controlling influence in the spread of beech on both sides of the Atlantic.

In the mountains of central Europe, beech is now a characteristic member of middle-altitude forests, in association with silver fir (*Abies alba*) and (at higher altitudes) spruce (*Picea abies*). In the Early and Mid Holocene, however, these trees were absent, and the middle-altitude forests were dominated instead by oak (Figure 8.8). The slow migration of beech, fir and spruce across Europe meant that the current forest zonation was established late, and in piecemeal fashion, in the Holocene. The major spread of beech occurred in the mid-Holocene [19]. In Switzerland the three trees entered the country by different migration routes (fir from the south, beech from the north-east, and spruce from the east), and rose to prominence in the middle-altitude forests at different times in different regions over at least a 1.5-ka period [40]. The extent to which forest clearance played a part in this scenario is still unclear.

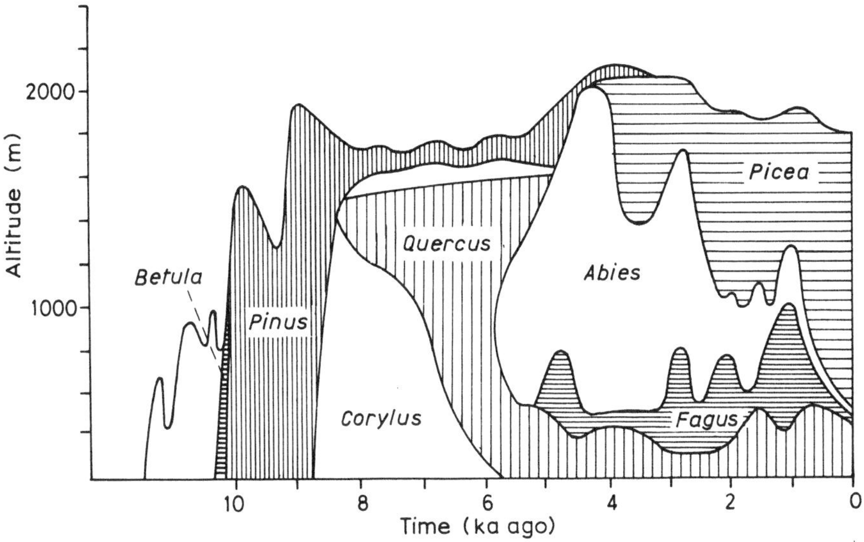

Figure 8.8 Change of dominance of forest trees with altitude in the Swiss Alps over the last 10 ka: an early diagram by Welten, based on pollen changes at a series of sites at different altitudes. Re-drawn from Faegri and Iversen [46].

8.4 SOME CONSEQUENCES OF PLANT MIGRATION

8.4.1 Compression and expansion of range

If climatic change is viewed simplistically as acting globally on a series of horizontally-disposed latitudinal or altitudinal zones (either climatic or vegetational), then its effect will be to displace these zones to new altitudes or latitudes. In the process, the width of these zones may be expanded or contracted; if a given plant population is viewed, also simplistically, as occupying a specified 'climatic space' that corresponds to a particular geographical region within which all the plant's climatic requirements are satisfied, then the climate space available to that population may also expand or contract as climate changes. The range of the plant population will accordingly not only be displaced, but may be varied in extent as well. Figure 8.9 shows two reconstructions that exemplify this process: the position and extent of vegetation zones in eastern North America over the last 20 ka (showing latitudinal displacements), and in the mountains of New Guinea over the last 15 ka (showing altitudinal displacements). The reconstructions demonstrate clearly the much greater 'inertia' of vegetation (i.e., resistance to change) both at low latitudes and at lower altitudes;

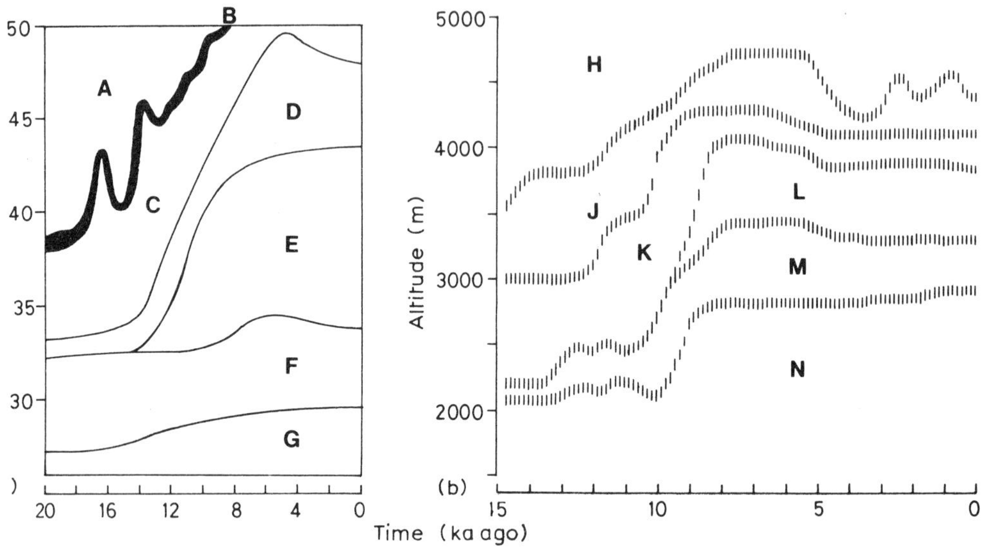

Figure 8.9 Expansion and contraction of climatic space. (a) Approximate positions of vegetation zones in eastern North America over the last 20 ka, based on studies at a series of sites close to the 85°W meridian (redrawn from Delcourt and Delcourt [23]); (b) reconstructed positions of major vegetation zone-boundaries in the mountains of New Guinea over the last 15 ka (redrawn from Hope [47]). A = Laurentide Ice Sheet, B = tundra, C = boreal forest, D = mixed conifer–hardwood forest, E = temperate deciduous forest, F = evergreen broadleaf forest, G = Gulf of Mexico; H = snow and ice, J = alpine zone, K = subalpine grassland L = subalpine forest, M = upper montane forest, N = lower montane forest.

although it would be unwise to generalize from such scanty data, nevertheless we might expect the most spectacular plant migrations in the past to have been in middle and higher latitudes and altitudes (where the greatest displacements of range were). Figure 8.9 indicates a marked compression of vegetational zones in middle latitudes and at middle altitudes during periods of glacial expansion, to the extent that some climatic space was almost entirely obliterated (notably the deciduous and subalpine forest 'spaces'). Shrinkage of climatic space must increase both inter- and intraspecific plant competition, and render extinction more probable. Rapid expansion of available climatic space, on the other hand, must place a premium on efficient seed-dispersal, and can be likened to a paper-chromatography system, separating out plant populations in space on the basis of their speeds of migration. In both Europe and North America, for example, pollen diagrams record a sequential spread of trees northwards in the Early Holocene. The sequence *Betula* → *Pinus* → *Corylus* → *Quercus, Ulmus* → *Alnus* → *Tilia* in Europe was undoubtedly in part a consequence of the relative cold-hardiness of these trees, but resulted also from their differing starting positions of migration as the ice-sheets melted, and from their differing efficiencies of seed-dispersal. A comparable situation pertained in eastern North America; but as the four trees featured in Figure 8.10 all have similar present-day northern limits (and hence may not differ markedly in cold-tolerance), their differing ranges 8 ka ago might have been the result of different migration histories during the preceding 2–4 ka [48].

Other processes (notably changing sea levels and mountain uplift) have also contributed to the expansion or contraction of climatic space over geological time. During the Pleistocene Epoch, for instance, changing sea levels were closely geared to the formation and melting of the major ice sheets, and must have offset to some extent the compression of vegetation zones during glacial periods by providing new habitable land immediately offshore of present-day coastlines.

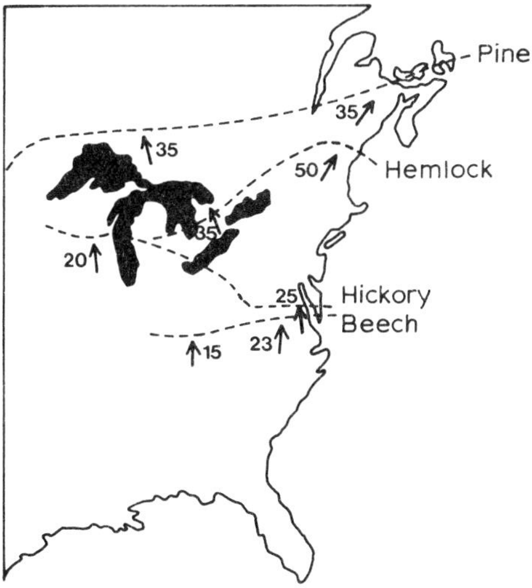

Figure 8.10 Northern limits of four trees in eastern North America 8 ka ago. Approximate migration rates (km per century) over the preceding 2–4 ka are calculated from the maps of Davis [48].

8.4.2 Fragmentation of range

Many plant ranges at the present-day are subdivided into two or more geographically separated ('disjunct') portions. Often there is a main range and outlying 'satellite populations' (Figure 8.11). Explanations of how such disjunct distributions have arisen are varied. At one extreme is the belief that long-distance dispersal at some time in the past was responsible for the establishment of populations remote from the parent one, and that the timing of these events was dictated by chance. Disjunct distribution is therefore evidence of an expanding population. At the other extreme is the belief that fragmentation of range occurred when a population was contracting under unfavourable conditions, and that the outlying populations are 'relicts' from a formerly more widespread and continuous distribution. A once-continuous population might have been fragmented not only by climatic change but also by destruction of habitats over portions of its range, by processes varying in scale from localized forest clearance to major continental displacements, mountain uplift and rising sea-levels. The climatic vicissitudes of the Pleistocene are often seen as a major cause of range fragmentation, ever since the classic exposition by Edward Forbes, and subsequently by Charles Darwin, in the middle of the last century, that the similarity of the alpine floras of Europe and Asia with the arctic floras far to the north was the result of the diffusion of arctic plants south over the lowlands during the glacial epoch, and their subsequent recession northwards or to higher altitudes at the termination of glaciation. However, we should be chary of seeking out a single process for all cases of range fragmentation, and of accepting facile explanations without supporting fossil evidence. The causes of disjunct distribution are almost certainly many and complex.

Two examples may serve to illustrate this complexity, and the necessity of adequate fossil evidence for deciphering the causal mechanisms. Both silver fir (*Abies alba*) and white spruce (*Picea glauca*) have distributions at the present-day (in Europe and North America, respectively) which comprise a main range and a number of outlying satellite populations (Figure 8.11). Silver fir is a dominant of middle-altitude forests in the mountains of central Europe, while white spruce dominates the boreal forests of Canada. Fossil records of white spruce from the time of the last glacial maximum

Figure 8.11 (a) Present-day range ≡ in Europe of silver fir (*Abies alba*), and approximate Hoxnian-interglacial |||||| and full-glacial ● distribution. Re-drawn from Frenzel [49] and Godwin [2]. (b) Present-day ≡ and full-glacial ||||| range of white spruce (*Picea glauca*) in North America. ⊥⊥⊥⊥, full-glacial limit of ice sheet. Re-drawn from Seddon [50].

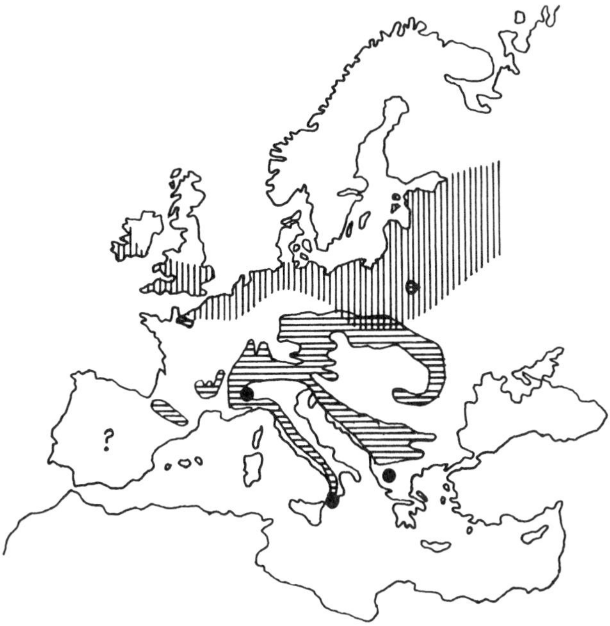

(a)

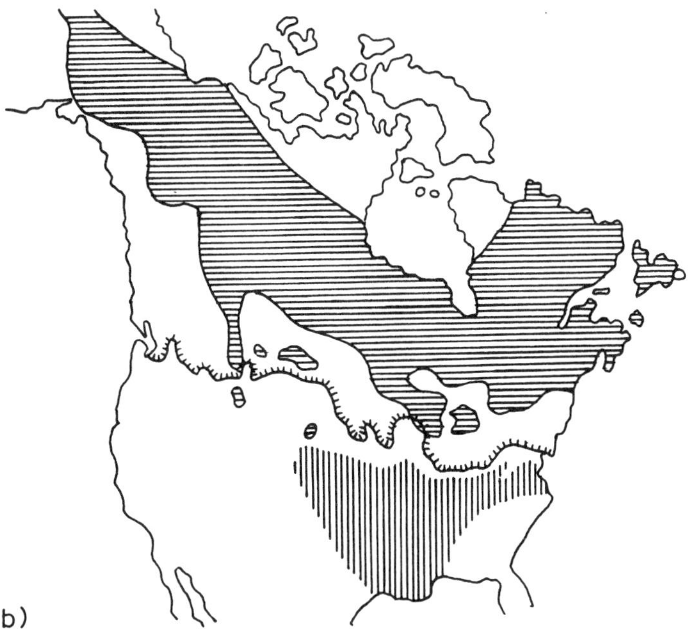

(b)

occur widely in the eastern half of the United States, where spruce was the major woody member of an open-canopy spruce parkland community. Almost all the present-day range lay then beneath the Laurentide and Cordilleran Ice Sheets. When these ice sheets began to melt, spruce moved northwards again, colonizing the recently-uncovered ground; however, in the south, prairie and forest invaded the former spruce parkland, so that spruce was forced to retreat northwards, reaching the position of its present southern limits about 7 ka ago [51]. It might be thought that the disjunct occurrences of spruce today are a memory of its full-glacial range. However, none of the present-day satellite populations occurs within this range, and three occupy locations that must have been ice-covered then. More probably, the satellite populations were 'pinched off' from the retreating range margin as spruce moved north in the Early Holocene.

In marked contrast to the extensive full-glacial range of white spruce, only three European sites are known where *Abies* pollen has been recorded in more than minimal amounts from deposits of latest Pleistocene age – from southern Italy, northern Italy and northern Greece [19]. The full-glacial range of silver fir must have comprised a mere handful of splinter populations. These were slow to expand subsequently, and the spread of silver fir in the mid-Holocene involved insertion into already-established middle-altitude forest communities. It is unlikely that a continuous lowland range was ever present. Hence the present-day satellite populations of silver fir in the Pyrenees and the Massif Central probably represent long-distance colonization in relation to an advancing range margin. Two further satellite populations, in Normandy and eastern Poland, occur within the area occupied by silver fir in the last (Hoxnian) interglacial, and might thus be envisaged as relicts from that time. However, there is no pollen evidence that *Abies* survived so far to the north during the last glaciation, and these populations too are best regarded as of Holocene origin (by long-distance dispersal).

The ranges of whole plant communities must also have been subject to contraction and fragmentation in the past. Four contrasting examples illustrate the process. Thus tree remains preserved north of the present treeline in Siberia, and radiocarbon-dated to 9220–4880 yr ago [52], show that the southern limit of the Eurasian tundra zone must have been displaced by up to 400 km northwards during this period, so that the tundra zone was broken up into separated blocks between the great gulfs and estuaries opening to the arctic coast of Siberia. In mountain regions throughout the world the alpine tundra zone has similarly expanded and contracted in response to changing temperature; in the northern Andes, for example, areas of treeless vegetation (paramo), now segregated among the higher mountain summits by a high-altitude treeline, were formerly part of a more extensive and nearly continuous alpine zone when the treeline was lowered to about 2000 m altitude during the last glaciation [53]. In tropical Africa it has been suggested that the evergreen rainforest was markedly contracted during periods of aridity, and survived only in a number of widely-separated localities, some of which are areas of high rainfall today [54]. During these dry periods savanna vegetation may have been much more widespread, but with the re-expansion of rainforest in the Holocene this savanna in its turn was greatly reduced in extent. Similar suggestions have been made for tropical South America, and one explanation of the 'islands' of lowland savanna that occur throughout the Amazonian rainforest at the present day is that they are 'relicts' from such a former dry period, when savanna was more widespread [55].

8.4.3 Refugia and reservoirs

Fluctuating climates over the last 65 Ma must have subjected all plant populations to periodic stress, when contractions and even fragmentations of range occurred; they must also have permitted periodic expansions of range, when inherent speeds of migration may have been a governing factor. The white spruce and silver fir examples illustrate two different types of plant survival under unfavourable conditions: (a) with a displaced range, but still as an essentially continuous population occupying a wide geographical area, and (b) as small populations segregated into locally favourable, and often geographically separated, sites. These latter sites are customarily referred to as **refuges** or **refugia**, a refugium being defined as 'a comparatively restricted area where a particular biome such as forest or savanna remained more or less constant during periods of vegetational shifts, thus . . . serving as a region of survival for at least part of the . . . animal and plant populations' [56]. A refuge shelters populations 'left behind' by migrating plants (and animals), or 'concentrated' by range contraction. No generally accepted terminology exists for the white spruce type of situation, where much greater numbers of individuals persist as a single wide-ranging population. The term **reservoir** is proposed, and will be used subsequently.

A special type of range-fragmentation in response to prolonged periods of unfavourable climate occurred in certain geographical regions. McGlone [57] argues that many members of the New Zealand flora never expanded or contracted their ranges by more than 50 km throughout the Pleistocene; instead many forest tree species probably survived during glacial periods in a large number of isolated sites throughout New Zealand. As a result, forest communities were reconstituted very quickly over a wide geographical area when climatic conditions improved again at the start of each interglacial stage. Measured rates of spread of *Nothofagus* (southern beech) populations at the present day of only 10 m per century certainly support this contention.

Detection of reservoirs from fossil records is generally not difficult; detection of refugia, by contrast, is extremely difficult, and very few sites have been investigated which can convincingly be claimed to be true refugia. Detection of refugia from fossil records involves examination of migration patterns of a particular taxon through the Holocene, and extrapolation of the pathways back to a likely **source area.** The most detailed analysis of survival patterns through the last glaciation is probably that by Tsukada [58–60] for the Japanese cedar, *Cryptomeria japonica.* Tsukada examined pollen diagrams from 163 sites in Japan, and found that at twenty-one of the sites *Cryptomeria* pollen was present 20 ka ago – though at only three was the quantity of pollen sufficient to indicate local presence of the cedar. Migration patterns of *Cryptomeria* in the Early Holocene showed at least five separate advancing fronts, apparently originating from a series of perglacial refugia located immediately offshore of the present coastline (Figure 8.12). Huntley and Birks [19] mapped the Holocene migration patterns of twenty-two tree pollen taxa in Europe, and identified seventeen separate areas from which migration pathways appeared to originate (Figure 8.12). As the areas are of considerable size, and as no attempt was made by Huntley and Birks to locate specific refugia within them, they are best regarded as source areas rather than as refugia. Comparable source areas for fourteen tree taxa in eastern North America, derived from the migration maps in Delcourt and Delcourt [20], are also shown in Figure 8.12.

Most of the discussion about the location and significance of refugia is based not on fossil evidence, however, but on the *present-day* distributions of plants and animals. Refugia are sought where the ranges of several or many taxa of similar behaviour coincide. Three types of **coincident range-pattern** (discussed in more detail below) have been used to locate possible refugia: clustering of rare species, centres of species diversity, and centres of endemism. Certain assumptions are commonly made in locating refugia by this method: that a given refugium will provide shelter for small populations of not just one but of several or many species; that subsequently under ameliorating conditions some of these populations will expand their ranges again, while others may remain localized; and that if populations of the same species are isolated from each other for long enough in separate refugia, then evolutionary change will occur and genetic and chromosomal differences between these populations will arise. Refugia thus figure prominently in many discussions about modes of speciation in plants. This aspect will be considered briefly later, but the major concern here is with the location of refugia.

Clustering of rare species has been especially studied in northern

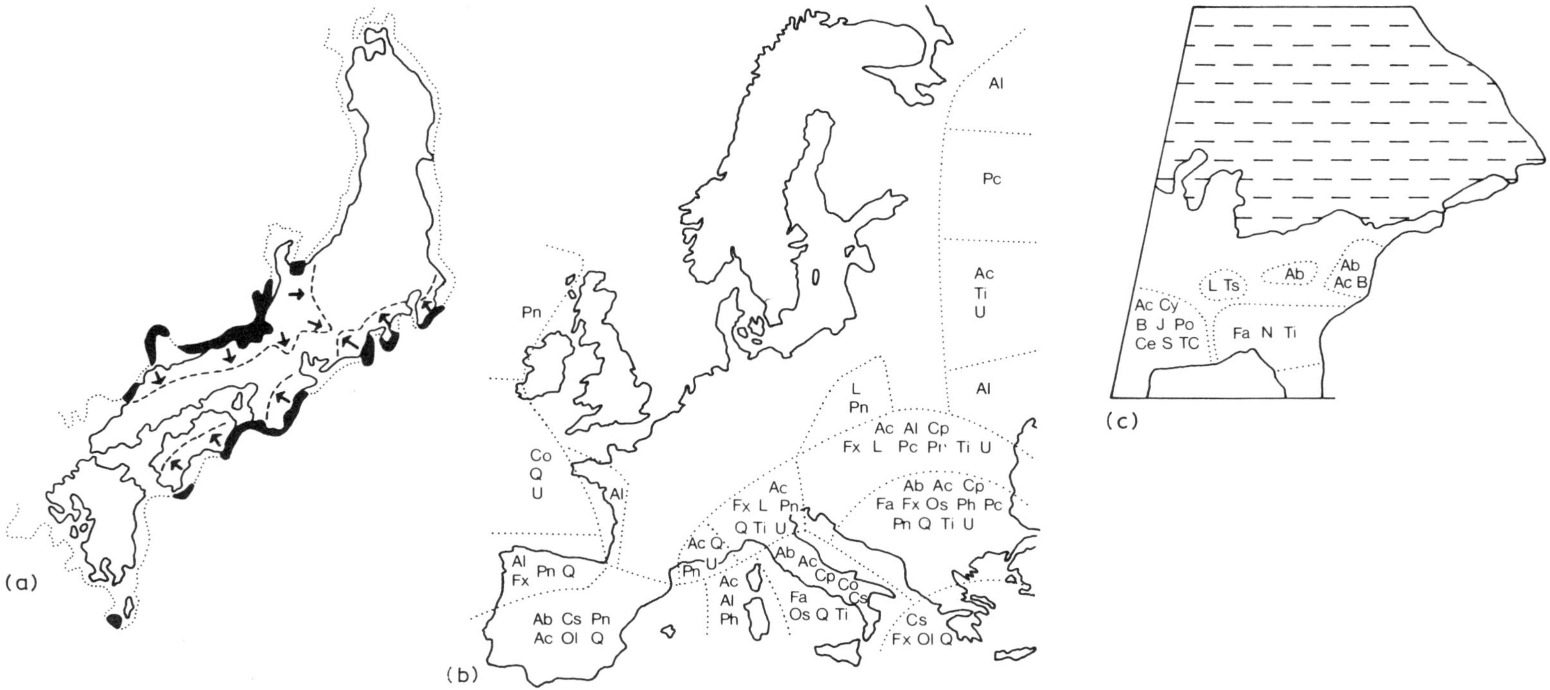

Figure 8.12 Perglacial survival. (a) Reconstructed locations of refugia of the Japanese cedar (*Cryptomeria japonica*) during the last cold stage in Honshu, Japan. ········, coastline at 20 ka ago; ●, postulated refugium; →, migration pathways in the Early Holocene; ––––, range margin at 8 ka ago. Re-drawn from Tsukada [58]. (b) Source areas of twenty-two tree taxa in Europe (re-drawn from Huntley and Birks [19]). (c) Source areas of fourteen tree taxa in eastern North America (based on maps in Delcourt and Delcourt [20]). Each source area is delimited by a dotted line, and includes the names of tree taxa presumed to have survived the last cold stage within it (Ab, *Abies*; Ac, *Acer*; Al, *Alnus*; B, *Betula*; Ce, *Celtis*; Co, *Corylus*; Cp, *Carpinus*; Cs, *Castanea*; Cy, *Carya*; Fa, *Fagus*; Fx, *Fraxinus*; J, *Juglans*; L, *Larix*; N, *Nyssa*; Ol, *Olea*; Os, *Ostrya*; Ph, *Phillyrea*; Pc, *Picea*; Pn, *Pinus*; Po, *Populus*; Q, *Quercus*; S, *Salix*; TC, Taxodiaceae + Cupressaceae; Ti, *Tilia*; Ts, *Tsuga*; U, *Ulmus*).

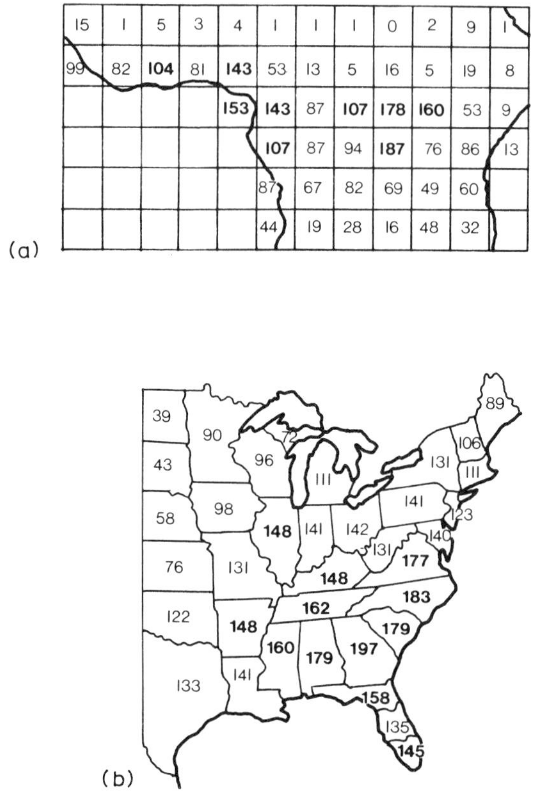

Figure 8.13 Species-diversity mosaics. (a) Numbers of passerine bird species in tropical Africa (re-drawn from Hamilton [54]); (b) numbers of tree species in the eastern United States (re-drawn from Watts [63]). Compare the positions of the high-diversity patches with the maps in Figures 10.12 and 10.11.

Europe, where certain famous localities, such as Teesdale in northern England and the Burren in western Ireland, have long been renowned among botanists for their collections of rare and localized species [61]. Such localities were formerly regarded as **perglacial refugia**, where cold-hardy plants survived in a generally glaciated landscape in local ice-free areas ('nunataks', discussed also in Chapter 14). Even if this interpretation were true, of more significance is the subsequent survival of these plants in the densely forested and highly competitive landscape of the early and mid-Holocene, and it is now widely recognized that the species-assemblages of Teesdale and the Burren persist today because of peculiarities in the Holocene history of these sites (the local failure of a forest cover to develop). The fossil evidence, moreover, suggests that the species assemblages were more probably formed in the late-glacial than in the full-glacial period, and hence that the localities are unlikely to have been perglacial refugia [2].

On a larger geographical scale, many regions have been shown to have non-random distributions of their component plant and animal populations. Each region consists of a mosaic of patches of high and low species-diversity, as is illustrated by the two examples in Figure 8.13: the numbers of species of forest birds in different parts of tropical Africa (in relation to a grid of 5 × 5° latitude/longitude 'squares'), and the numbers of tree species in eastern North America (per state). It has been suggested that the high-diversity patches are centred on refugia, where varied collections of plants and animals survived during periods of unfavourable climate. Subsequently, most populations expanded their ranges again, but to varying distances away from each refugium, so that gradients of diversity were produced, with the greatest numbers of species still close to the refugia. It must be remembered, however, that spatial variations in species-diversity can be produced in many ways – as a result of environmental heterogeneity, competition and predation, and differences in environmental stability and productivity [62]. Consequently, the demonstration of a diversity mosaic in a particular region does not necessarily require the presence of refugia to explain that mosaic, and many biogeographers now accept that centres of species-diversity are no guide to the locations of refugia.

There is more agreement about the relationship of centres of endemism to refugia, as a result particularly of recent work on the distributions of plants and animals in tropical South America [64,65]. In many groups of neotropical plants and animals there are wide-ranging taxa that each consist of a series of closely-related **vicariant** (i.e., geographically segregated) populations, variously categorized as species, subspecies or geographical races. Each population in a series typically has a restricted geographical range (so that it is endemic to a particular small part of Amazonia) and does not overlap with neighbouring populations; yet the combined ranges of all the populations may cover much of Amazonia (Figure 8.14). Within a given group of animals or plants many such taxa may occur, and if the distributions of all the vicariant populations of these taxa are compared, **centres of endemism** can be recognized – geographically localized 'hotspots' where the distributions of several unrelated endemic populations coincide. Such range-patterns have been analysed most fully for nymphalid butterflies [66–68], but good data also exist for flowering plants [64,69,70,71], birds [72–74] and frogs [75,76]. Thirty-seven centres of endemism in tropical South America have been recognized for butterflies, 26 for flowering plants, 11 for birds and 17 for tree

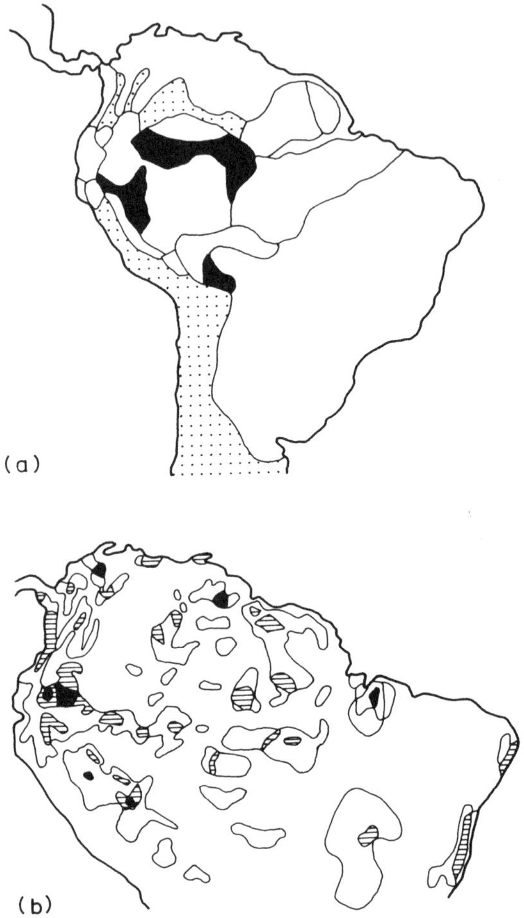

Figure 8.14 Centres of endemism. (a) The ranges of the fifteen geographical races of the butterfly *Heliconius erato* cover much of Amazonia (the butterfly is absent only from the stippled areas), yet there is little overlap (the black areas) between the different ranges (from Brown [66]). (b) The degree of coincidence of the postulated centres of endemism for woody angiosperms [64], nymphalid butterflies [66] and forest birds [68]; centres of endemism postulated for one group only , two groups , and all three groups .

frogs [65]. When all groups are considered there is approximate coincidence of some at least of the centres of endemism (Figure 8.14). It has accordingly been suggested that such shared locations represent refugia where varied assemblages of forest taxa survived through 'climatic crises' when the Amazonian rainforest was markedly contracted in extent. Many workers see this contraction as occurring during drier periods of the Pleistocene (the glacial stages of higher latitudes), and would accordingly interpret centres of endemism as equivalent to perglacial refugia. Whilst there are probably sufficient data now to validate the existence of centres of endemism in particular groups, there is no general agreement as to their biogeographical and evolutionary significance [64,77]. Thus centres of endemism occur in areas with high present-day rainfall and more fertile soils [66], and so could in part reflect current environmental heteroigeneity; conversely, there is as yet no unambiguous fossil evidence that the Amazonian rainforest was substantially reduced in extent during drier periods of the Pleistocene. Protagonists of the refuge theory assert that the current patterns of specific and subspecific differentiation in many rainforest taxa are best interpreted as the result of speciation in small isolated populations ('allopatric speciation'), followed by range expansion; where such expanding populations come into contact again, hybridization may occur along 'zones of secondary contact'. Opponents claim that similar patterns can be produced in continuous populations dispersed over a spatially heterogeneous environment [78]. Nor is there agreement over the distinctness of the proposed refugia. For Africa, attempts to delineate refugia by objective criteria have met with some measure of success [79]; however, for South America the degree of congruence of the refugia proposed for different plant and animal groups has been claimed not to be significantly different from random overlap [80]. Finally, as explained below, known rates of speciation in plant and animal taxa may not be consistent with the Pleistocene origin for some at least of the taxa that have been linked with the refuge theory. The refuge theory, although an attractive one, must as yet be regarded as unproven [84].

In recent years attempts have been made to 'test' the refuge theory by determining the ages of vicariant populations by means of so-called 'albumin clocks'. Animal and plant proteins are complex assemblages of amino-acids (about 500 different ones in albumins), which in closely related taxa differ in only a few respects. It has been claimed that the extent to which the precise amino-acid sequences in proteins from two related taxa differ is proportional to the time elapsing since the two taxa last shared a common ancestor. This difference can be measured by standard immunological techniques, using antigen/antibody reactions [81,82]. For albumins, the rate of change in the amino-acid sequences has been calibrated with some consensus of agreement (one amino-acid substitution every 0.6 Ma). Heyer and Maxson [76] found that the amino-acid sequences in albumin from eleven species of frogs of the genus *Leptodactylus* (an Amazonian genus) were sufficiently different to render it unlikely that any of the species had originated more recently than the Pliocene Epoch. In frogs of the genus *Heleioporus*, where western Australian species are believed to have evolved from eastern Australian species by isolation during dry periods of the Pleistocene, albumin similarities again indicated that speciation had occurred considerably prior to that time (4–12 Ma ago [83]).

Although these examples relate to animals, there is no reason to suppose that speciation has been more rapid in plants. The alignment of present-day distribution patterns of plants around supposed Pleistocene refugia may consequently be a secondary feature, overlain on a more ancient one that originated prior to the climatic fluctuations of the Pleistocene.

8.4.4 Assemblages and communities

The fossil record of plant responses to the climatic vicissitudes of the Pleistocene and Holocene Epochs contrasts markedly with that for the Tertiary Period. The evidence for the last 2 Ma is dominated by individuality not communalism. Populations of different plant taxa appear to have migrated independently of each other in response to climatic changes, and plant communities have arisen fortuitously at **meeting points** in time and space of a series of different migration pathways. The two records are not mutually incompatible, once it is realized that the Tertiary record is a coarse-resolution one, documenting the taxa found at various meeting points, and not the ways by which these taxa had reached there. The Quaternary record suggests that meeting points are characteristic of periods of relative climatic stability, when some plant migrations are slowing down or have ceased altogether, and populations of slower-moving taxa are enabled to catch up. Evidence at this level of resolution is lacking for the Tertiary Period, but that does not mean that the same process did not occur then as in the Quaternary.

The Quaternary record, moreover, is a very limited one, restricted to a few temperate tree taxa. There is no evidence for the behaviour of subordinate taxa – the herbaceous plants forming the ground layer of present-day temperate broadleaf forests, for example. It is difficult to believe that these migrated completely independently of the tree taxa. Nor is there evidence for the vastly more complex plant communities of tropical regions, where many more taxa of closely similar ecological requirements are thought to occur than in temperate regions. The regional plant assemblages characteristic of higher latitudes at the present day originated later in the Tertiary Period than modern lower-latitude assemblages, as specialized and limited subsets, and were subjected to more extreme climatic fluctuations. Their patterns of response to these fluctuations may accordingly have been simplified because of their impoverished character. Consequently, we are not yet in a position to speculate meaningfully on the persistence of plant communities as entities over long periods of time (section 15.2.1).

REFERENCES

1. Iversen, J. (1944) *Viscum, Hedera* and *Ilex* as climate indicators. *Geol. För. Stockh. Förh.*, **66**, 463–83.
2. Godwin, H. (1975) *History of the British Flora*, 2nd edn, Cambridge University Press, Cambridge.
3. Crocker, R. L. and Major, J. (1955) Soil development in relation to vegetation and surface age at Glacier Bay, Alaska. *J. Ecol.*, **43**, 427–48.
4. Andrews, J. T. (1973) The Wisconsin Laurentide ice sheet: dispersal centers, problems of rates of retreat, and climatic implications. *Arctic Alpine Res.*, **5**, 185–99.
5. Atkinson, T. C., Briffa, K. R. and Coope, G. R. (1987) Seasonal temperatures in Britain during the last 22,000 years, reconstructed using beetle remains. *Nature*, **325**, 587–92.
6. Lamb, H. H. (1985) Climate and landscape in the British Isles, in *The English Landscape: Past, Present and Future* (ed. S. R. J. Woodell), Oxford University Press, Oxford, pp. 148–167.
7. Chappell, J. (1978) Chronological methods and the ranges and rates of Quaternary physical changes, in *Biology and Quaternary Environments* (eds D. Walker and J. C. Guppy), Australian Academy of Science, Canberra, pp. 1–34.
8. Cooper, W. S. (1939) A fourth expedition to Glacier Bay, Alaska. *Ecology*, **20**, 130–55.
9. Reiners, W. A., Worley, I. A. and Lawrence, D. B. (1971) Plant diversity in a chronosequence at Glacier Bay, Alaska. *Ecology*, **52**, 55–69.
10. Birks, H. J. B. (1986) Late-Quaternary biotic changes in terrestrial and lacustrine environments, with particular reference to north-west Europe, in *Handbook of Holocene Palaeoecology and Palaeohydrology* (ed. B. E. Berglund), Wiley, Chichester, pp. 3–65.
11. Bennett, K. D. (1983) Postglacial population expansion of forest trees in Norfolk. *Nature*, **303**, 164–7.
12. Watts, W. A. (1973) Rates of change and stability in vegetation in the perspective of long periods of time, in *Quaternary Plant Ecology* (eds H. J. B. Birks and R. G. West), Blackwell, Oxford, pp. 195–206.
13. Bennett, K. D. (1986) The rate of spread and population increase of forest trees during the postglacial. *Phil. Trans. R. Soc. Lond. B*. **314**, 523–31.
14. Chen, Y. (1988) Early Holocene population expansion of some rainforest trees at Lake Barrine basin, Queensland. *Aust. J. Ecol.*, **13**, 225–33.
15. Davis, M. B. (1978) Climatic interpretation of pollen in Quaternary sediments, in *Biology and Quaternary Environments* (eds D. Walker and J. C. Guppy), Australian Academy of Science, Canberra, pp. 35–51.
16. Jacobson, G. L. (1979) The palaeoecology of white pine (*Pinus strobus*) in Minnesota. *J. Ecol.*, **67**, 697–726.
17. Tallantire, P. A. (1972) Spread of spruce (*Picea abies* (L.) Karst) in Fennoscandia and possible climatic implications. *Nature*, **236**, 64–5.

18. Deacon, J. (1974) The location of refugia of *Corylus avellana* L. during the Weichselian glaciation. *New Phytol.*, **73**, 1055–63.
19. Huntley, B. and Birks, H. J. B. (1983) *An Atlas of Past and Present Pollen Maps for Europe: 0–13,000 Years Ago*, Cambridge University Press, Cambridge.
20. Delcourt, P. A. and Delcourt, H. R. (1987) *Long Term Forest Dynamics of the Temperate Zone*, Springer Verlag, New York.
21. Jacobson, G. L., Jr, Webb, T., III, and Grimm, E. C. (1987) Patterns and rates of vegetation change during the deglaciation of eastern North America, in *North America and Adjacent Oceans during the Last Deglaciation* (eds W. F. Ruddiman and H. E. Wright Jr), Geological Society of America, Boulder, CO, pp. 277–88.
22. Wells, P. V. (1983) Paleobiogeography of montane islands in the Great Basin since the last glaciopluvial. *Ecol. Monogr.*, **53**, 341–82.
23. Delcourt, P. A. and Delcourt, H. R. (1983) Late-Quaternary vegetational dynamics and community stability reconsidered. *Quat. Res.*, **19**, 265–71.
24. Ritchie, J. C. and MacDonald, G. M. (1986) The patterns of post-glacial spread of white spruce. *J. Biogeogr.*, **13**, 527–40.
25. White, J. M., Mathewes, R. W. and Mathews, W. H. (1985) Late Pleistocene chronology and environment of the 'ice-free corridor' of northwestern Alberta. *Quat. Res.*, **24**, 173–86.
26. Green, D. G. (1982) Fire and stability in the postglacial forests of southwest Novia Scotia. *J. Biogeogr.*, **9**, 29–40.
27. Davis, M. B. (1984) Holocene vegetational history of the eastern United States, in *Late-Quaternary Environments of the United States. Volume 2. The Holocene* (ed. H. E. Wright Jr), Longman, London, pp. 166–81.
28. Turner, J. (1962) The *Tilia* decline: an anthropogenic interpretation. *New Phytol.*, **61**, 328–41.
29. Payette, S. and Gagnon, R. (1985) Late Holocene deforestation and tree regeneration in the forest-tundra of Quebec. *Nature*, **313**, 570–2.
30. Crawford, R. M. M. (1988) *Studies in Plant Survival. Ecological Case Histories of Plant Adaptation to Adversity*, Blackwell, Oxford.
31. Tranquillini, W. (1979) *Physiological Ecology of the Alpine Timberline: Tree Existence at High Altitudes with Special Reference to the European Alps*, Springer Verlag, Berlin.
32. Kullman, L. (1986) Late Holocene reproductional patterns of *Pinus sylvestris* and *Picea abies* at the forest limit in central Sweden. *Can. J. Bot.*, **64**, 1682–90.
33. Kullman, L. (1987) Long-term dynamics of high-altitude populations of *Pinus sylvestris* in the Swedish Scandes. *J. Biogeogr.*, **14**, 1–8.
34. Gorchakovsky, P. L. and Shiyatov, S. G. (1978) The upper forest limit in the mountains of the boreal zone of the USSR. *Arctic Alpine Res.*, **10**, 349–63.
35. Bryson, R. A., Irving, W. N. and Larsen, J. A. (1965) Radiocarbon and soils evidence of former forest in the southern Canadian tundra. *Science*, **147**, 46–8.
36. Nichols, H. (1967) The postglacial history of vegetation and climate at Ennadai Lake, Keewatin, and Lynn Lake, Manitoba. *Eiszeitalter Gegenw.* **18**, 176–97.
37. Nichols, H. (1970) Late Quaternary pollen diagrams from the Canadian Arctic barren grounds at Pelly Lake, Northern Keewatin, N.W.T. *Arctic Alpine Res.*, **2**, 43–61.

38. Nichols, H. (1968) Chronology of peat growth in Canada. *Palaeogeogr. Palaeoclimatol. Palaeoecol.*, **6**, 61–5.
39. Nichols, H. (1975) The time perspective in northern ecology: palynology and the history of the Canadian boreal forest. *Proceedings of the Circumpolar Conference on Northern Ecology, Ottawa 1975*. National Research Council of Canada, Ottawa.
40. Burga, C. A. (1988) Swiss vegetation history during the last 18,000 years. *New Phytol.*, **110**, 581–602.
41. Lamarche, V. C. (1973) Holocene climatic variations inferred from treeline fluctuations in the White Mountains, California. *Quat. Res.*, **3**, 632–60.
42. Kullman, L. (1988) Holocene history of the forest–alpine tundra ecotone in the Scandes Mountains (central Sweden). *New Phytol.*, **108**, 101–10.
43. Kubitzki, K. (1961) Zur Synchronisierung der nordwesteuropäischen Pollendiagramme. *Flora, Jena*, **150**, 43–72.
44. Van Zeist, W. (1959) Studies on the Post-boreal vegetational history of south-eastern Dienthe (Netherlands). *Acta bot. neerl.*, **8**, 156–84.
45. Huntley, B. and Webb, T., III (1989) Migration: species' response to climatic variations caused by changes in the earth's orbit. *J. Biogeogr.*, **16**, 5–19.
46. Faegri, K. and Iversen, J. (1964) *Textbook of Pollen Analysis*, 2nd edn, Blackwell, Oxford.
47. Hope, G. S. (1976) The vegetational history of Mt Wilhelm, Papua New Guinea. *J. Ecol.*, **64**, 627–63.
48. Davis, M. B. (1976) Pleistocene biogeography of temperate deciduous forests. *Geoscience & Man*, **13**, 13–26.
49. Frenzel, B. (1973) *Climatic Fluctuations of the Ice Age* (transl. by A. E. M. Nairn), Case Western Reserve University Press, Cleveland.
50. Seddon, B. (1971) *Introduction to Biogeography*, Duckworth, London.
51. Bernabo, J. C. and Webb, T. (1977) Changing patterns in the Holocene pollen record of northeastern North America: a mapped summary. *Quat. Res.*, **8**, 64–96.
52. Khotinskiy, N. A. (1984) Holocene vegetation history, in *Late Quaternary Environments of the Soviet Union* (ed. A. A. Velichko), Longman, London, pp. 179–200.
53. Vuilleumier, B. S. (1971) Pleistocene changes in the fauna and flora of South America. *Science*, **173**, 771–80.
54. Hamilton, A. C. (1982) *Environmental History of East Africa. A Study of the Quaternary*, Academic Press, London.
55. Huber, O. (1982) Significance of savanna vegetation in the Amazon Territory of Venezuela, in *Biological Diversification in the Tropics* (ed. G. T. Prance), Columbia University Press, New York, pp. 221–44.
56. Haffer, J. (1982) General aspects of the refuge theory, in *Biological Diversification in the Tropics* (ed. G. T. Prance), Columbia University Press, New York, pp. 6–24.
57. McGlone, M. S. (1985) Plant biogeography and the Late Cenozoic history of New Zealand. *N.Z. J. Bot.*, **23**, 723–49.
58. Tsukada, M. (1982) *Cryptomeria japonica*: glacial refugia and late-glacial and postglacial migration. *Ecology*, **63**, 1091–1105.

59. Tsukada, M. (1985) Map of vegetation during the last glacial maximum in Japan. *Quat. Res.*, **23**, 369–81.
60. Tsukada, M. (1986) Altitudinal and latitudinal migration of *Cryptomeria japonica* for the past 20,00 years in Japan. *Quat. Res.*, **26**, 135–152.
61. Pigott, C. D. (1956) The vegetation of Upper Teesdale in the North Pennines. *J. Ecol.*, **44**, 545–86.
62. Krebs, C. J. (1985) *Ecology. The Experimental Analysis of Distribution and Abundance*, 3rd edn, Harper and Row, New York.
63. Watts, W. A. (1983) Vegetational history of the eastern United States 25,000 to 10,000 years ago, in *Late-Quaternary Environments of the United States*, Vol. 1, *The Late Pleistocene* (ed. S. C. Porter), Longman, London, pp. 294–310.
64. Prance, G. T. (1982) Forest refuges: evidence from woody angiosperms, in *Biological Diversification in the Tropics* (ed. G. T. Prance), Columbia University Press, New York, pp. 137–57.
65. Whitmore, T. C. and Prance, G. T. (1987) *Biogeography and Quaternary History in Tropical Latin America*, Oxford University Press, New York.
66. Brown, K. S., Jr (1982) Paleoecology and regional patterns of evolution in neotropical forest butterflies, in *Biological Diversification in the Tropics* (ed. G. T. Prance), Columbia University Press, New York, pp. 255–308.
67. Brown, K. S., Jr (1987) Soils and vegetation, in *Biogeography and Quaternary History in Tropical Latin America* (eds T. C. Whitmore and G. T. Prance), Oxford University Press, Oxford, pp. 19–45.
68. Turner, J. R. G. (1982) How do refuges produce biological diversity? Allopatry and parapatry, extinction and gene flow in mimetic butterflies, in *Biological Diversification in the Tropics* (ed. G. T. Prance), Columbia University Press, New York, pp. 309–34.
69. Gentry, A. H. (1982) Phytogeographic patterns as evidence for a Chocó Refuge, in *Biological Diversification in the Tropics* (ed. G. T. Prance), Columbia University Press, New York, pp. 112–36.
70. Prance, G. T. (1987) Biogeography of neotropical plants, in *Biogeography and Quaternary History in Tropical Latin America* (eds T. C. Whitmore and G. T. Prance), Oxford University Press, New York, pp. 46–65.
71. Steyermark, J. A. (1982) Relationships of some Venezuelan forest refuges with lowland tropical floras, in *Biological Diversification in the Tropics* (ed. G. T. Prance), Columbia University Press, New York, pp. 182–220.
72. Haffer, J. (1969) Speciation in Amazonia forest birds. *Science*, **165**, 131–7.
73. Haffer, J. (1987) Quaternary history of tropical America, in *Biogeography and Quaternary History in Tropical Latin America* (eds T. C. Whitmore and G. T. Prance), Oxford University Press, Oxford, pp. 1–18.
74. Pearson, D. L. (1982) Historical factors and bird species richness, in *Biological Diversification in the Tropics* (ed. G. T. Prance), Columbia University Press, New York, pp. 441–52.
75. Duellman, W. E. (1982) Quaternary climatic–ecological fluctuations in the lowland tropics: frogs and forests, in *Biological Diversification in the Tropics* (ed. G. T. Prance), Columbia University Press, New York, pp. 389–402.
76. Heyer, W. R. and Maxson, L. R. (1982) Distributions, relationships, and zoogeography of lowland frogs: the *Leptodactylus* complex in South America,

with special reference to Amazonia, in *Biological Diversification in the Tropics* (ed. G. T. Prance), Columbia University Press, New York, pp. 375–88.

77. Simpson, B. and Haffer, J. (1978) Speciation patterns in the Amazonian forest biota. *Ann. Rev. Ecol. Syst.*, **9**, 497–518.
78. Endler, J. (1982) Pleistocene forest refuges: fact or fancy? in *Biological Diversification in the Tropics* (ed. G. T. Prance), Columbia University Press, New York, pp. 641–57.
79. Mayr, E. and O'Hara, R. J. (1986) The biogeographic evidence supporting the Pleistocene forest refuge hypothesis. *Evolution*, **40**, 55–67.
80. Beven, S., Conner, E. F. and Beven, K. (1984) Avian biogeography in the Amazon basin and the biological model of diversification. *J. Biogeogr.*, **11**, 383–99.
81. Maxson, L. R., Sarich, V. M. and Wilson, A. C. (1975) Continental drift and the use of albumin as an evolutionary clock. *Nature*, **255**, 397–400.
82. Sarich, V. M. and Wilson, A. C. (1966) Quantitative immunochemistry and the evolution of primate albumins: micro-complement fixation. *Science*, **154**, 1563–6.
83. Maxson, L. R. and Roberts, J. D. (1984) Albumin and Australian frogs: molecular data a challenge to speciation model. *Science*, **225**, 957–8.
84. Nelson, B. W., Ferreira, C. A. C., da Silva, M. F. and Kawasaki, M. (1990) Endemism centres, refugia and botanical collection density in Brazilian Amazonia. *Nature*, **345**, 714–6.

PART FOUR

The History of Forest, Dryland and Coldland Communities

CHAPTER NINE

The forested world of the Tertiary

9.1 INTRODUCTION

As we saw in Chapter 7, the forests that had re-formed by the Eocene Epoch after the probable global catastrophe that ushered in the Cenozoic Era, and particularly those in the northern hemisphere, represent the true progenitors of modern plant communities, rather than the earlier Cretaceous plant cover. The accumulated global biomass of flowering plants in these Eocene forests was probably greater than at any time before or since. In the forests of 50 Ma ago we can recognize for the first time plants that can be assigned with some confidence to extant families or even to extant genera, so that we can begin to make direct floristic comparisons between these forests and their possible modern counterparts. They thus form a convenient baseline for a consideration of the development of modern floras and plant communities.

However, in an evolutionary sense these forests represent an essentially low-diversity response to a low-diversity environment (Chapter 4): their component taxa were adapted to only a limited range of environmental conditions. Subsequently in the Tertiary Period the plant cover was subjected to increasing environmental heterogeneity. As surface relief was accentuated by mountain uplift, so a whole new suite of local and regional climates was generated to which the existing plant taxa were poorly adapted. The spread of seasonally cold and seasonally dry climates, and then of year-round cold and arid climates, produced an evolutionary challenge to which the surviving gymnosperms and pteridophytes (still important components of the Eocene forests) were largely unable to respond, but which the angiosperms readily accepted. The successful evolution over the long time-span of the Tertiary Period of varied herbaceous floras adapted to cold or arid environments, coupled with the progressive diversification of woody taxa adapted to seasonal environments, was the essential prelude to the ability of the angiosperms to cope with the subsequent rigours of the Pleistocene Ice Age. The success of the angiosperms in mastering the vicissitudes of the Quaternary Period was made possible only because of their previous responses to the upheavals of the Tertiary Period.

9.2 EOCENE FLORAS

9.2.1 Imperfections of the floristic data

Although numerous floras of Eocene age have been described from many parts of the world, and floristic lists published, various problems exist with the records before the lists can be used for vegetation reconstructions. These problems are not unique to Eocene floras – they are encountered in all palaeoecological studies involving preserved plant remains (section 2.2). One problem that is particularly relevant to Eocene floras, however, is that some fossil floras have undergone repeated revisions over the years, so that the original floristic lists are no longer reliable. Thus the various identifications by Oswald Heer last century of tropical plants in Arctic Eocene floras are no longer believed valid. In the case of two of the richest Eocene floras (the Wilcox Flora* from the south-eastern United States and the London Clay Flora from South-east England), complete floristic lists incorporating the various revisions have not been published (and for other floras, too, for one reason or another, complete floristic lists have never been published).

Different fossil floras, moreover, may be dominated either by leaf remains, by preserved seeds and fruits, or by pollen grains. The specificity of identification possible, the environment of deposition, and the likely vegetation 'catchment' are all different for these three types of flora [1], so that valid comparisons between them may be difficult. For northern-hemisphere floras of Eocene age these problems have not been considered of major importance, but for southern Australian floras, for example, the data from pollen and leaf remains show few common taxa: *Nothofagus*, one of the chief pollen producers in the Eocene there, is totally absent from some leaf floras [2].

The geographical coverage of described Eocene floras is very uneven, with the best-worked examples in North America and Europe, and a dearth of records from most tropical and arid regions. Reconstructions of global vegetation patterns in the Eocene must accordingly involve considerable conjecture and extrapolation – a situation which is exacerbated by the occurrence in all floras of combinations of taxa not found in any current vegetation types. This pattern is so consistent that any explanation based merely on adulteration of the autochthonous (local) component with a strong allochthonous (non-local) presence (derived, for example, from adjacent high-altitude vegetation) must be regarded with suspicion. Sus-

* Some published floristic lists may relate to a single site only ('floras'), whereas others may be pooled lists from several sites where the same rock formation is present ('Floras').

pect too must be any direct floristic comparisons of Eocene floras with modern floras – as the presence in all investigated Eocene floras of many plant remains that cannot be related to surviving taxa necessarily means that attempts to utilize the floristic lists as the basis of comparison must involve part only of the total data set (those plant remains that can be related to modern genera or families). The total data set of angiosperm records is pertinent only to physiognomic comparisons, based for example on leaf criteria. For such comparisons, however, the simple classification of major forest types given in Chapter 1 is clearly inadequate. More elaborate classifications have been devised (see next section), but even these are applicable primarily to northern hemisphere floras.

The Eocene, finally, was a period of considerable climatic change, but many floras cannot be dated with sufficient precision to show whether they relate to the temperature maximum of about 50 Ma ago, or to the cooler episodes before and after. Comparisons between imprecisely dated floras may consequently lead to invalid conclusions about the distribution of particular vegetation types.

9.2.2 General character of the floras

All described Eocene floras from about 50 Ma ago are dominated by woody plants. In the northern hemisphere a forest flora has been inferred for Grinnell Land in the Canadian Arctic (currently at 81°N latitude) and for Spitsbergen, both close to the shores of the then-landlocked Arctic Ocean [3]. In the southern hemisphere, the presence of *Nothofagus*-type macrofossils in Antarctica later in the Tertiary Period [4] suggests that southern polar regions too were forested in the Eocene. Fossilized wood of Eocene age in the North African deserts [5] demonstrates that these areas were tree-covered also. The Eocene must thus have been an almost totally forested world. In both hemispheres gymnosperms were an important component of the forests at high latitudes, though at middle and low latitudes evergreen angiosperms were predominant. Leaf sizes of these evergreen angiosperms were typically large (lengths in excess of 12 cm – so-called 'mesophylls' and 'megaphylls'), and the general character of the fossil leaf floras below about latitudes 55–60° resembled that of present-day tropical and subtropical forests. At high latitudes in the northern hemisphere (>70°), leaf sizes were also large, but angiosperms there were mainly deciduous – as were many of the conifers (*Ginkgo, Glyptostrobus, Larix, Metasequoia, Pseudolarix* and *Taxodium*). Comparable forests in the southern hemisphere had a very different aspect, as a result of the absence of deciduous angiosperms and needle-leaved gymnosperms [6] (section 9.2.4). At intervening latitudes (60–70°) in both hemispheres, most broadleaf trees were evergreen and 'notophyllous' (with leaf lengths of 8–12 cm).

A small number of Eocene floras (some not precisely dated and probably postdating the thermal maximum) have been interpreted as derived from scrubby woodland adapted to seasonally dry subtropical climates [3,7–14]. Component plants of these floras typically had small evergreen or deciduous leaves ('microphylls'), often of leathery texture (sclerophyllous). The occurrence of more extreme drought conditions in southern South America and south-central Australia at this time, leading perhaps to the development of savanna woodland, has been inferred from the fossil presence of grazing faunas [13] and of small amounts of grass pollen [2], whilst there is purely geological evidence for wide-ranging drier climates in central and eastern Asia, North America, South America and North Africa in the earlier Eocene [15] (Figure 4.2). Consequently, the inference must be that drought-adapted plants were widespread at this time in latitudes 30–45° in both hemispheres, and that this zone represented the subtropical high-pressure belt displaced polewards by 10–15° latitude from its present position. Component plants of this zone in the northern hemisphere have been termed the **Madrean–Tethyan Geoflora** by Axelrod [8]; they were characteristically small deep-rooted trees or shrubs with small evergreen leaves, often with a thick cuticle. Plants comprising the more mesic forest vegetation at higher latitudes in the northern hemisphere have been termed the **Arctotertiary Geoflora**; those at lower latitudes constitute the **Tropical–Tertiary Geoflora** [3].

9.2.3 Physiognomic comparisons

Various attempts have been made to relate the early Eocene forests to modern forest types. As pointed out earlier, direct floristic comparisons are probably not valid because of the 'botanical anonymity' (i.e., lack of present-day homologues) of much of the fossil material. Comparisons using leaf characteristics (section 4.3.1), on the other hand, have the merit of utilizing data for all identified taxa, whether they have surviving relatives or not. Such comparisons require an equivalent and wide-ranging data-set from modern forests, and that has recently been provided by Wolfe's comprehensive study [16] of modern eastern Asian forests – which have been much less impoverished by glaciation than the corresponding forests of North America and Europe. Wolfe recognizes ten mesic forest types (Figure 9.1), characterized thermally (by mean annual temperature and temperature-range), physiognomically (by the percentage of hardwood taxa with entire-margined leaves), and floristically (by typical famil-

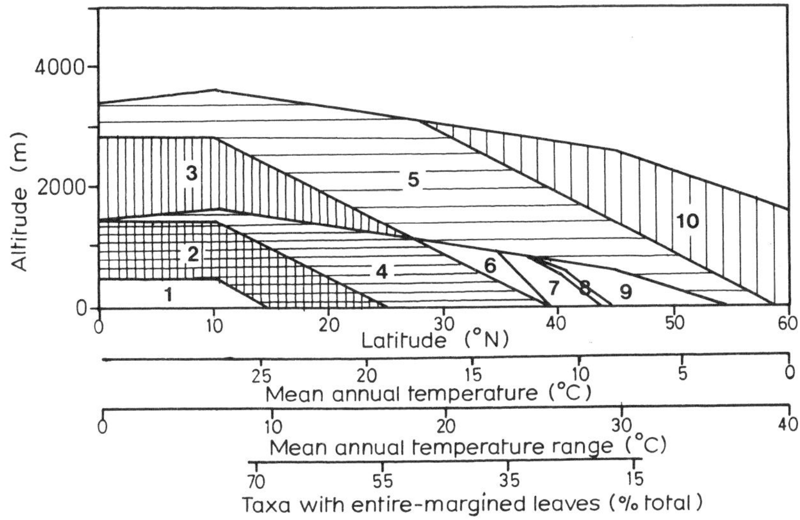

Figure 9.1 Schematic distribution of modern forest-types in the northern hemisphere, based on East Asian forests, and some thermal and physiognomic properties (re-drawn from Wolfe [16]). 1 = tropical rain forest; 2 = paratropical rain forest; 3 = microphyllous broadleaved evergreen forest; 4 = notophyllous broadleaved evergreen forest; 5 = mixed coniferous forest; 6 = mixed broadleaved evergreen and coniferous forest; 7 = mixed mesophytic forest; 8 = mixed broadleaved deciduous forest; 9 = mixed northern hardwood forest; 10 = taiga.

ies and genera). His classification can be applied with some modifications to forest-types elsewhere in the world also.

Physiognomic characterizations of Eocene floras have concentrated on leaf-margin analysis. Typical published values of the proportion of hardwood taxa with entire-margined leaves are: Alaska 65–82% [17,18]; Rocky Mountains 55% [12]; south-eastern United States 82% [19]; and the Rhineland 75% [20]. These values fall predominantly within the range shown by modern paratropical forests (Figure 9.1). On the basis of such values, and on general physiognomic features, Wolfe [14] has attempted a reconstruction of global forest types at the Eocene thermal maximum of about 50 Ma ago (Figure 9.2). He recognizes four major forest types only: tropical rainforest, extending to about 50° latitude in both hemispheres; paratropical rainforest, extending to 60–65° latitude in both hemispheres; notophyllous broadleaved evergreen forest, between latitudes 60° and 70°; and polar deciduous forest, in high polar regions. The evidence presented in section 9.2.2 suggests that a further forest type, megathermal semi-deciduous woodland, must also have been present between latitudes 30° and 45° in continental interiors.

Wolfe's map, however, is highly generalized, and his positioning of

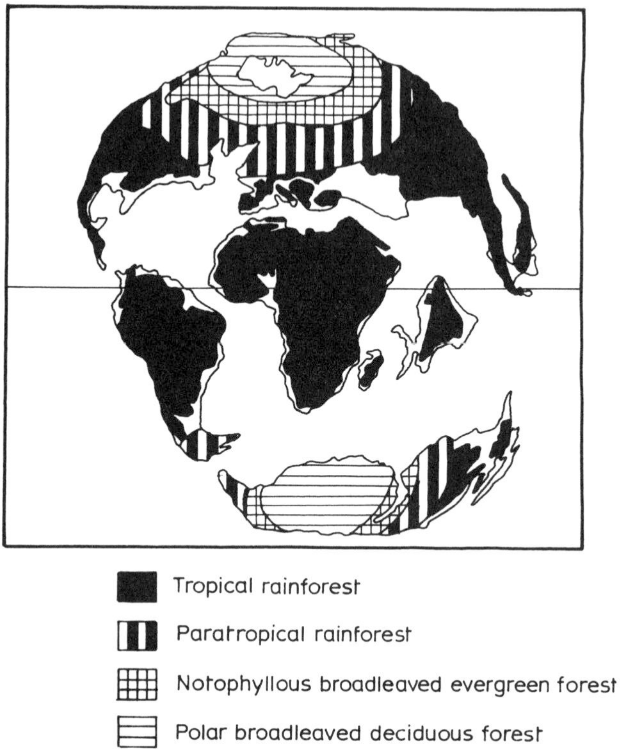

Figure 9.2 Generalized distribution of forest types in the Early Eocene: data of Wolfe [14], re-drawn on a base-map of Barron *et al.* [21]. The extent of forests is shown to continental shelf margins at higher latitudes, but only to actual coastlines at lower latitudes.

forest zones in the southern hemisphere in particular is very speculative. There are no fossil records to support the idea of an Antarctic Continent covered with deciduous forest similar to that in arctic regions. A rather different reconstruction of African Eocene forests too is given by Axelrod and Raven [9]. Nevertheless, Wolfe's map brings out clearly the enormous global extent of tropical and subtropical forests 50 Ma ago.

9.2.4 Floristic comparisons

All investigators of Eocene fossil floras from middle and high latitudes, particularly in the northern hemisphere, have emphasized three features: the presence in them of many taxa now characteristic of lower latitudes;

the presence in floras from one continent of taxa now restricted to another continent; and the presence of assemblages of taxa not now found together in any single extant vegetation type. Thus high-latitude Eocene floras, it is claimed, contain many warm-temperate taxa currently found 40–50° latitude nearer to the equator, while middle-latitude floras contain many present-day tropical taxa; North American Eocene floras are reported to contain taxa now found only in eastern and south-eastern Asia, while European floras contain some taxa now found only in the New World; and middle-latitude Eocene floras are reported to contain taxa now segregated among both tropical and temperate forests types. Although such claims are based on floristic comparisons between modern and fossil floras where up to two-thirds of the fossil taxa may have no living relatives, nevertheless these features of Eocene floras seem so clear-cut that they must have real significance.

The three features can be illustrated by floristic comparisons between six northern-hemisphere Eocene floras: the Wilcox Flora of the southeastern United States, the Lower Ravenian Flora of southern Alaska, the Tipperary and Kisinger Lakes Floras of Wyoming, the London Clay Flora of south-east England, and composite lists for Arctic regions and for Japan. Sources of floristic data (almost certainly incomplete) are given in Table 9.1 (columns 1–6), which lists extant angiosperm families recorded in two or more of these floras. These families are grouped into five categories, based on their present geographical ranges (as given by Willis [22]): wide-ranging, cool-temperate, warm-temperate, tropical and subtropical, and confined to tropical regions. The lists comprise 54 families out of a total of 85 extant families represented in the six floras (64%). In view of the disparateness of the data (the floras containing very different proportions of preserved leaves, seeds, fruits and pollen grains), the number of common families is surprisingly high. Twenty-three of the 54 wider-ranging families (43%) have tropical/subtropical distributions at the present day, though these families are poorly represented in the Arctic floras from the Eocene. In the five middle-latitude floras, the percentage of tropical/subtropical families ranges from 36 to 52%.

Many of these tropical/subtropical families are also represented in Eocene floras from low latitudes. Such low-latitude floras are floristically very rich, but they also show greater geographical differences between continents than do middle- and high-latitude floras – because of the early development of ocean barriers between the low-latitude land masses. Consequently, the number of wide-ranging extant families in low-latitude Eocene floras is rather small (Table 9.1, column 7).

At the generic level also there is a strong presence in middle-latitude floras of taxa with present-day ranges centred in tropical and subtropical regions. Numbers of such genera recorded in the five middle-latitude

Table 9.1 Extant families represented by fossil taxa in two or more of the following six Eocene floras: 1 = Arctic floras [3,23]; 2 = Japanese floras [24]; 3 = London Clay Flora [25,26]; 4 = Wyoming floras [12]; 5 = Wilcox Flora [27,28]; 6 = southern Alaskan floras [17]. Column 7 lists wide-ranging families in low-latitude floras [9,29,30]

	1	2	3	4	5	6	7
Wide-ranging families							
Altingiaceae	+		+	+	+		
Apocynaceae		+		+			+
Aquifoliaceae		+		+	+		
Celastraceae					+	+	
Euphorbiaceae		+	+	+	+		+
Fagaceae	+	+	+		+		
Juglandaceae	+	+	+	+	+	+	
Leguminosae		+		+	+	+	+
Magnoliaceae	+		+		+		
Myricaceae	+	+			+		
Oleaceae	+	+			+		
Rhamnaceae		+			+	+	
Rosaceae			+	+	+		
Rutaceae					+	+	+
Smilacaceae		+			+		
Tiliaceae	+	+	+	+			
Ulmaceae	+	+		+		+	
Cool–temperate families							
Aceraceae	+	+		+	+		
Betulaceae	+	+	+	+	+	+	
Caprifoliaceae		+	+				
Cercidiphyllaceae	+	+	+		+		
Cornaceae	+		+	+		+	
Corylaceae	+	+	+				
Nyssaceae			+		+		
Platanaceae	+	+	+	+	+		
Salicaceae	+			+			
Warm–temperate families							
Eucommiaceae		+		+			
Hamamelidaceae	+	+	+				
Meliaceae		+	+	+	+	+	
Meliosmaceae			+	+			+
Menispermaceae			+	+			+
Tropical and subtropical families							
Actinidiaceae				+		+	
Aristolochiaceae				+	+		
Ehretiaceae		+	+		+		
Flacourtiaceae		+	+		+		
Lauraceae	+	+	+	+	+	+	+
Moraceae		+			+		+
Myrsinaceae		+			+		+
Myrtaceae		+		+	+	+	+
Palmae		+	+	+		+	+
Sapindaceae		+		+	+	+	+
Symplocaceae		+	+	+	+		
Verbenaceae					+	+	
Vitidaceae	+		+			+	
Exclusively tropical families							
Alangiaceae		+		+		+	
Anacardiaceae		+	+	+	+	+	+
Annonaceae			+		+	+	
Araliaceae				+	+		
Ebenaceae		+		+	+		+
Icacinaceae			+	+		+	
Rhizophoraceae			+			+	
Sapotaceae		+			+		+
Sterculiaceae				+	+		+
Zygophyllaceae				+	+		

Eocene floras listed in Table 9.1 are as follows: Japanese 17 (30% of total extent genera); London Clay 26 (52%); Wyoming 23 (48%); Wilcox 46 (49%); and Alaskan 31 (74%).

Wolfe [16] recognizes a number of 'floristic components' (families or genera with geographically restricted distributions) that are present in modern Asian forests. The more important of these are listed in Table 9.2. The components can also be recognized in the Eocene floras. All the floras contain a greater number of them than are typically present in any single modern forest-type (and it must be remembered that this conclusion is based on a consideration of part only of the total floristic list). Nevertheless,

Table 9.2 Characterization of six Eocene Floras in terms of floristic components* represented in modern forest types of eastern Asia [16]; numbers are the numbers of appropriate taxa present in the fossil flora

	Arctic	Japan	London Clay	Wyoming	Wilcox	Alaska
Tropical/subtropical families	1	7	7	9	7	5
Evergreen 'laurels'	2	2	3	1	3	1
Temperate genera (hardwoods)	4	7	5	2	5	2
Hardy gymnosperms	3	0	1	0	0	0

* Extant taxa comprising the floristic components are:

1. Tropical/subtropical families: Anacardiaceae, Annonaceae, Apocynaceae, Araliaceae, Burseraceae, Dipterocarpaceae, Ebenaceae, Guttiferae, Icacinaceae, Meliaceae, Menispermaceae, Myristicaceae, Myrtaceae, Palmae, Sterculiaceae, Vitidaceae. 2. Evergreen 'laurels': Fagaceae (evergreen), Lauraceae, Magnoliaceae, Theaceae. 3. Temperate hardwoods: *Acer, Alnus, Ampelopsis, Berchemia, Celastrus, Celtis, Euonymus, Hydrangea, Koelreuteria, Myrica, Parthenocissus, Photinia, Pyracaentha, Rhamnus, Rhododendron, Rhus, Rosa, Rubus, Salix, Styrax, Ulmus, Viburnum, Zelkova*. 4. Hardy gymnosperms: *Abies, Larix, Picea, Pinus, Pseudotsuga, Tsuga*.

the floristic data support the leaf-margin analyses in showing that the closest resemblances of the middle-latitude Eocene forests are to the modern paratropical forests. The Arctic forests, however, have no close modern parallel.

Eocene floras from the seasonally dry climates of the southern United States, southern Europe and North Africa, and the southern Soviet Union are rather different in character. These floras are entirely included within the region occupied then by tropical and paratropical rainforests, and were probably derived from them as a subset of plants that evolved xeromorphic features. Axelrod [7,8] gives a long list of such characteristic Madrean –Tethyan taxa. Because of the early separation of neotropical and palaeotropical floras, however, different xeromorphic assemblages developed from them in North America and in Eurasia, and the number of taxa common to the two regions is rather small. Genera represented by sclerophyllous species in both regions include *Acacia, Arbutus, Myrica, Persea, Pistacia, Quercus, Rhamnus, Rhus, Sabal, Sapindus, Zanthoxylum* and *Zizyphus* [7,8,31].

Much less is known about Eocene floras from the southern hemisphere. Floristic lists for high- and middle-latitude floras contain, like their northern-hemisphere counterparts, assemblages of taxa now segregated among several different modern plant communities [9,11,32,33]. The main taxa are of Gondwanic origin, and largely confined to the southern hemisphere: *Nothofagus*, Myrtaceae, Podocarpaceae and Proteaceae, for example. At higher latitudes (Patagonia, Tasmania, New Zealand and Antarctica), evergreen beeches (*Nothofagus*) and the 'southern conifers' (*Araucaria, Dacrydium, Libocedrus, Microcachrys* and *Podocarpus*) dominate Early Eocene pollen floras, and may have formed forests similar to the cool-temperate rainforests found today in New Zealand and southern Chile. Equatorwards, in southern Australia and northern Patagonia, tropical/subtropical elements predominate. Notable Australian pollen types related to extant subtropical taxa are *Anacolosa* (Olacaceae), *Beauprea* (Proteaceae), *Bombax* (Bombacaceae), *Cupania* (Sapindaceae), *Nypa* (Palmae) and *Santalum* (Santalaceae). These tropical/subtropical elements are generally not present in the higher-latitude floras.

The southern Australian Eocene floras, in particular, show a mingling of taxa not found together today [34] – though the marked discrepancies between pollen and leaf floras from the same or nearby localities [2] suggest that the forests may have consisted of a fine mosaic of different communities rather than being one uniform type. The character of the leaf floras suggests a diverse subtropical rainforest vegetation resembling certain rainforest types still persisting today in north-eastern Queensland [35]. Subtropical elements in these Eocene forests that are no longer found on the Australian mainland include *Beauprea* (now confined to New

Caledonia), *Microcachrys* (now confined to Tasmania), and species of the 'brassii' group of *Nothofagus* (now found only in New Guinea, New Britain and New Caledonia), for example.

Almost nothing is known about the origin and character of sclerophyllous floras in the southern hemisphere in the Early Tertiary Period. Sclerophylly is a distinctive feature particularly of the contemporary Australian flora, especially in the families Epacridaceae, Myrtaceae, Papilionaceae and Proteaceae, and may have evolved in response not only to seasonally dry climates but also to phosphate deficiency in the soil [36]. It is unlikely, however, that sclerophylly was widespread in Australian Eocene floras; for two major present-day genera at least, *Eucalyptus* and *Acacia*, fossil documentation begins later – in the Oligocene and Miocene, respectively [2].

9.3 DIVERSIFICATION OF VEGETATION THROUGH THE TERTIARY

9.3.1 The general pattern of change

Physiognomically, as we have seen, the fossil biomes of the Eocene seemingly relate to a limited number only of modern vegetation types. This restricted diversity can be illustrated schematically by reference to Figure 9.3, which is an idealized chart of the distribution of present-day biomes in relation to mean annual temperature and precipitation. The chart utilizes data of Whittaker [37], Walter [38] and Wolfe [16], and follows Wolfe's categorization of annual temperature regimes into tropical, paratropical, subtropical, temperate, paratemperate and subtemperate. The chart depicts what might be called **global ecospace** and its subdivision. The current extent of global ecospace encompasses the full range of temperature and precipitation conditions to the right of the line ABC on the chart. However, during the thermal maximum of the Eocene Epoch, global ecospace was far less diverse: its extent can perhaps be represented by the portion of the chart to the right of the line DEF. Subsequently global ecospace expanded. The line GHJ defines its likely extent during the Middle Miocene, about 15 Ma ago. The main cause of this expansion of global ecospace was the development of new seasonal climates, whereby part of each year became increasingly unfavourable for active plant growth – because of either low temperature or low rainfall or both. Initially woody plants remained dominant throughout the global ecospace, though new strategies were evolved to cope with the lengthening stress period each year. In some habitats leaf size was reduced, though an evergreen habit was retained; in other habitats a deciduous habit was adopted. Ultimately new

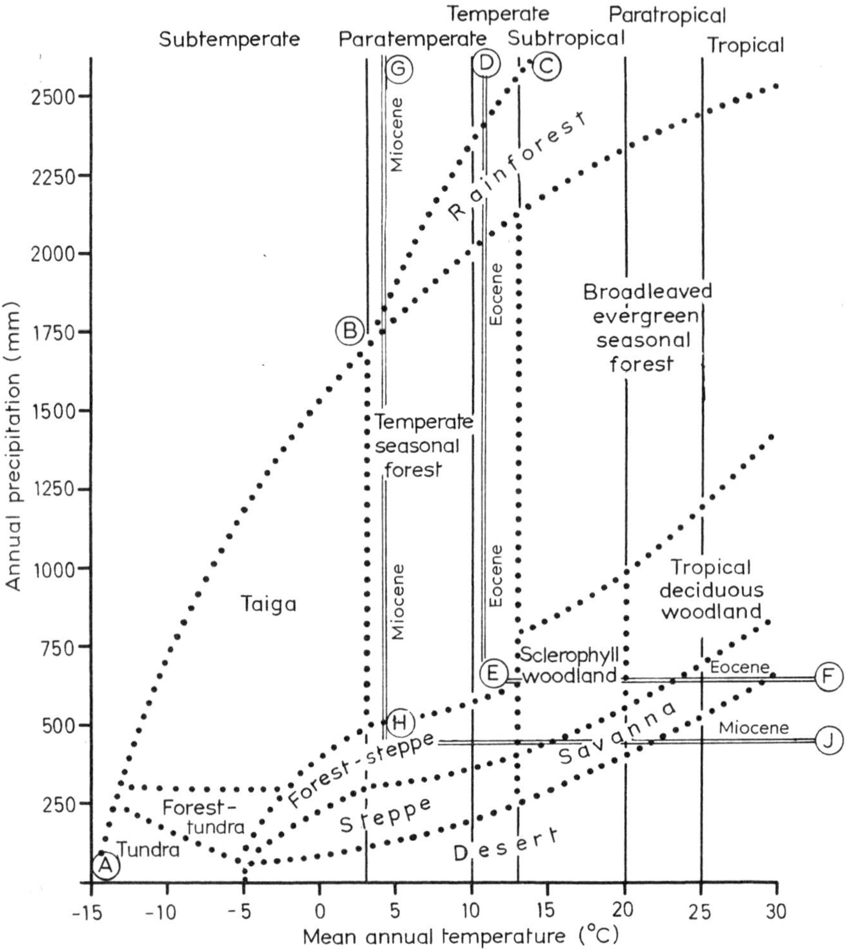

Figure 9.3 A schematic representation of global ecospace.

ecospace was created with conditions too severe for the survival of woody plants, and biomes dominated by herbaceous plants developed. By the close of the Tertiary Period, global ecospace had diversified to nearly its present-day extent: on the chart (Figure 9.3), its extent and subdivision then differed little from the contemporary situation. Increasing seasonality of climate through the Tertiary was thus expressed in a shift from the equable conditions of the Eocene towards accentuated annual variability of both temperature and rainfall, which were superimposed upon the general trend of declining mean temperatures.

The new seasonal ecospace was largely the outcome of continental displacement (Chapter 4). Part at least of the declining temperature trend through the Tertiary can be explained by changing global dispositions of land and sea affecting the global patchwork of albedo (section 4.7). By the Early Miocene (17–24 Ma ago), moreover, the realignment of land masses was beginning to produce critical changes in the patterns of ocean currents. The elimination of the Tethys Sea 17–18 Ma ago by the northward drift of the African Plate, followed by the closure of the 'Panamanian portal' as North and South America drifted into contact, and the enhanced 'blocking effect' of the Malesian archipelago by the northward drift of the Australian Plate, led to the replacement of the vast ocean gyres of the Early Tertiary (Figure 4.2) by more circumscribed gyres within each of the major oceans. The heating effect of warm currents at high latitudes was thus diminished, and temperature gradients between equator and poles accentuated [9]. Cold waters from polar latitudes, brought south by these ocean gyres, began increasingly to affect the western sides of the major land masses by the Late Tertiary (cf. Figure 3.3); this effect was particularly marked in the southern hemisphere, where refrigeration of the polar regions was encouraged by the development of the cold transglobal Circum-Antarctic Current as early as 22 Ma ago.

Plate contacts in the Miocene led both to general continental uplift and to more localized orogenesis. These fundamental changes in topography produced fundamental changes in climatic patterns. In addition to the rainshadow effects created by orogenesis, there was a general intensification of the mid-latitude high-pressure areas, a strengthening of atmospheric circulation systems (section 3.3), and a progressive enhancement of the subtropical arid zone [9]. All these processes resulted in diversification of the global ecospace.

9.3.2 The Miocene world

The deep-ocean oxygen-isotope record shows a major fall in temperature at the end of the Eocene Epoch (Figure 4.3), which led to the elimination of many tropical and some warm-temperate elements from high-latitude floras [26] (Table 7.7). The ensuing Oligocene Epoch was cooler and drier globally, with falling sea levels (Figure 4.5) and probably the first development of permanent ice over the Antarctic continent by its close (Table 4.1). Plants adapted to seasonally dry climates spread [7,9,39]. Ocean temperatures at higher latitudes rose again temporarily in the Middle Miocene, 15–17 Ma ago, when land temperatures in Europe and North America (though not at lower latitudes) were probably about 7°C higher than they are at present [20,40]. Associated high sea levels resulted in generally moist

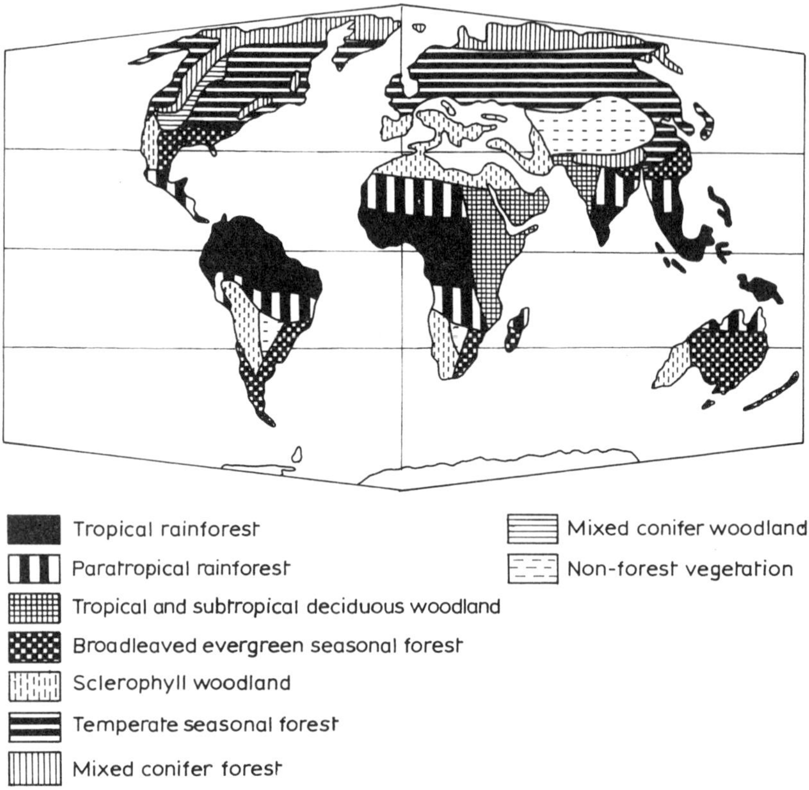

Figure 9.4 Generalized vegetation map for the Middle Miocene Epoch. Redrawn with modifications from Olson [42], on a base map of continental outlines by Parrish *et al.* [15].

genial climates, and the development of a rich and diversified forest cover, which extended to the shores of the Arctic Ocean and to the periphery at least of Antarctica [41]. Wolfe [14] and Olson [42] have produced generalized vegetation maps for this period. Figure 9.4 is based on these maps, and shows clearly the vegetation changes that are presumed to have taken place since the Middle Eocene some 30 Ma earlier (Figure 9.2). The wide-ranging, large-leaved, evergreen, tropical and paratropical rainforests of the Eocene had contracted through about 35° of latitude, and been replaced at higher latitudes by smaller-leaved seasonal forests and woodlands with a variable (but often high) proportion of deciduous taxa. The greater diversity of vegetation types in the Middle Miocene (nine, as against five Eocene types) was largely the result of adaptation to strongly seasonal climates at all latitudes: deciduous woodlands in the tropics and subtropics of Africa and India, sclerophyll woodlands between latitudes 20° and 40° in both hemispheres, and mixed coniferous forest at higher latitudes and altitudes in the northern hemisphere. Nevertheless, the tenuousness of the evidence on which this Miocene map is based must not be overlooked. The vegetation reconstructions for almost the whole of the southern hemisphere rely on a very small number of fossil sites, and rather different reconstructions, based on the same very limited data, have been proposed for South America [43], Africa [9] and Australia [44]. The extent to which non-forest types (savanna, grassland, or even desert), in particular, had developed in continental interiors in the subtropics is quite unclear.

The best documented feature of the Middle Miocene vegetation is the wide-ranging occurrence of temperate broadleaved forests in the northern hemisphere. These forests contained taxa now characteristic of the present-day forest vegetation of western and eastern North America, non-Mediterranean Europe and East Asia, but were more diverse than any single modern forest-type in these regions. Parallels have been drawn with the modern 'mixed mesophytic forests' of China [45], but a better analogy might be to regard them as an amalgam of at least four of Wolfe's forest-types [16] (types 6–9 in Figure 9.1). Although there must have been regional variation in these Miocene forests, many taxa apparently occurred over almost the entire range. Among these taxa were: cold–temperate hardwoods such as *Alnus, Betula, Corylus, Nyssa, Salix* and *Ulmus*; cold-tolerant conifers such as *Picea, Pinus* and *Tsuga*; warm–temperate trees such as *Carya* and *Liquidambar* (still widespread), *Ceridiphyllum* and *Glyptostobus* (now restricted to eastern Asia), and *Sequoia* (now restricted to western North America); and subtropical taxa such as *Engelhardtia*. All these (and many other) taxa had also been widespread earlier on, in the Middle Eocene Epoch, but in rather different forest-types. Consequently, to regard the middle-latitude Miocene forests as derived solely from southward displacement of an Eocene Arctotertiary Geoflora would be erroneous. The middle-latitude Miocene flora certainly contained many temperately-adapted taxa of that geoflora that had migrated southward with deteriorating climate, but it had also drawn in taxa from the Early Tertiary paratropical forests – taxa that had been 'left behind', and had evolved *in situ* to produce new lineages adapted to colder conditions than those favoured by their Eocene ancestors [45].

Similar considerations must also apply to the flora of the sclerophyll woodlands at lower latitudes (the Madrean–Tethyan Geoflora), associated particularly with the subtropical high-pressure belt and its margins in both hemispheres. That belt, with its seasonal or more prolonged drought conditions, remained much more nearly stationary as vegetation zones were displaced progressively equatorwards through the Tertiary Period. Consequently, new lineages were continually being drawn in as temperate taxa migrated equatorwards. Some adapted to the drier climates, and persisted, so that recruitment to the Madrean–Tethyan Geoflora was continually occurring. The concept of geofloras as originally proposed by Chaney (section 1.3) can thus have only limited significance in the context of Tertiary vegetation history [45,46].

9.3.3 The spread of non-forest communities

Following the thermal optimum of the Middle Miocene Epoch (section 4.4), renewed orogenesis affected most parts of the globe, and continued, often with increasing intensity, into the subsequent Pliocene Epoch (section 4.2). As a result, the range of climatic and topographic conditions over the earth's surface was enhanced, and global ecospace was further diversified. Ocean temperatures (and presumably land temperatures also) fell sharply between about 15 and 13 Ma ago, and again between 7 and 5 Ma ago, leading first to enlargement of the East Antarctic Ice Sheet and then to icing-up of West Antarctica [47]. The ensuing fall in global sea levels, coupled with intensification of high pressure in the subtropics and increasing rainshadow effects in the lee of newly-uplifted mountain ranges, combined to make the Late Miocene the driest part of the Tertiary Period. This arid episode was manifested most spectacularly in the almost complete drying-up of the Mediterranean Sea [48]. In northern Israel as a result [49], pollen spectra with 50% tree pollen in the Middle Miocene (and including such mesophytic taxa as *Alnus, Corylus, Engelhardtia, Juglans, Platycarya* and *Pterocarya*) were replaced in the Late Miocene by spectra with almost no tree pollen, and with 'arid' indicators such as *Artemisia* and Chenopodiaceae prominent.

There have been speculations that a wide range of dryland communities had developed earlier, in the drier conditions of the Oligocene 25–35 Ma ago [9,50], with centres of aridity in South America in the rainshadow of the rising Andean Cordillera, in the low-latitude Sahara and Kalahari Deserts, and in the remote interior of Middle Asia [51]. However, the major evidence of diversifying dryland vegetation comes from the Late Miocene and Pliocene. Certainly, the environmental conditions necessary for the creation of some present-day desert regions did not exist before the Miocene: the Middle Eastern desert region was largely submerged below the Tethys Sea [52], the American Southwest was still open to the influence of rain-bearing westerly winds, and parts of Middle Asia were not yet isolated from the south-west monsoons by the mountain barrier of the

Himalayas. All that changed with the frenzied tectonic activity of the Miocene and Pliocene, and during the last 5–10 Ma of the Tertiary Period non-forest communities became widely established over the same general areas occupied today by desert, steppe and savanna vegetation.

However, although floras containing non-woody taxa similar to those predominant today in desert, steppe and savanna areas were undoubtedly present in the Late Tertiary, the component taxa might have been assembled into communities very different from modern dry-climate ones. Thus pollen floras of Late Miocene and Pliocene age, even from present-day desert areas, almost invariably contain tree pollen, often of mesophytic taxa. For example, pollen of Betulaceae, Fagaceae, Juglandaceae and Ulmaceae accompanies that of *Pinus*, *Artemisia* and Chenopodiaceae in Miocene pollen spectra from the Gobi Desert region of north-west China [53]. Whether this pollen was transported from faraway deciduous forest sources to the south, or whether it was derived from local woodland patches growing in damper situations within the region, is not clear. Most palaeoecologists would probably favour an explanation based on local sources of pollen in arid regions – either from continuous scrubby woodland with an understorey of desert plants, or from local woodland patches intermixed with more open vegetation.

Nevertheless, the macrofossil, pollen and faunal evidence for the Late Tertiary supports the existence of a wide range of dryland communities. True desert conditions may have been present in the Sahara and the Namib [54], and semi-desert in southern Africa and the Near East [9,49]. The extensive saline deposits that formed as the remnants of the Tethys Sea evaporated could have triggered the evolution of many desert halophytes [15]. Some sort of sclerophyllous woodland was almost certainly present in the American Southwest, whilst fossil floras of Pliocene age from the Great Plains region of the United States attest to a gradual restriction of forest to river valleys there as grassland communities developed on the drier interfluves. The existence of extensive areas of open grassland is suggested also by the presence of grazing and browsing faunas (horses, camels and osteodonts) in Kansas and Oklahoma after 10 Ma ago [55]. The presence of grazing faunas has similarly been used as evidence of grassland conditions in South America, but here aridification occurred rather earlier – by 15 Ma ago at least [13] – and by the Pliocene thorn scrub and grassland probably occupied much of southern South America east of the Andes [13,43]. More open conditions probably developed within the Amazonian forest too in the Pliocene, to form a major 'savanna avenue' extending north from the dry woodlands of the Brazilian Chaco, through the Mato Grosso of Brazil, to the highlands of Guyana and Venezuela [13,56].

In East Africa, grazing faunas diversified later, about 2.5 Ma ago, at a time when pollen evidence indicates a contraction of rainforest [57]. Savanna and grassland may have spread sufficiently then to separate the formerly continuous rainforest into eastern and western segments [9]. In India the dominant rainforest of the Middle Tertiary retreated gradually eastwards and southwards in the Pliocene, as arid conditions developed in Pakistan and to the south of the rapidly-rising Himalayas [8,58]. Arid conditions almost certainly developed too in the interior of the Australian continent, but adequate fossil documentation is lacking [2,59,60].

In areas outside the tropics where seasonally dry climates developed in the Late Tertiary, the major dry period initially occurred during the cool season, and 'mediterranean' climates with summer drought did not develop until the accumulation of polar ice allowed the formation of cold ocean currents affecting the western sides of the major land masses [61]. For some areas, full summer drought conditions may not have been realized until the glacial stages of the Pleistocene [9], but in the northwest Mediterranean Basin at least there is some evidence that dry summers developed in the Late Pliocene. About 2.8 Ma ago here, pollen of woody taxa requiring year-round moisture (e.g., *Engelhardtia, Hamamelis, Nyssa* and *Symplocos*) declined to near-extinction, and pollen of taxa characteristic of present-day shrublands around the Mediterranean Basin (*Cistus, Olea, Phillyrea, Pistacia, Quercis ilex*-type) became a constant component of the sediments [62]. However, it must not be forgotten that adaptation to summer drought is a secondary evolutionary feature that developed in plants already adapted to seasonal drought; the adaptation necessitated a shift in the time of seedling germination and establishment from late spring/early summer to late winter/early spring, together with concomitant adjustments in the timing of flowering and seed maturation [8]. Hence the presence of 'mediterranean' genera in the fossil record does not necessarily imply the existence of a mediterranean climate. *Arbutus, Chamaerops, Juniperus, Pistacia* and *Quercus coccifera*, for example, characteristic members of present-day shrublands in southern Europe, are recorded from European sites back to the Early Oligocene, when they were part of the developing Madrean–Tethyan Geoflora [63].

Spread of non-forest plant communities late in the Tertiary Period occurred also at high altitudes, as mountain ranges were thrust up above the attainable treeline. Documentation of this spread of herbaceous plants relies largely on the pollen record, and is consequently rather non-specific. However, it is clear both from the pollen record and from present-day distribution patterns that the colonizing species of these newly-created habitats were drawn both from local sources and from more distant ones. Thus pollen analyses from the high northern Andes [43] show developing montane and alpine floras in the Pliocene Epoch that were derived mainly from neotropical elements, but drew in some North American taxa as well (*Alnus, Aster*, Cruciferae, Ericaceae, *Erigeron, Sambucus* and *Viburnum*) and at least one Antarctic one (*Lagenophora*). Present distribution patterns of alpine plants on the East African mountains [9] similarly indicate an input of exotic genera in the past – from southern Africa and from temperate regions of the northern hemisphere.

Adaptation to conditions of extreme cold almost certainly occurred in these habitats and not in high-latitude (arctic) ones. Tundra vegetation dominated by herbaceous plants was almost non-existent before the Pleistocene Epoch, with boreal forest and spruce parkland reaching north to the shores of the Arctic Ocean [64]. At a few sites only is there pollen evidence of dwarf-shrub tundra in the Late Pliocene: on Seward Peninsula, Alaska [39], and in the Kolyma lowlands of north-east Siberia [65].

9.4 THE TERTIARY VEGETATION OF WESTERN NORTH AMERICA

9.4.1 Modern vegetation types

In previous pages we have discussed some of the major trends in vegetation history through the Tertiary Period, and in particular the diversification of global ecospace, and the gradual break-up of the complex plant assemblages of the Early Tertiary into less species-rich ones that were more nearly related to modern plant communities. These trends are especially well shown in western North America, as a result of the investigation over many years of numerous fossil floras, both macrofossil and pollen, and the publication of a series of synoptic accounts dealing with the vegetation histories of localized areas*. The patterns of change that these accounts reveal to have taken place in western North America can perhaps be assumed to have been followed, albeit modified, in other parts of the world where documentation is less good.

We consider first the modern vegetation of western North America [10,72]. In the lowlands below 50°N (except along the coast north of San Francisco) there is now a range of dry woodland, scrub, grassland and desert communities. Grassland is dominant over the Great Plains to the east of the Rocky Mountains, and occurs also as 'relict' patches along with chaparral and evergreen (live) oak–madrone woodland in central California west of the Sierra Nevada. Cold and warm desert, respectively, covers the floor of the Great Basin, and the Mojave and Sonoran Deserts to the south (Figure 9.5(a)), and passes upwards into desert grassland and pinyon–juniper woodland at intermediate altitudes in the Great Basin and

* Recent accounts include those of Axelrod [55,66–68], Axelrod and Raven [9], Leopold and MacGinitie [12], Raven and Axelrod [69], Robichaux and Taylor [70] and Wolfe [45,71].

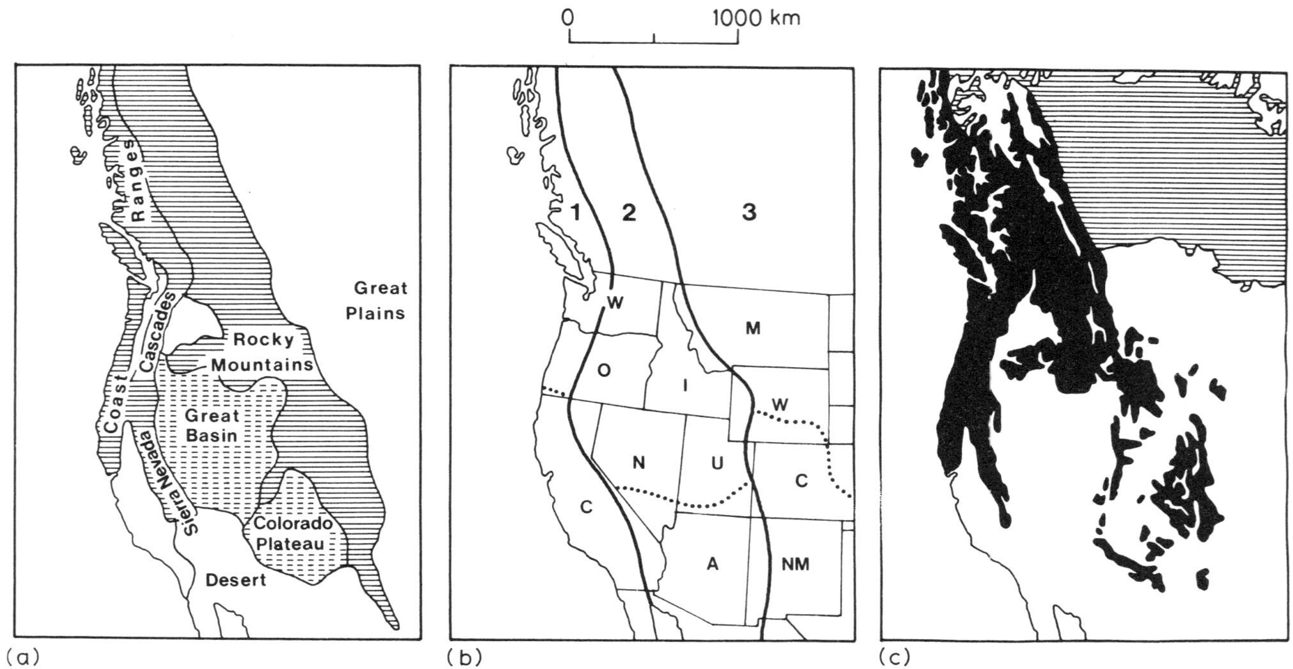

Figure 9.5 Western North America. (a) Major physiographic regions (major mountain ranges shaded, intermontane areas hatched); (b) rainfall regions: 1 =summer drought; 2 = warm-season maximum plus winter rain; 3 = warm-season maximum. State boundaries shown; regions to the south of the dotted line experience pronounced seasonal drought; (c) major forest areas: black = mixed and subalpine coniferous forest; shaded = boreal coniferous forest.

Table 9.3 Major vegetation types of western North America, and some characteristic dominant plants [10,72]

Boreal coniferous forest	*Picea glauca, P. mariana, Abies balsamea*
Subalpine forest	*Abies lasiocarpa, Picea engelmannii, Tsuga mertensiana*
Mixed coniferous forest:	
Coastal	*Tsuga heterophylla, Sequoia sempervirens*
Sierran	*Pseudotsuga menziesii, Abies concolor*
Cordilleran	*Pseudotsuga menziesii*
Pinyon–juniper woodland	*Juniperus* spp. (e.g., *J. monosperma*), *Pinus* spp. (e.g., *P. monophylla*)
Oak–madrone woodland	*Quercus* spp. (evergreen), *Arbutus menziesii*
Chaparral	*Adenostoma fasciculatum, Arctostaphylos* spp., *Ceanothus* spp.
Prairie grassland	Various Gramineae
California grassland	*Stipa pulchra* and introduced grasses (*Avena* spp., *Bromus* spp.)
Cold desert	*Artemisia tridentata, Atriplex confertifolia*
Warm desert	*Franseria dumosa, Larrea divaricata*
Desert grassland	*Bouteloua* spp., *Prosopis juliflora*

Rocky Mountains. Surviving forest in western North America (Figure 9.5(c)) is dominated by coniferous taxa, and occurs at low altitudes along the west coast northwards from San Francisco to southern Alaska and widely throughout Canada, and also at higher altitudes in all the main mountain ranges. The most luxuriant forest occurs along the west coast and along the western flanks of the Coast Ranges, the Sierra Nevada and the Cascades; the montane forests of the Rocky Mountains, the subalpine forests near timberline on all the mountains, and the Canadian boreal forest are floristically much more impoverished. Different assemblages of conifers are now characteristic of the coastal, Sierran, Cordilleran (Rocky Mountain), subalpine and boreal forests (Table 9.3). In all these forests woody angiosperms play only a subordinate role.

Climate is a major determinant of these modern vegetation types. The high mountain ranges of western North America exert a powerful blocking effect on rain-bearing westerly winds, and pronounced rainshadow conditions are produced on the eastern sides of the Sierra–Cascades and the Rockies. Rain-bearing westerlies are largely confined to the winter months, and summer drought affects the whole west-coast region (zone 1 in Figure 9.5(b)); that drought is most pronounced in California. Rainfall increases progressively northwards, from <400 mm p.a. in central California to well over 1000 mm p.a. in Washington and British Columbia; the higher rainfalls in the north are associated with lower summer temperatures. To the east of the Rockies precipitation is much less (<300 mm p.a.), and is mainly in the spring or summer months (zone 3 in Figure

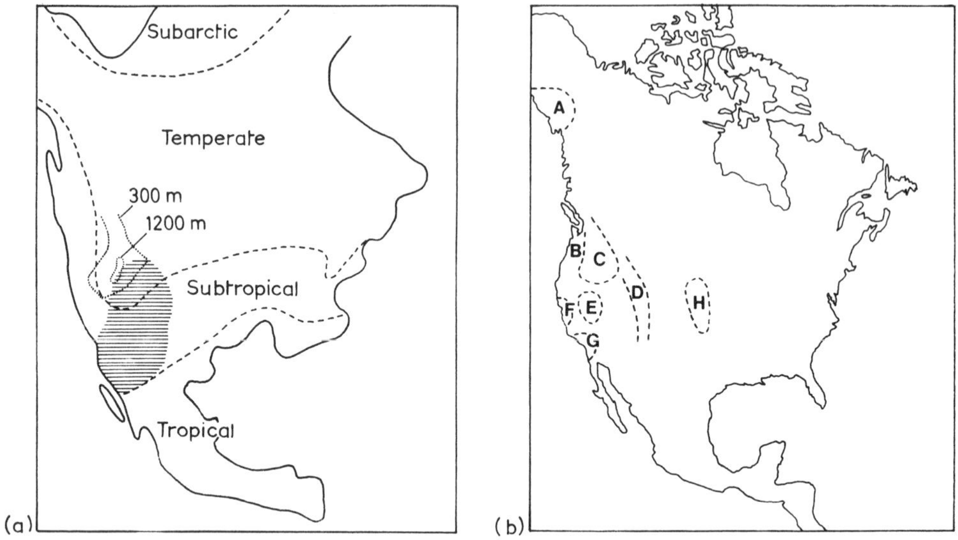

Figure 9.6 The Tertiary vegetation of North America. (a) Approximate forest zones in the Eocene. Reconstructed continental outlines are shown. The shaded area was occupied by a complex of forest, woodland and scrub in the Oligocene. Reconstructed contour lines (300 m and 1200 m altitude) for the Central Rocky Mountains are shown as close-dotted lines. Based on maps in Axelrod [55,67], Daubenmire [10] and Vankat [72]. (b) Site-groupings of Miocene and Pliocene floras. See text for explanation. Modern continental outlines are shown.

9.5(b)). In the intervening Rocky Mountain–Intermontane region (zone 2), precipitation varies generally from 200 to 500 mm p.a., and typically shows a warm-season maximum and often a subsidiary winter peak as well. Drought conditions are alleviated at higher altitudes on the mountains, but there the winters are correspondingly more severe and the growing season shorter. Winters generally also become more severe eastwards and northwards from the west coast.

9.4.2 Palaeogene vegetation*

Western North America was almost entirely forested during the Early Tertiary Period. Approximate boundaries of floristic/physiognomic zones in North America for the Eocene Epoch are shown in Figure 9.6. Much of Canada was covered with broadleaved forest ('temperate mesic' forest),

* The terms 'Palaeogene' and 'Neogene' are used here to denote the Early Tertiary (Palaeocene, Eocene and Oligocene) and Late Tertiary (Miocene and Pliocene) Periods, respectively. Several published American works include in the Pliocene Epoch what is defined as Late Miocene (12–5 Ma ago) in Chapter 2.

composed of both evergreen and deciduous tree taxa of the Arctotertiary Geoflora. Among these taxa were trees closely related to those currently characteristic of the western North America forests; others with relatives characteristic of the eastern North American forests today; and yet others with modern relatives found only in the forests of eastern Asia (Table 9.4). At high latitudes this forest was considerably impoverished, with only *Alnus, Betula, Corylus, Pinus, Populus, Salix* and *Taxodium* recorded at some sites [10]. Southwards throughout the United States, and extending northwards along the west coast into southern Alaska, tropical and subtropical elements were predominant. Thus 32 of the 36 extant dicotyledonous genera recorded in floras from the Cook Inlet area of Alaska at 60°N latitude have modern representatives in lowland Malaysia, and six are now restricted to that region (*Anamirta, Hypserpa, Limacia, Luvunga, Mastixia* and *Stemonurus* [45]). This west-coast forest probably grew in an ever-wet subtropical environment, which was maintained into the Oligocene – though a cooling trend is apparent from the floristic composition of floras in the Late Eocene. Further south in the western United States, however (Figure 9.6), there was an increasing proportion through the Palaeogene of small-leaved, evergreen, sclerophyllous taxa apparently adapted to drier climates (e.g., *Berberis, Cercocarpus, Rhus, Vauquelinia* and *Zizyphus*), together with neotropical taxa not recorded from the Alaskan forests (*Astronium, Bursera, Cardiospermum, Colubrina, Oreopanax, Persea* and *Thouinia*). The sclerophyllous element in this region was intensified by a southward shift of the North American Plate in the Eocene Epoch, which brought much of the southern United States into the influence of the subtropical high-pressure belt – so leading to a spread of the formerly localized Madrean–Tethyan Geoflora of the Cretaceous [7,73]. In the northern part of this region particularly, a diversified relief was present, with the Central Rocky Mountains in Idaho and eastern Nevada rising in places to 1500–1800 m altitude [67,69]. Here elements of the Arctotertiary Geoflora formed montane forests in more mesic situations, and probably interdigitated in the Oligocene with sclerophyllous vegetation on drier

Table 9.4 Floristic elements in North American mixed conifer–hardwood forests in the Eocene [68,69]

Characteristic of modern western American forests: *Abies, Acer, Amelanchier, Chamaecyparis, Cornus, Crataegus, Picea, Pinus, Populus, Pseudotsuga, Rhododendron, Sequoia, Thuja*
Characteristic of modern eastern North American forests: *Acer, Aesculus, Carpinus, Diospyros, Fagus, Halesia, Ilex, Liquidambar, Nyssa, Ostrya, Sassafras, Ulmus*
Characteristic of (or confined to) modern East Asian forests: *Cercidiphyllum, Ginkgo, Keteleeria, Koelreuteria, Metasequoia, Pseudolarix, Zelkova*

slopes. By 25 Ma ago these montane forests had acquired a distinctive flora containing limited subsets from both sources – notably *Abies, Picea, Pinus* and *Pseudotsuga*, with *Acer, Cercocarpus, Philadelphus, Ptelea, Ribes* and *Salix* [69]. Locally, parts of the Rocky Mountains were of sufficient elevation for conifer-dominated forests to form. Here *Abies, Chamaecyparis, Larix, Picea, Pinus, Pseudotsuga, Pseudolarix, Sequoiadendron, Thuja* and *Tsuga* were the dominant taxa [67].

9.4.3 Neogene vegetation

Except for the American Southwest (where hot arid conditions are the result of its position within the subtropical high-pressure belt), the climatic differentiation present in western North America today is primarily dependent upon topography. The modern topography of this region developed gradually through the Tertiary Period as the major mountain ranges were uplifted. This gradual uplift governed the onset of climatic dryness in the interior. Climatic dryness was manifested in seasonal drought, with drought conditions initially most pronounced probably in winter; only late in the Tertiary is there evidence of summer drought. Superimposed upon the development of local seasonally dry climates was a general lowering of temperature, consequent upon the global trend of climatic deterioration through the Tertiary. This decline was accentuated as a high-relief topography formed. The Rocky Mountains were the first of the three major mountain chains to form. A low-relief nucleus was present in the Early Tertiary (Figure 9.6(a)), and subsequent intermittent volcanic activity and uplift culminated in the Miocene, when the Cordilleran Axis may have been raised by 1500–1800 m. At this time the Sierra Nevada was still a low ridge only 500–800 m high. Final shaping of the Sierras, and full uplift of the Cascades and the Coast Ranges, was delayed until the Pliocene. Seasonally dry climates thus developed first in the east of the region, in the Rocky Mountains, where fossil floras were already modified by the Middle Oligocene. Thence dry conditions spread gradually westwards. Drought was prevalent at lower altitudes in the Great Basin by the Late Miocene (Early Pliocene of some American workers), and extended to coastal California by the Pliocene. Modernization of floras in western North America was thus achieved at different times in different areas.

Forest was still predominant throughout the region in the Middle Miocene. Derivative assemblages of the Arctotertiary and Madrean–Tethyan Geofloras had by then almost entirely replaced the neotropical and palaeotropical elements of the Eocene. Regional differentiation of Miocene floras is apparent, with both 'native' and 'exotic' taxa present in assemblages from particular geographical areas ('native' taxa being defined as those with close relatives still present in the area, and 'exotic' taxa as those with close relatives no longer present). The positions of the more important site-groupings (based on information in Wolfe [71], Axelrod [68], Robichaux and Taylor [70], and other sources) are shown in Figure 9.6(b).

In southern Alaska (assemblage A in Figure 9.6(b)) conifers were dominant, with fossil species similar to the present-day dominants of both the coastal and the boreal forests occurring together in the same floras; among the subordinate hardwoods were several exotics. A rich mixed conifer–hardwood forest (assemblage C) occurred to the south, on the low uplands that were to form the Cascades and the Columbia Plateau, with a coniferous component that included taxa characteristic of both the Sierran and coastal forests today, and a varied collection of exotic dicots. Dicotyledonous trees were predominant to the west, in the coastal lowlands running south from Puget Sound to central California (and including the low ridge of the northern Sierras); here the broadleaved forest (assemblage B) contained a mixture of evergreen and deciduous taxa no longer found in the region. Away to the east on the Rocky Mountains, by contrast, was coniferous forest of largely modern aspect, with few exotics (assemblage D). In between, on the eastern side of the low Sierran ridge and extending over parts of the Great Basin, was further mixed conifer–hardwood forest (assemblage E), of rather different composition from that to the north. This forest, although ancestral to the modern Sierran ones, still retained much of the complex character of the Early Palaeogene forests, with taxa referable to a variety of different floristic elements (Table 9.5). Even areas that today are grassland, desert or scrub probably supported tree growth in the Middle Miocene. Thus fossil floras of that age from the central Great Plains (assemblage H) contain a variety of tree taxa as well as grasses, whilst coastal California (assemblage G) was covered by oak–laurel woodland similar in character to the present live oak–madrone woodlands. Only

Table 9.5 Taxa recorded as fossils from the Miocene mixed conifer–hardwood forest of Nevada (assemblage E, Figure 9.6) which are now segregated into geographically-separate forest types*

1. Components of the Arctotertiary Geoflora
   Sierran forest: *Abies/concolor/, Pinus/lambertiana/*
   Coastal forest: *Thuja/plicata/, Tsuga/heterophylla/*
   Cordilleran forest: *Populus/angustifolia/*
   East American–East Asian: *Ailanthus, Platanus, Ulmus*
2. Components of the Madrean–Tethyan Geoflora
   Californian woodland and chaparral: *Arbutus, Cercocarpus, Chrysolepis*
   Mexican woodland: *Bumelia, Persea*
   Pinyon–juniper woodland: *Cercocarpus, Juniperus*

* / . . . / signifies the modern species closest to the fossil one [66].

from the Mojave and Sonoran Desert regions of the American Southwest are there Miocene floras with a significant proportion of scrubby taxa, indicative of some sort of thorn scrub.

Modernization of floras proceeded rapidly in the Late Miocene. Intensifying drought conditions and rapidly declining temperatures after 15 Ma ago led to the elimination of most exotic taxa from the interior by 10 Ma ago, and the spread of oak–pinyon–juniper woodland over much of the Great Basin and the lower slopes of the Rockies. Exotic taxa survived rather longer in the coastal lowlands, but even there had largely disappeared by 8 Ma ago. The conifer forests retreated westwards as drought intensified in the interior, gradually occupying the coastal lowlands and the upper slopes of the mountain ranges as they were uplifted. As the Sierra Nevada was elevated in the Pliocene, Sierran elements from the mixed conifer–hardwood forest of the Middle Miocene (assemblage E, Figure 9.6) crossed the divide and became established on the moister western slopes. With progressive restriction of many of the conifer taxa to the west-coast lowlands, the colder lowlands of the interior at higher latitudes were occupied by an impoverished boreal flora. At lower altitudes within the subtropical arid zone, sclerophyllous woodland and scrub expanded, to occupy the whole of the present-day desert region from California eastwards to New Mexico and Texas. Fossil floras of Pliocene age from west-central California (assemblage F) indicate the development of chaparral-type vegetation there.

Figure 9.7, adapted from two diagrams by Axelrod [68,74], illustrates

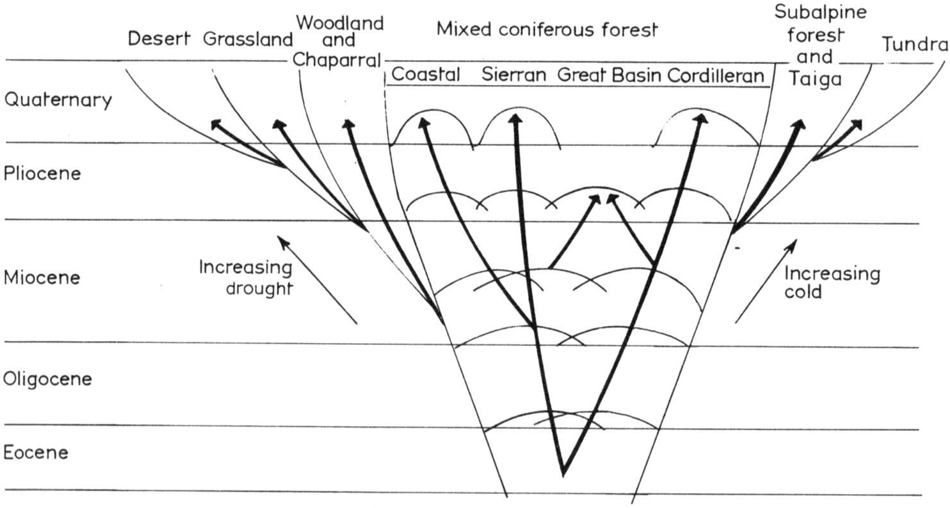

Figure 9.7 Schematic representation of vegetation diversification in western North America in the Cenozoic Era (adapted from Axelrod [68,74]).

schematically the course of plant community history in western North America over the last 50 Ma. It depicts the progressive radiation of new vegetation types through the Tertiary Period from the relatively undifferentiated forest cover of the Eocene, both by segregation of specialized assemblages of certain of the forest taxa themselves and by formation of assemblages composed of entirely new taxa. Climatic change is seen as the driving force in the evolution of both forest and non-forest assemblages. The gradual differentiation, commencing early in the Tertiary, of distinct Coastal, Sierran, Great Basin and Cordilleran forest-types, for example, occurred in response to the development of different climatic regimes [68]: oceanic, with high precipitation, mostly as rain; suboceanic, with colder winters, but with adequate rain; continental and semi-arid, with severe winters and hot summers; and continental with summer rain. Comparable trends but with lower overall rainfall generated the various non-forest communities later in the Tertiary.

9.4.4 Patterns of survival and extinction

The Tertiary floras of western North America occupy a central place in discussions of phytogeographical diversification. They provide evidence that is pertinent to the origins of the modern plant assemblages not only of this particular region but also of the eastern United States, eastern Asia, Mexico and Europe.

For part at least of the Tertiary Period these land areas were all linked, allowing interchange of plant taxa between them. Interchange was not unrestricted, however, but selective, with developing climatic and topographic barriers acting as 'filters', on suites of taxa that were also changing as vegetation zones were forced southwards by deteriorating climate at higher latitudes throughout the Tertiary. Until late in the Tertiary (about 6 Ma ago) Central America north of Panama was a cul-de-sac for southward-migrating taxa, because the Panamanian Isthmus was submerged, but it can never have offered 'open house' to all such taxa because of the steadily intensifying aridity in the subtropical high-pressure belt to the north. This 'drought filter' may already have been present in the Eocene [8,75], when it could have prevented drought-intolerant thermophilic plants in southern Alaska from moving south into Central America as climate worsened in the Early Oligocene; on the other hand it could well have allowed certain Rocky Mountain taxa, already partly adapted to seasonally dry climates, to pass through. Because some of the Alaskan plants had continuous ranges across the Bering land bridge into eastern Asia, they were able to move south successfully on the far side of the Pacific Ocean. Modern derivatives of these plants are thus palaeotropical in distribution, and the 'palaeotropical element' is accordingly well represented in Alaskan Eocene floras. The 'neotropical element' is much more evident in Rocky Mountain Eocene floras [12].

It has been suggested that by the later Tertiary Period temperatures had fallen sufficiently over the southern United States to allow temperate tree taxa to migrate south into the Mexican uplands, where they contributed to the developing montane forests. At the present day at least twenty-three genera of trees characteristic of the broadleaved forests of the eastern United States occur also in the Mexican uplands, as the same or as closely related species [76]. They first appear in the Mexican pollen record in the Middle Miocene. Eight genera are common to these Mexican Miocene floras and to the earlier Eocene floras of the south-eastern United States (*Abies, Alnus, Celtis, Fagus, Juglans, Liquidambar, Myrica* and *Ulmus* [76]), where they occupy a subordinate role in subtropical forest. (Axelrod [8] has argued, however, that temperate tree taxa could have been present in Mexico in similar assemblages much earlier than the Miocene, and spread within the region, rather than migrated into it, as climate worsened.)

The dominant conifers of the modern temperate forests of western North America, on the other hand, do not occur in Mexico. They were almost certainly prevented from migrating south by developing aridity in the American Southwest, accompanied by high summer temperatures.

The land connection between western North America and eastern Asia across the Bering land bridge survived for much of the Tertiary (until about 15 Ma ago [39], and was reconstituted periodically at times of low sea-level during the Pleistocene. For much of its existence the Bering land bridge lay within the zone of temperate broadleaved forest, and allowed free interchange of the elements of the Arctotertiary Geoflora. The close floristic resemblances between the forests of eastern Asia and of North America are testimony to this interchange, and are manifested almost entirely within taxa of this geoflora [46] (Table 7.4). The present-day richness of the eastern Asian forests is the result of much greater survival of Arctotertiary elements there than in either eastern or western North America. This diversity is displayed both in the variety of taxa present and in the variety of forest types. The distribution of modern forest types in eastern Asia has been shown by Wolfe [16] to be closely governed by temperature – such that a plot of forest distribution in relation to mean annual temperature and mean annual temperature range achieves almost complete separation of the ranges of the major types. Figure 9.8, based on temperature data for 427 stations in eastern Asia adjacent to modern forest, shows the non-overlapping ecospace of broadleaved evergreen (A) and deciduous (B) forests, and mixed (C) and boreal (D) coniferous forests. Hot summers appear to restrict the distribution of conifer-dominated forests, as these forests are absent from ecospace with mean temperatures above 20°C in

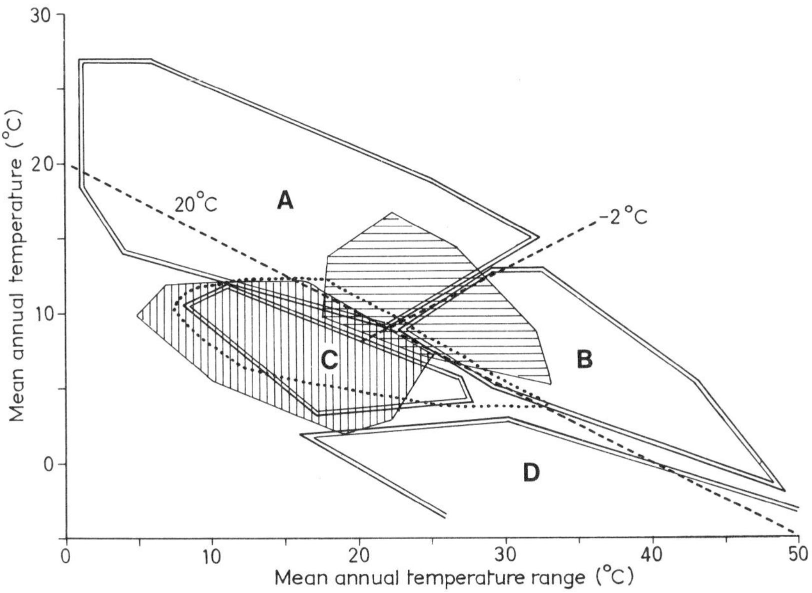

Figure 9.8 Schematic representation of the ranges ▭ of four major forest-types in eastern Asia, based on recent temperature data for 427 stations adjacent to one or more of these forest-types. A = broadleaved evergreen forest, B = broadleaved deciduous forest, C = mixed coniferous forest, D = boreal coniferous forest. Also shown are the ranges of the mixed coniferous forests of western North America |||||, the broadleaved deciduous forests of the eastern United States ☰, and the broadleaved deciduous forests of Europe ·······. The two interrupted lines are the theoretical plots of the 20°C mean temperature of the warmest month and the −2°C mean temperature of the coldest month. Adapted from Wolfe [16].

the warmest month. North American forests follow a similar pattern (the shaded areas on Figure 9.8), so that the present-day contrast between the mixed coniferous forests of the western United States and the broadleaved deciduous forests of the eastern United States can be seen as resulting from the different summer climates of the two regions. Many woody angiosperms were eliminated from the west as declining summer temperatures in the Late Tertiary, accompanied by increasing summer drought southwards, placed these trees at a competitive disadvantage compared with the taller-growing conifers. Coniferous taxa, on the other hand, may never have been common in the central and eastern United States because of high summer temperatures – as witness the almost complete absence of conifers from the Miocene and Pliocene floras of the Great Plains. Ample summer rainfall in the east favoured the growth of broadleaved forests, with conifers restricted to the higher elevations of the Appalachian Range. Diversity of taxa here was low compared with the west.

Both Axelrod [8] and Wolfe [16] have pointed out that broadleaved evergreen trees are poorly represented in the eastern American forests, as compared with the eastern Asian ones. As Figure 9.8 shows, broadleaved deciduous forest in the eastern United States spreads into ecospace occupied by evergreen forest in eastern Asia. There the climatic separation of deciduous and evergreen forest appears to be achieved by winter cold – with evergreen forests being absent from areas where the mean temperature of the coldest month falls below −2°C. The Atlantic coasts of the United States came under the influence of the cold southward-flowing Labrador Current in the Late Tertiary as Arctic regions froze over, and this could have produced prolonged cold spells in winter inimical to the survival of evergreen taxa.

Broadleaved deciduous forest is also characteristic of much of Europe between about latitudes 45° and 60°N – the same latitudinal zone as is occupied by mixed coniferous forest in western North America, and well to the north of comparable forests in the eastern United States. As Figure 9.8 shows, the ecospace filled by European broadleaved forest (the area enclosed by the dotted line) is more nearly comparable with that of the western American conifer forests than with that of the eastern American deciduous ones. Fossil floras from the Late Tertiary in central Europe [77] show clearly that conifers were an important component of the forests then. Their subsequent disappearance is one of the consequences of the events of the Pleistocene Ice Age that are discussed in the next chapter.

REFERENCES

1. Krasilov, V. A. (1975) *Paleoecology of Terrestrial Plants. Basic Principles and Techniques* (transl. by H. Hardin), Wiley, New York.
2. Lange, R. T. (1982) Australian Tertiary vegetation. Evidence and interpretation, in *A History of Australasian Vegetation* (ed. J. M. B. Smith), McGraw-Hill, Sydney, pp. 44–89.
3. Takhtajan, A. (1969) *Flowering Plants. Origin and Dispersal* (transl. by C. Jeffrey), Oliver and Boyd, Edinburgh.
4. Thomson, M. R. A. and Burn, R. W. (1977) Angiosperm fossils from latitude 70°S. *Nature*, **269**, 139–41.
5. Aubréville, A. (1970) La flore tropicale tertiaire du Sahara. *Adansonia* Series 2, **10**, 9–14.
6. Box, E. O. (1986) Some climatic relationships of the vegetation of Argentina in global perspective. *Veröff. Geobot. Inst., Zürich*, **91**, 181–216.
7. Axelrod, D. I. (1958) Evolution of the Madro–Tertiary Geoflora. *Bot. Rev.*, **24**, 433–509.
8. Axelrod, D. I. (1975) Evolution and biogeography of Madrean–Tethyan sclerophyll vegetation. *Ann. Miss. Bot. Gard.*, **62**, 280–334.

9. Axelrod, D. I. and Raven, P. H. (1978) Late Cretaceous and Tertiary vegetation history of Africa. *Monogr. Biol.*, **31**, 77–130.
10. Daubenmire, R. (1978) *Plant Geography, with Special Reference to North America*, Academic Press, New York.
11. Kemp, E. M. (1978) Tertiary climatic evolution and vegetation history in the Southeast Indian Ocean region. *Palaeogeogr. Palaeoclimatol. Palaeoecol.*, **24**, 169–208.
12. Leopold, E. M. and MacGinitie, H. D. (1972) Development and affinities of Tertiary floras in the Rocky Mountains, in *Floristics and Palaeofloristics of Asia and Eastern North North America* (ed. A. Graham), Elsevier, Amsterdam, pp. 147–200.
13. Webb, S. D. (1978) A history of savanna vertebrates in the New World. Part II: South America and the Great Interchange. *Ann. Rev. Ecol. Syst.*, **9**, 393–426.
14. Wolfe, J. A. (1985) Distribution of major vegetational types during the Tertiary, in *The Carbon Cycle and Atmospheric CO_2: Natural Variations Archean to Present* (eds E. T. Sundquist and W. S. Broecker), American Geophysical Union, Washington, pp. 357–75.
15. Parrish, J. T., Ziegler, A. M. and Scotese, C. R. (1982) Rainfall patterns and the distribution of coals and evaporites in the Mesozoic and Cenozoic. *Palaeogeogr. Palaeoclimatol. Palaeoecol.*, **40**, 67–101.
16. Wolfe, J. A. (1979) Temperature parameters of humid to mesic forests of eastern Asia and relation to forests of other regions of the Northern Hemisphere and Australasia. *US Geol. Survey Prof. Pap.*, **1106**, 1–37.
17. Wolfe, J. A. (1977) Palaeogene floras from the Gulf of Alaska region. *US Geol. Survey Prof. Pap.*, **997**, 1–108.
18. Wolfe, J. A. (1980) Tertiary climates and floristic relationships at high latitudes in the northern hemisphere. *Palaeogeogr. Palaeoclimatol. Palaeoecol.*, **30**, 313 –23.
19. Dorf, E. (1964) The use of fossil plants in palaeoclimatic interpretations, in *Descriptive Palaeoclimatology* (ed. A. E. M. Nairn), Interscience, New York, pp. 13–31.
20. Schwarzbach, M. (1968) Das Klima des rheinischen Tertiärs. *Z. dt. geol. Ges.*, **118**, 33–88.
21. Barron, E. J., Sloan, J. L. II, and Harrison, C. G. A. (1980) Potential significance of land–sea distribution and surface albedo variations as a climatic forcing factor: 180 m.y. to the present. *Palaeogeogr. Palaeoclimatol. Palaeoecol.*, **30**, 17–40.
22. Willis, J. C. (1966) *A Dictionary of the Flowering Plants and Ferns*, 7th edn, Cambridge University Press, Cambridge.
23. Schweitzer, H.-J. (1980) Environment and climate in the early Tertiary of Spitsbergen. *Palaeogeogr. Palaeoclimatol. Palaeoecol.*, **30**, 297–311.
24. Tanai, T. (1972) Tertiary history of vegetation in Japan, in *Floristics and Palaeofloristics of Asia and Eastern North America* (ed. A. Graham), Elsevier, Amsterdam, pp. 235–55.
25. Collinson, M. E. (1983) *Fossil Plants of the London Clay*, Palaeontological Association, London.

26. Daley, B. (1972) Some problems concerning the early Tertiary climate of southern Britain. *Palaeogeogr. Palaeoclimatol. Palaeoecol.*, **11**, 177–90.
27. Jones, E. L. (1961) Environmental significance of palynomorphs from Lower Eocene sediments of Arkansas. *Science*, **134**, 1366.
28. Sharp, A. J. (1951) The relation of the Eocene Wilcox Flora to some modern floras. *Evolution, Lancaster, Pa*, **5**, 1–5.
29. Bande, M. B. and Prakash, U. (1986) The Tertiary flora of Southeast Asia with remarks on its palaeoenvironment and phytogeography of the Indo-Malayan region. *Rev. Palaeobot. Palynol.*, **49**, 203–33.
30. Menéndez, C. A. (1969) Die fossilen Floren Südamerikas. *Monogr. Biol.*, **19**, 519–61.
31. Arènes, J. and Depape, G. (1956) La Flore Burdigalienne des Iles Baléares (Majorque). *Revue Gén. Bot.*, **63**, 347–90.
32. Martin, H. A. (1978) Evolution of the Australian flora and vegetation through the Tertiary: evidence from pollen. *Alcheringa*, **2**, 181–202.
33. Mildenhall, D. C. (1980) New Zealand Late Cretaceous and Cenozoic plant biogeography: a contribution. *Palaeogeogr. Palaeoclimatol. Palaeoecol.*, **31**, 197–233.
34. Hill, R. S. (1987) Discovery of *Nothofagus* fruits corresponding to an important Tertiary pollen type. *Nature*, **327**, 56–8.
35. Christophel, D. C. and Greenwood, D. R. (1989) Changes in climate and vegetation in Australia during the Tertiary. *Rev. Palaeobot. Palynol.*, **58**, 95–109.
36. Beadle, N. C. W. (1966) Soil phosphate and its role in molding segments of the Australian flora and vegetation, with special reference to xeromorphy and schlerophyll. *Ecology*, **47**, 992–1007.
37. Whittaker, R. H. (1970) *Communities and Ecosystems*, Collier-Macmillan, Toronto.
38. Walter, H. (1973) *Vegetation of the Earth*, 2nd edn (transl. by J. Wieser), Springer Verlag, Berlin.
39. Wolfe, J. A. and Leopold, E. B. (1967) Neogene and Early Quaternary vegetation of northwestern North America and northeastern Asia, in *The Bering Land Bridge* (ed. D. M. Hopkins), Stanford University Press, CA, pp. 193–206.
40. Axelrod, D. I. and Bailey, H. P. (1969) Paleotemperature analysis of Tertiary floras. *Palaeogeogr. Palaeoclimatol. Palaeoecol.*, **6**, 163–95.
41. Brady, H. and Martin, H. (1979) Ross Sea region in the Middle Miocene: a glimpse into the past. *Science*, **203**, 437–8.
42. Olson, J. S. (1985) Cenozoic fluctuations in biotic parts of the global carbon cycle, in *The Carbon Cycle and Atmospheric CO_2: Natural Variations Archean to Present* (eds E. T. Sundquist and W. S. Broecker), American Geophysical Union, Washington, pp. 377–96.
43. Solbrig, O. T. (1976) The origin and floristic affinities of the South American desert and semidesert regions, in *Evolution of Desert Biota* (ed. D. W. Goodall), Texas University Press, Austin, pp. 7–49.
44. Beard, J. S. (1977) Tertiary evolution of the Australian flora in the light of latitudinal movements of the continent. *J. Biogeogr.*, **4**, 111–8.
45. Wolfe, J. A. (1972) An interpretation of Alaskan Tertiary floras, in *Floristics and*

Palaeofloristics of Asia and Eastern North America (ed. A. Graham), Elsevier, Amsterdam, pp. 201–33.

46. Graham, A. (ed.) (1972) *Floristics and Palaeofloristics of Asia and Eastern North America*, Elsevier, Amsterdam.
47. Brain, C. K. (1984) The Terminal Miocene Event: a critical environmental and evolutionary episode? in *Late Cainozoic Palaeoclimates of the Southern Hemisphere* (ed. J. C. Vogel), Balkema, Rotterdam, pp. 491–8.
48. Benson, R. H. (1976) Testing the Messinian crisis biodynamically: an introduction. *Palaeogeogr. Palaeoclimatol. Palaeoecol.*, **20**, 3–11.
49. Horowitz, A. and Horowitz, M. (1985) Subsurface Late Cenozoic palynostratigraphy of the Hula Basin, Israel. *Pollen Spores*, **27**, 365–90.
50. Lancaster, N. (1984) Aridity in southern Africa: age, origins and expression in landforms and sediments, in *Late Cainozoic Palaeoclimates of the Southern Hemisphere* (ed. J. C. Vogel), Balkema, Rotterdam, pp. 433–44.
51. Walter, H. and Box, E. O. (1983) The deserts of Central Asia, in *Ecosystems of the World 5. Temperate Deserts and Semi-Deserts* (ed. N. E. West), Elsevier, Amsterdam, pp. 193–236.
52. Windley, B. F. (1977) *The Evolving Continents*, Wiley, London.
53. Song, Z., Li, H., Zheng, Y. and Liu, G. (1979) Miocene floristic regions of China, in *Paleontology in China 1979* (eds C. Teichart, L. Lu and C. Pei-Ji), Geological Society of America, Boulder, CO, pp. 249–54.
54. Van Zinderen Bakker, E. M. (1980) Comparison of late-Quaternary climatic conditions in the Sahara and the Namib–Kalahari region. *Palaeoecol. Africa*, **12**, 381–94.
55. Axelrod, D. I. (1985) Rise of the grassland biome, central North America. *Bot. Rev.*, **51**, 163–201.
56. Van der Hammen, T. (1983) The palaeoecology and palaeogeography of savannas, in *Ecosystems of the World 13. Tropical Savannas* (ed. F. Bourlière), Elsevier, Amsterdam, pp. 19–35.
57. Cooke, H. B. S. (1984) Horses, elephants and pigs as clues in the African later Cenozoic, in *Late Cainozoic Palaeoclimates of the Southern Hemisphere* (ed. J. C. Vogel), Balkema, Rotterdam, pp. 473–82.
58. Sahni, A. and Mitra, H. C. (1980) Neogene palaeobiogeography of the Indian subcontinent with special reference to fossil vertebrates. *Palaeogeogr. Palaeoclimatol. Palaeoecol.*, **31**, 39–62.
59. Beadle, N. C. W. (1981) The vegetation of the arid zone, in *Ecological Biogeography of Australia*, Vol. 1 (ed. A. Keast), Junk, The Hague, pp. 695–716.
60. Martin, H. A. (1981) The Tertiary flora, in *Ecological Biogeography of Australia*, Vol. 1 (ed. A. Keast), Junk, The Hague, pp. 391–406.
61. Di Castri, F. (1981) Mediterranean-type shrublands of the world, in *Ecosystems of the World 11. Mediterranean-type Shrublands* (eds F. di Castri, D. W. Goodall and R. L. Specht), Elsevier, Amsterdam, pp. 1–52.
62. Suc, J.-P. (1984) Origin and evolution of the Mediterranean vegetation and climate in Europe. *Nature*, **307**, 429–32.
63. Pons, A. (1981) The history of the Mediterranean shrublands, in *Ecosystems of the World 11. Mediterranean-type Shrublands* (eds F. di Castri, D. W. Goodall and R. L. Specht), Elsevier, Amsterdam, pp. 131–8.

64. Nelson, R. E. and Carter, L. D. (1985) Pollen analysis of a Late Pliocene and early Pleistocene section from the Gubik formation of arctic Alaska. *Quat. Res.*, **24**, 295–306.
65. Petrov, O. M. (1967) Paleogeography of Chukotka during Late Neogene and Quaternary time, in *The Bering Land Bridge* (ed. D. M. Hopkins), Stanford University Press, CA, pp. 144–71.
66. Axelrod, D. I. (1959) Late Cenozoic evolution of the Sierran bigtree forest. *Evolution, Lancaster, Pa*, **13**, 9–23.
67. Axelrod, D. I. (1968) Tertiary floras and topographic history of the Snake River basin, Idaho. *Geol. Soc. Am. Bull.*, **79**, 713–34.
68. Axelrod, D. I. (1976) History of the coniferous forests, California and Nevada. *Univ. Calif. Publs Bot.*, **70**, 1–62.
69. Raven, P. H. and Axelrod, D. I. (1977) Origin and relationships of the California flora. *Univ. Calif. Publs Bot.*, **72**, 1–134.
70. Robichaux, R. H. and Taylor, D. W. (1977) Vegetation-analysis techniques applied to Late Tertiary fossil floras from the western United States. *J. Ecol.*, **65**, 643–60.
71. Wolfe, J. A. (1969) Neogene floristic and vegetational history of the Pacific Northwest. *Madroño*, **20**, 83–110.
72. Vankat, J. L. (1979) *The Natural Vegetation of North America. An Introduction*, Wiley, New York.
73. Axelrod, D. I. and Raven, P. H. (1985) Origins of the Cordilleran flora. *J. Biogeogr.*, **12**, 21–47.
74. Axelrod, D. I. (1957) Late Tertiary floras and the Sierra Nevadan uplift. *Geol. Soc. Am. Bull.*, **68**, 19–46.
75. Dilcher, D. L. (1973) A paleoclimatic interpretation of the Eocene floras of southeastern North America, in *Vegetation and Vegetational History of Northern Latin America* (ed. A. Graham), Elsevier, Amsterdam, pp. 39–59.
76. Graham, A. (ed.) (1973) *Vegetation and Vegetational History of Northern Latin America*, Elsevier, Amsterdam.
77. Szafer, W. (1954) Pliocene flora from the vicinity of Czorsztyn (West Carpathians) and its relationship to the Pleistocene. *Instytut Geologiczny Prace*, **11**, 1–238.

CHAPTER TEN

Forests in decline

10.1 INTRODUCTION

The environmental instability of the Pleistocene Epoch, manifested most clearly in its repeated high-amplitude climatic cycles, generated a corresponding instability in the global vegetation cover. The picture revealed by pollen and macrofossil analyses is of vegetation in a ceaseless state of flux, with continual shifts in composition. The times when vegetation and climate were more nearly in equilibrium were few, and probably during the colder parts of each climatic cycle, when forests were in decline worldwide. Episodes of forest expansion, under interglacial conditions, were of shorter duration and greater flux. The fossil evidence for that vegetation flux is most complete for the warm-stage forest communities of Europe, and it is clear from this evidence that current European forests (of the Holocene or Flandrian warm stage) differ in important respects from those of earlier warm stages. We cannot assume, in fact, that any of the plant communities of previous interglacials were similar in composition (or range) to those surviving at the present day. Nevertheless this is what is commonly done, especially for biomes other than forest, and for geographical regions other than Europe, where there are few pollen data to provide concrete data. Some aspects of the vegetation tableau through the full duration of the Pleistocene are accordingly considered first, in section 10.2, to illustrate this flux, with particular reference to the European fossil record.

The current world vegetation provides some approximation to that characteristic of one set of climatic extremes in the Pleistocene. The vegetation at the other set of climatic extremes (the 'glacial stages') epitomizes more clearly the challenge of the Pleistocene for plant survival. Evidence about glacial-stage vegetation is widely available only for the most recent glacial stage, and it is this evidence that is reviewed in the later part of the chapter. The picture is for a narrow time-band around 20 ka ago, and its relevance to the total Pleistocene tableau has to be gauged by extrapolation from the limited data outlined in section 10.2 on pattern and process in European interglacial forests, and from the review of environmental changes and plant responses in Chapters 5 to 8. To view each glacial/interglacial cycle as a simple alternation between two climatic extremes, however, is a gross oversimplification. The complex character of even the last glacial stage is apparent from fine-resolution analyses of both ocean sediments (Figure 5.3) and polleniferous land deposits (Figure 13.6). The oxygen-isotope values show the glacial part of the cycle to consist of a series of alternating cold stadials and warmer interstadials, with corresponding shifts in the composition of the plant cover revealed in the pollen counts. Because of the marked asymmetry of the glacial/interglacial cycle, moreover, conditions of maximum cold were delayed until the final stadial in the cycle, centred around 18–20 ka ago. In those regions where alternating wet and dry periods through the glacial stage can also be recognized, then maximum cold apparently coincided with a shift towards greater aridity (cf. Figure 10.2). Vegetation reconstructions for this narrow 18–20-ka time-band are thus not representative of the glacial stage as a whole, and the vegetation instability of the entire Pleistocene Epoch must be seen as continuing, albeit in muted form, through the cold stages as well as the warm ones.

10.2 PATTERNS AND PROCESSES OF FLORISTIC CHANGE

10.2.1 Overall trends

The harshness of the Pleistocene Epoch for plant life is encapsulated in its record of extinction. Extinction marks the failure of a plant taxon to return to a particular geographical area after being driven thence by changing climate at some stage in the Pleistocene – a failure determined in part at least by topographic considerations such as the effectiveness of barriers to migration (the east/west-trending mountain ranges of Europe, for example). In only a few parts of the world, however, is there a sufficiently full fossil record spanning the whole Pleistocene to document this process adequately. The record, moreover, extends only to forest vegetation, and is thus concerned with extinctions resulting from cold; information on the effects of warmer conditions on cold-adapted vegetation is almost entirely lacking.

As a result of differential extinctions in the Pleistocene, forest types of common ancestry in the Tertiary came to differ markedly in their floristic composition. Thus the present-day differences between the temperate forest vegetation of Europe and eastern Asia are assumed to be the consequence of wholescale extinction of most of the Tertiary conifers

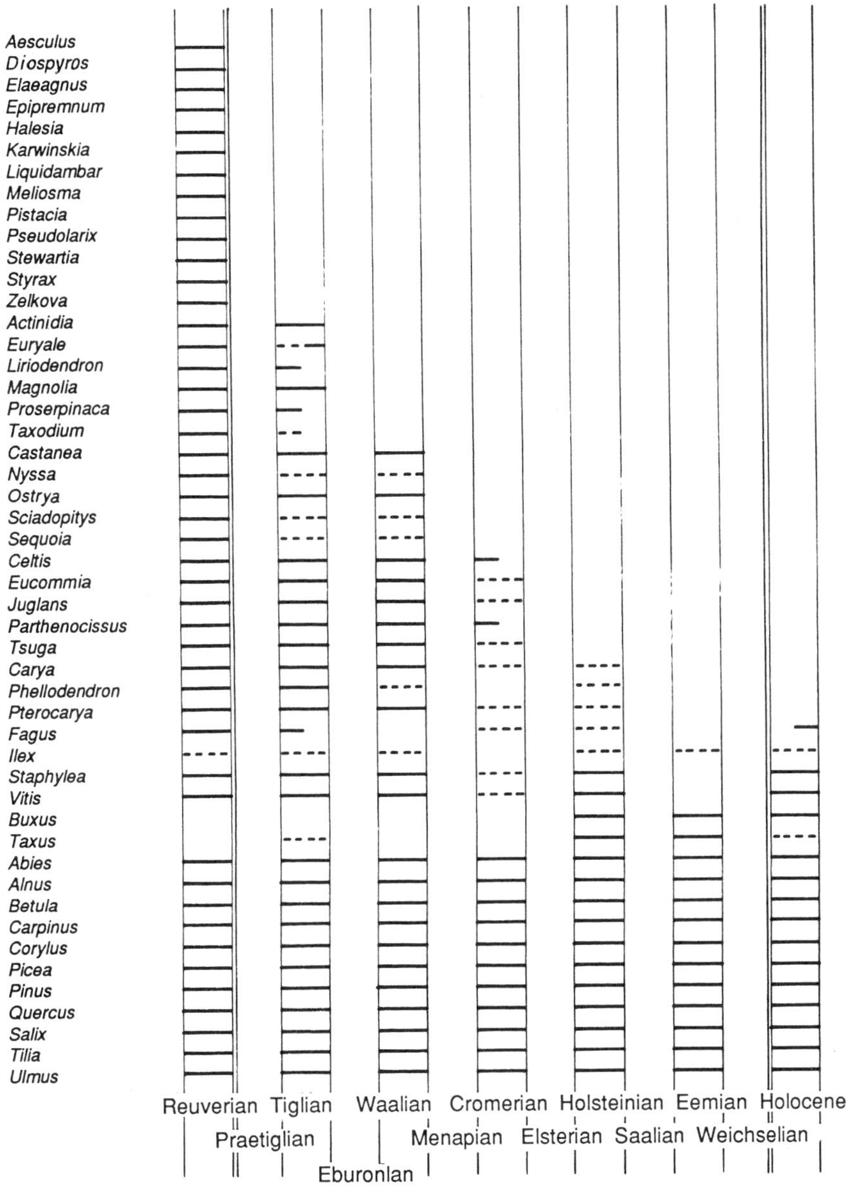

Figure 10.1 The occurrence of tree taxa in warm stages (Table 5.2) of the Pleistocene in northwest Europe. Adapted from Van der Hammen, Wijmstra and Zagwijn [3] and Straka [4].

during the Pleistocene Epoch in Europe and their survival in eastern Asia (and to a lesser extent also in the Caucasus). The European extinctions are well documented, but eastern Asian survival is still largely conjectural (though see Tanai [1]). Figure 10.1 shows the pattern of extinction of forest trees in north-west Europe in the Pleistocene. The record of extinctions starts in fact with the Praetiglian cold stage (about 2.4 Ma ago), which is of Late Pliocene age *sensu stricto*. Many frost-sensitive trees disappeared from north-west Europe at this time, in what may have been 'a biological catastrophe of far greater consequence' than the events associated with any subsequent cold stage [2]. Nevertheless, each subsequent cold stage resulted in change; after each of the stages it seems likely that some forest trees failed to re-establish their former ranges. In north-west Europe, extinctions of forest trees were concentrated into the Lower Pleistocene, and already by the Middle Pleistocene the forests were beginning to assume a modern aspect. In eastern Europe, however, the process was more delayed, with Tertiary 'relicts' surviving well into the Middle Pleistocene in Hungary [5], and into the Upper Pleistocene in the Caucasus [6].

The picture of floristic change in Figure 10.1 is one of progressive impoverishment, with taxa being lost stepwise from the diverse assemblages of the Late Tertiary, yet leaving behind a persistent nucleus of the more hardy forest trees (*Abies, Alnus, Betula, Carpinus, Corylus, Picea, Pinus, Quercus, Salix, Ulmus,* and perhaps *Tilia*) that returned successfully after each cold stage. There is little evidence of recruitment of new taxa. However, the picture should not be regarded as necessarily typical of all biomes in the Pleistocene, as it relates to one that was adversely affected by the prevailing cold. The gradual decline of temperature through the Pleistocene inevitably led to an increase in the extent of ground available for colonization by non-woody plants, as treelines were lowered altitudinally and latitudinally. We can surmise that in these situations there was an active expansion of certain opportunistic plant taxa into the new ecospace, and the development of progressively richer herbaceous assemblages. Thus in mountain regions the long-term picture may have been of increasingly more diverse alpine vegetation through the Pleistocene, maintained despite regular setbacks in the warmer stages. The data for the high Andes of Colombia [7] certainly support his scenario. At rather lower altitudes there, continued uplift of the Andes during the Pleistocene led to an expanded montane forest zone also, where progressive diversification similarly occurred.

A third scenario, particularly relevant to tropical and subtropical regions, has been enthusiastically championed by some workers [8]. In this scenario (not universally accepted), environmental changes are thought not to have been sufficiently stringent to bring about widespread extinction of plant taxa. Instead, plant populations repeatedly contracted and expanded in range, alternately undergoing fragmentation and re-mixing. As a result, diversity was enhanced, either from speciation in isolated sub-populations during periods of range-contraction, or from hybridization following range-expansion (section 8.5). In summary then, long-term floristic trends in different parts of the world through the Pleistocene may have involved both impoverishment and enrichment, and the oft-cited impoverishment of temperate forests is only one of at least three alternative patterns of biome development.

10.2.2 The glacial/interglacial cycle

The repetitive character of Pleistocene climatic change is clearly seen in the succession of cold/warm cycles (at least seventeen) revealed in the ocean oxygen-isotope record (section 5.3). Where sufficient pollen documentation is available for a particular geographical region, then broadly similar patterns of change in the pollen assemblages can be recognized through each cycle in the sequence. Accordingly, the concept of a generalized cycle of environmental and vegetational change occupying 90–120 ka has been widely used as an aid to understanding the ecological processes operating repetitively through Pleistocene time. The concept, originally proposed by Iversen [9] for the later-Pleistocene high-amplitude glacial/interglacial cycles, subdivides each cycle into **cryocratic**, **protocratic**, **mediocratic** and **telocratic** phases. In north-west Europe these phases encompass the changes represented in Figure 10.2:

Cold, dry, continental climate → increasing warmth → warm moist climate → increasing cold.
Base-rich, skeletal mineral soil → unleached soil with low humus content → brown-earth soil → leached humus-rich podsol or peat → disruption of soil stratification by frost-action and solifluction.
Sparse herb assemblages → open forest, scrub, or grassland → temperate deciduous forest → conifer-dominated forest, heath, or bog.

The glacial/interglacial cycle thus incorporates three concepts: the cycle of increase and decrease in mean temperature, the progressive closure and then regression of the forest cover, and the unidirectional soil leaching [2]. Typical trees of the protocratic phase in north-west Europe are *Betula*, *Populus* and *Salix*; of the mediocratic phase, *Quercus*, *Ulmus* and *Tilia*; and of the telocratic phase, *Picea*, *Abies* and *Fagus*. More detailed characterizations of the associated floristic assemblages are given by Birks [11].

Soil changes in the interglacial part of the cycle are envisaged as proceeding independently of climatic change, with climatic deterioration

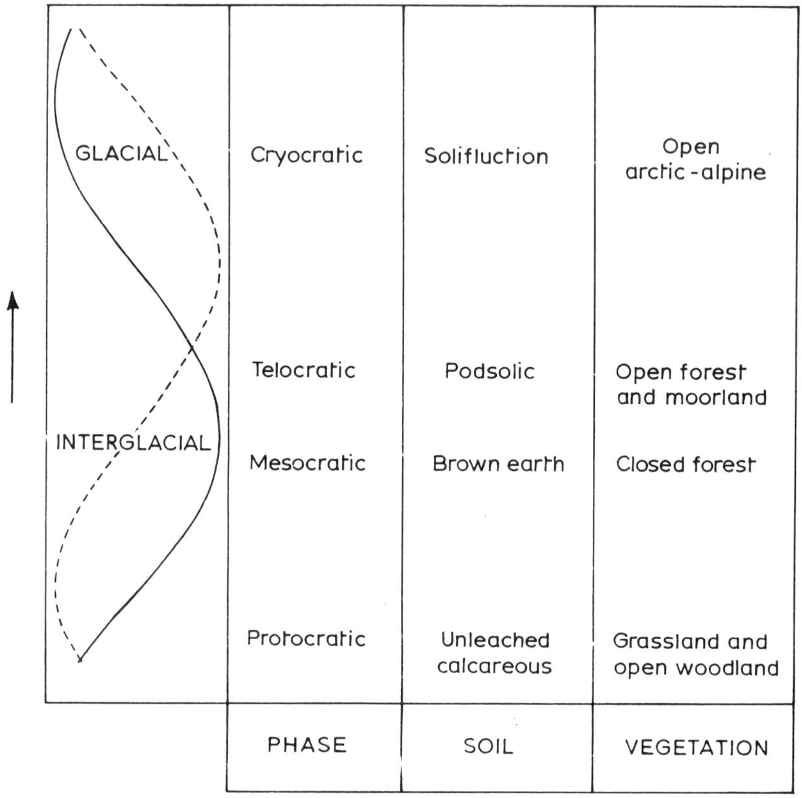

Figure 10.2 Schematic representation of changes through a glacial/interglacial cycle in relation to warmth (——) and moisture (– – –). Modified from Frenzel [10].

apparent only after soil leaching and acidification have occurred. Andersen [12] accordingly interpolates an additional **oligocratic** phase into the cycle after the mesocratic phase, to emphasize the non-synchroneity of edaphic and climatic change.

Reconstructions of the pollen assemblages found in the various phases of a typical glacial/interglacial cycle have also been attempted for other geographical regions besides north-west Europe [11]. For many biomes, however, there are insufficient data to reconstruct the full cycle, and only the extreme vegetation types are known. These are discussed more fully later in the chapter.

It is important to recognize that the cryocratic, protocratic, mediocratic and telocratic phases typically do not occupy the same lengths of time within the cycle. The cryocratic phase was the most protracted (section 5.4), and involved a long decline to maximum cold, interrupted by several warmer interstadials. The protocratic phase, by contrast, was of short duration, and constituted an abrupt termination to the more stable conditions of the cryocratic phase. The mediocratic and telocratic phases together probably occupied less than one-fifth of the total cycle.

A supplementary model favoured by some workers has a moisture cycle superimposed on the temperature cycle, but out of step with it (Figure 10.2). Accordingly, the cryocratic phase is initially wet, but subsequently becomes increasingly dry, whilst the protocratic and early mediocratic phases are dry and the telocratic phase is wet. In view of the complexities of global climate it is unlikely that the model has universal applicability, though it may be relevant for lower latitudes, where moisture changes during the cycle were probably more pronounced than temperature changes.

10.2.3 Interglacial forests

The concept of common patterns and processes of change in all glacial/interglacial cycles is probably true in only a very general way. The very limited evidence currently available suggests rather that each cycle may have been unique climatically and vegetationally [2]. This evidence includes detailed oxygen-isotope analyses of successive cycles in ocean cores (cf. Figure 5.1), theoretical calculations based on Milankovitch theory (cf. section 5.5), and pollen diagrams from the various warm stages of the European Pleistocene [10]. All three lines of evidence support the idea of continuous vegetation flux throughout the Pleistocene.

The pollen evidence is largely restricted to temperate forest communities in Europe. Nevertheless, it indicates the sorts of variability that may have occurred in other biomes also during the Pleistocene. A typical pollen diagram for one glacial/interglacial cycle (arboreal taxa only) is shown in Figure 10.3. The pollen sequence can be regarded as encompassing much or all of the protocratic phase of the cycle (zone I), the mediocratic phase (zones II and III*), and the earlier part of the telocratic phase (zone IV). Turner and West [13] refer to these zones as the Pre-temperate, Early-temperate, Late-temperate and Post-temperate, respectively. In pollen diagrams from different warm stages of Europe, three main patterns of variability are apparent: some taxa are present only in the earlier Pleistocene warm stages (notably *Tsuga*, *Carya* and *Pterocarya*); other taxa are consistently present, but vary in the timing of their expansion (for example, *Corylus* and *Alnus*); yet others vary markedly in their abundance in the various warm stages (for example, *Abies*, *Carpinus*, *Fagus*, *Tilia*). The differences in behaviour of the major tree pollen taxa that survived into or

* Zone III is Andersen's [12] oligocratic phase.

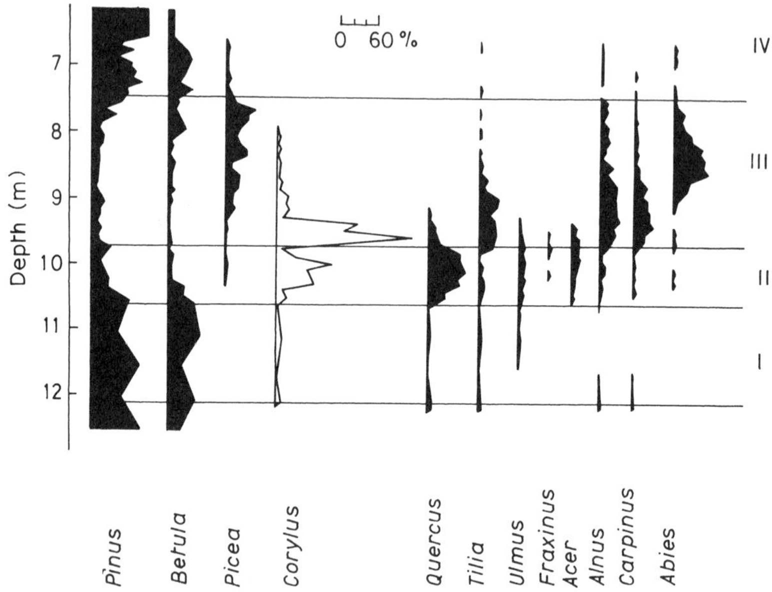

Figure 10.3 Pollen diagram from interglacial deposits of Eemian (= Ipswichian) age at Bedlno, Poland (arboreal taxa only, % total tree pollen). Re-drawn from Turner and West [13].

through the Pleistocene produce a constantly shifting composition of the warm-stage forests [14], and must result from the action of a complex of factors related in part at least to the migratory history of the taxon: the distance to be covered from its cold-stage refugia, its speed of migration, and its ability to compete effectively with other taxa, for example. The differing patterns of *Corylus* in the Hoxnian and Ipswichian interglacials and the Holocene in Britain [14,15] might be explained on this basis. Other behavioural differences could have resulted from the varying lengths of time in each warm stage when barriers to migration were effective: the detailed timing of the isolation of Britain from the continent of Europe by rising sea levels, for example, might have been a contributory factor in the failure of *Abies* and *Picea* to enter Britain in the Ipswichian interglacial. Furthermore, the various warm stages may also have differed climatically, with the Ipswichian interglacial, for example, characterized by a more continental climate in Britain than either the Hoxnian or the Holocene [14].

The effects of such climatic differences are more apparent in eastern Europe, where cold/warm oscillations through the Pleistocene were combined with dry/moist fluctuations that periodically allowed the spread of steppe communities. The continual flux in the composition of forest communities through the Pleistocene is clearly shown in Ronai's pollen analyses of Lower and Middle Pleistocene sediments from the Jászladány borehole in Hungary. The six pollen assemblages shown in Figure 10.4 are typical of the sixteen assemblages recognized by Ronai in these sediments, and are characterized by differing proportions of a range of tree and non-tree taxa. In addition to the cold/warm and dry/moist oscillations accompanying the assemblages is a general climatic trend through the Lower and Middle Pleistocene towards lower temperatures, so that the uppermost assemblage in Figure 10.4 probably represents an almost treeless steppe community.

The examples discussed above relate to periods of time when forest communities were in the ascendancy worldwide. During the colder stages of the Pleistocene, by contrast, the global range of forest and woodland was markedly reduced. For the maximum of the last cold stage, 18–20 ka ago, a rather detailed picture of the extent of that reduction is becoming available. This picture is considered in the following pages.

10.3 EURASIAN FULL-GLACIAL VEGETATION

Most pollen diagrams from unglaciated land surfaces in Eurasia north of latitude 45°N show minimum amounts of tree and shrub pollen around 18–20 ka ago (Figure 13.6(c)). Less than 10% tree and shrub pollen in this time-interval is characteristic of such different and widely separated sites as Grande Pile Bog in the southern Vosges, France [16], and Duvannyy Yar in the Kolyma lowlands of northeastern Siberia [17]. The tree and shrub pollen was derived entirely from cold-hardy taxa such as *Betula, Juniperus, Larix, Pinus* and *Salix.* Even at more southerly sites such as Tenaghi Philippon in northern Greece [18] and the Ghab valley in northwestern Syria [19], tree and shrub pollen does not rise above 15% of total pollen. Such low counts of pollen of woody taxa are found in present-day surface samples only from non-forest vegetation types such as tundra, desert and steppe (and then not always) [20]. A reconstruction of vegetation types in Eurasia 18–20 ka ago accordingly shows a broad belt of non-forest vegetation more than 1500 km wide stretching from the southern margins of the Scandinavian Ice Sheet and the ice-covered Arctic Ocean to the shores of the Mediterranean and Black Seas and the high mountains of central Asia (Figure 10.5). The local ice-caps that formed on the high mountain ranges to the south of the main Scandinavian Ice Sheet (the Pyrenees, Alps, Carpathians, Tien Shan, Altai and Sayan Mountains) undoubtedly reinforced its climatic effects, and extended periglacial conditions over a vast area. Permanently frozen subsoils were widespread, with the southern limit of permafrost displaced by up to 2000 km from its present position. The low temperatures were accompanied by enhanced aridity, affecting

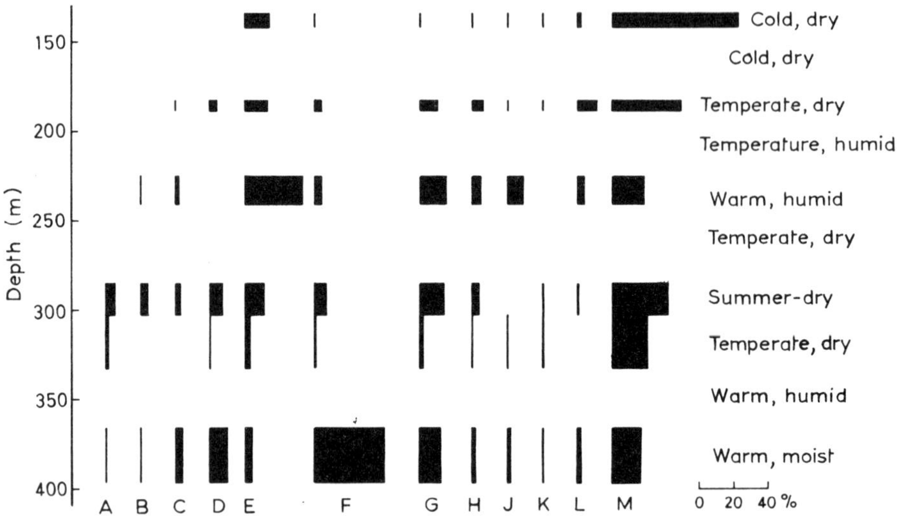

Figure 10.4 Pollen analyses from the Lower and Middle Pleistocene of the Jászladány borehole in Hungary. Mean pollen frequencies (% total pollen) of groups of tree taxa over selected depth-intervals are shown. The assumed moisture status of each depth-interval is indicated on the right. Data of Ronai [5]. A = *Gingko, Keteleeria, Podocarpus*; B = *Cedrus*, Palmae; C = *Ilex, Rhus*; D = *Carya, Engelhardtia, Nyssa, Pterocarya*; E = *Tsuga*, Taxodiaceae; F = *Alnus, Fraxinus, Ulmus*; G = *Castanea, Quercus, Tilia*; H = *Acer, Juglans*; J = *Abies*; K = *Carpinus, Fagus, Ostrya*; L = *Betula, Salix*; M = *Larix, Picea, Pinus*. The groupings of pollen taxa represent increasing tolerance to cold.

even the northern coasts of the Mediterranean Sea [23,24], and by increasing continentality of climate [25]. The combined effects of these factors must be seen as the cause of the almost total demise at full glacial of the deciduous and coniferous forests that cover much of Eurasia today (Figure 10.5).

Forest communities were clearly very restricted at full glacial. In only a few of the several hundred pollen spectra used in the vegetation reconstructions in Figure 10.5 are there appreciable quantities of tree pollen 18–20 ka ago. Most of these spectra are from sites in or close to areas of high relief: for example, the Sierra Nevada of southern Spain [26], Macedonia [3], the Caucasus [6], the Levant [19] and the Mongolian Altai [10]. In order to understand the distribution of forest communities in Eurasia at full glacial, palaeoecologists have sought present-day analogues in regions of similarly arid climate – for example, the American Southwest and central Asia. In these regions forest communities currently form a well-defined belt at mid-altitudes on the higher ground, above the desert, scrub, or grassland of the lowlands, and below the high-altitude alpine tundra. Increased moisture availability coupled with still moderate temperatures at

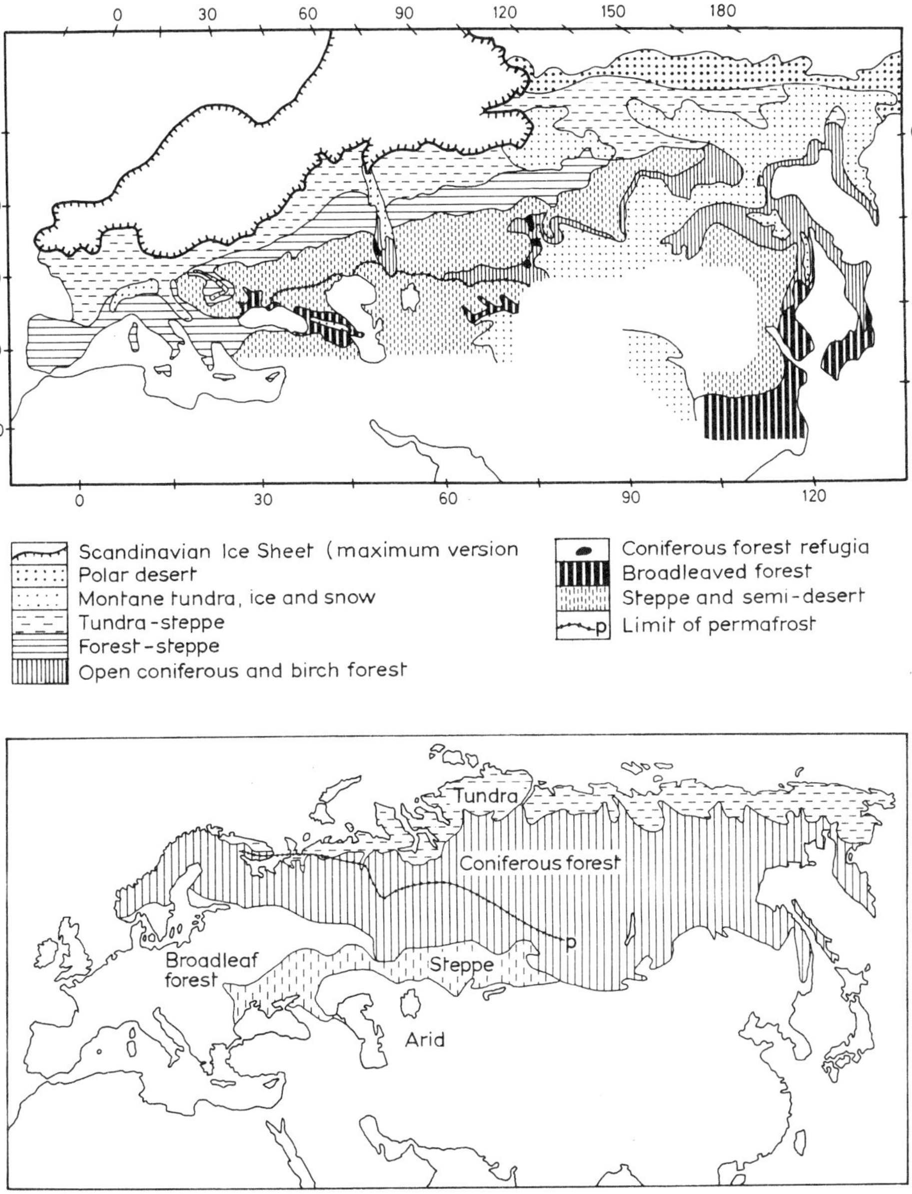

Figure 10.5 Vegetation types of Eurasia 18–20 ka ago (above) and at the present day (below). Based on Frenzel [21] and Grichuk [22].

middle altitudes is seen as the controlling factor allowing tree growth in the prevailing arid climate. A similar localized distribution pattern might have characterized the forest stands of the Eurasian full glacial.

The best corroborative evidence comes from two sites in northern Greece: Tenaghi Philippon at 48 m altitude and Ioannina at 500 m altitude. Tree pollen was entirely absent from the former site 18–20 ka ago, when the pollen spectrum was dominated by grasses, chenopods and *Artemisia*. At the higher-altitude site, on the other hand, appreciable quantities of tree pollen (oak, pine, and some beech) persisted through the full glacial. Van der Hammen *et al.* [3] accordingly present a reconstructed vegetational transect along a north–south line through Europe, which shows forest restricted to a narrow montane belt on south-facing slopes, intercalated between cool steppe in the lowlands and alpine tundra above (Figure 10.6). The most probable picture of forest survival at full glacial in Eurasia is thus of small widely-separated stands at middle altitudes in locally favourable situations in mountainous areas of southern Europe and southern Russia. The European refugia for forest trees postulated by Huntley and Birks [27] (Figure 8.12) must be regarded merely as geographical source areas within which such localized stands occurred. The differing migration patterns shown subsequently by individual tree taxa in the Early Holocene, moreover, indicate that such stands were of widely-varying composition.

Although such a reconstruction applies particularly to deciduous tree taxa, closed coniferous forest must also have been markedly restricted in distribution 18–20 ka ago. High values of conifer pollen at some sites (up to 70% at Padul in southern Spain [26], for example) are the basis for locating the main coniferous forest refugia shown in Figure 10.5: the southern Ural Mountains and the Altai and Sayan Mountains. Open coniferous forest was rather more widespread, and scattered groves of pine, birch, and

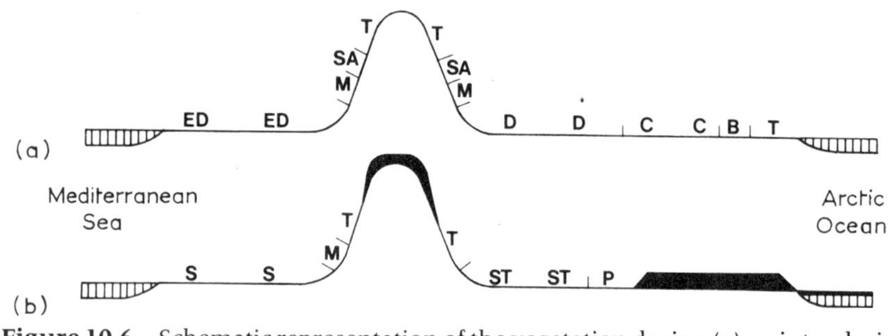

Figure 10.6 Schematic representation of the vegetation during (a) an interglacial and (b) a glacial stage along a north/south transect through Europe. Modified from Van der Hammen *et al.* [3]. B = birch forest; C = boreal coniferous forest; D = mixed deciduous forest; ED = mixed evergreen–deciduous forest; ■ = glacial ice; M = montane forest (beech/fir/spruce); P = polar desert; S = steppe; SA = subalpine forest (pine); ST = steppe–tundra; T = tundra.

perhaps spruce, probably occurred throughout the region of forest–steppe shown in Figure 10.5. The Colchis region of southern Russia, encompassing the eastern shores of the Black Sea and the high ranges of the Caucasus, is widely recognized to have been a major forest refugium during the glacial stages of the Pleistocene, but even there the main forests 18–20 ka ago were of pine and birch, and broadleaved trees were probably very localized in distribution [6]. More continuous broadleaved forest is presumed to have survived in eastern Asia, though this survival is poorly documented.

Climatic reconstructions for 18–20 ka ago (Figure 5.1) show the zone of temperate westerly winds, which governs the distribution of broadleaved deciduous forest in Europe today, displaced southwards to lie over southern Spain and North Africa [28]. There are claims of full-glacial tree remains as far south as latitude 23°N on the Hoggar massif in the central Sahara [29], but both the dating and the provenance of the tree pollen finds are uncertain [10]. More convincing is the occurrence of well-developed palaeosols at several sites between latitudes 32° and 36°N in the currently semi-arid climates of Tunisia and Morocco; these palaeosols have been dated to around 20 ka ago [30]. Pollen spectra associated with the palaeosols contain 20–30% tree pollen [31]. To what extent North Africa could have functioned as a refuge for European forest trees (and particularly Mediterranean taxa [32]) during glacial stages of the Pleistocene, however, is unclear, as there was probably never a dryland connection between Europe and Africa at the western end of the Mediterranean Sea.

At the vast majority of full-glacial sites in Eurasia that have been investigated, non-woody taxa predominate in the pollen spectra. The major components are grasses (Gramineae), chenopods (Chenopodiaceae) and *Artemisia*, but a wide variety of other pollen and spore taxa is also present. Some of these subordinate taxa are characteristic of particular vegetation types at the present day: for example, Cyperaceae (moss–lichen tundra); *Armeria*, Caryophyllaceae, *Cerastium*, Compositae, Cruciferae, *Plantago*, *Potentilla* and *Rumex* (dry fellfield); *Draba*, *Oxyria*, *Saxifraga*, *Silene* and *Lycopodium* spp. (arctic–alpine communities); *Eurotia* and *Kochia* (cool steppe); *Plantago maritima* and *Suaeda* (saline communities); and *Juniperus* and *Salix* (dwarf-shrub communities). A broad distinction can be drawn between more northerly sites, where *Artemisia* was a major pollen component, and more southerly sites, where chenopods predominated [10,11]. The more northerly sites have been categorized as tundra, tundra–steppe and forest–steppe in Figure 10.5, the more southerly ones (with reduced representation of the arctic and alpine elements) as steppe and desert. Pollen spectra of the first group differ from those of present-day tundra communities in a number of important respects. Cyperaceae values are typically lower, and Gramineae and *Artemisia* values higher; tree and shrub pollen values are considerably lower (except in forest–steppe), whilst Ericaceae is almost unrepresented; and steppe and saline elements are present [11,17,33]. Various attempts have been made to reconstruct the character of these open cryophilic full-glacial communities [10,34,35]. The closest modern analogues would appear to be the scattered patches of tundra–steppe in northeastern Siberia [36], but it is probable that a rather small-scale mosaic of plant communities (of which tundra–steppe was but one element) contributed to the pollen spectrum at any one site; Schweger [37], for example, postulates willow scrub, sedge tundra, fellfield and polar desert communities in eastern Beringia. The more southerly full-glacial steppe communities have been likened to the cool and arid montane steppes found today in the Pamirs of southern Russia [32]. At their maximum extent around 20 ka ago these steppes extended both northwards and southwards beyond their present boundaries [25], reaching southwards to the shores of the Black Sea and also westwards into the Hungarian Plain. Associated with them was a rich and varied fauna, which also occurred on the tundra–steppe further north: mammoth, steppe bison, woolly rhinoceros, reindeer, musk-ox, yak, arctic fox, horse, saiga antelope, cave lion, ground squirrel and lemming [38]. The presence of many large herbivores on the tundra–steppe has generated a lively controversy about their means of sustenance [35]. Claims of highly productive grassland communities at full glacial [39,40] have been rejected by most palaeoecologists in favour of a mosaic of tundra, steppe and scrub communities of lower productivity, but supplying a variety of food resources [37].

Full-glacial plant communities in Eurasia represent the nadir of productivity during the last glacial stage. Interstadials earlier in the stage were characterized by a spread of dwarf shrubs and conifers. The stratigraphy of the last glacial stage in Eurasia is now sufficiently well known for there to be little risk of confusing stadial and interstadial vegetation records. Nevertheless, problems do still exist. For example, radiocarbon dates of frozen carcasses of mammoths from Siberia show that their ages fall into two main groups – one from 30 to 45 ka ago and one from 11 to 14 ka ago [38]. These age-ranges correspond to interstadial rather than stadial conditions, and it is still not clear whether the apparent absence of carcasses from the full glacial is merely a reflection of failure of preservation, or whether the low productivity then did indeed limit the population sizes of the large herbivores. In other less well-documented parts of the world confusion between stadial and interstadial climates and vegetation is more likely, particularly where climatic differences were expressed through the moisture cycle rather than through the temperature cycle. Apparently conflicting vegetation reconstructions for the same region may be an outcome of this confusion.

10.4 NORTH AMERICAN FULL-GLACIAL VEGETATION

10.4.1 The general character of the vegetation

The complex character of the last glacial stage is also evident in North America, where a sequence of stadials and interstadials culminating in a period of maximum cold late in the stage has been described. Glaciation in this final cold period, the **Late Wisconsin**, commenced 24–25 ka ago, peaked at 18–20 ka ago, and was in full decline in most areas by 14 ka ago [41]. In regions close to the ice sheets, and also in the American Southwest, where high lake levels coincided with maximum glaciation, a fairly stable vegetation cover was probably maintained for much of the 10-ka period [42,43]. However, in the eastern Great Plains the composition of the plant cover changed markedly about 21 ka ago, whilst in the southeastern United States, for example, vegetation recovery was already apparent by 16.5 ka ago [44, 45]. Accordingly, for comparative purposes, studies on full-glacial vegetation in North America need ideally to be concentrated into the narrow time-band 21–17 ka ago.

Pollen diagrams extending back to this time-band are rather few in number, particularly in the west. Consequently, no comprehensive recent vegetation maps have been attempted for the full glacial – and earlier ones (by Frenzel [10] and Whitehead [46]) are largely hypothetical. Nevertheless, a general pattern is clearly discernible (Figure 10.7). At all sites except those in close proximity to the ice sheets the full-glacial pollen spectra contain abundant tree pollen – almost always in excess of 30%, and sometimes reaching 80%. This is true of such diverse sites as the dried-up Lake Bonneville in the salt flats of Utah [47], Muscotah Marsh in the prairies of Kansas [48], and Anderson Pond in the Appalachian Mountains of Tennessee [44]. Spruce pollen is predominant in the northwestern and northeastern United States and in the Great Plains, and pine in the Rocky Mountains, the American Southwest, and the eastern states from Tennessee and the Carolinas southwards. Macrofossil records suggest that the spruce was *Picea engelmannii* (Engelmann spruce) in the west, and *Picea glauca* (white spruce) in the east [49,50]. Jack Pine (*Pinus banksiana*) similarly was probably dominant in the southeast, and bristlecone pine (*Pinus longaeva*), limber pine (*Pinus flexilis*) and pinyon pines (*Pinus edulis*, *P. monophylla* and *P. remota*) were variously dominant in the west [45,51].

The much greater survival of trees in North America than in Eurasia at full glacial that the fossil record documents was almost certainly the result of different climatic and topographic conditions in the two land masses. The absence of lower-latitude mountain ranges aligned parallel to the ice front in North America, coupled with the more southerly position reached by the ice sheets, meant that extensive permafrost did not form to the south

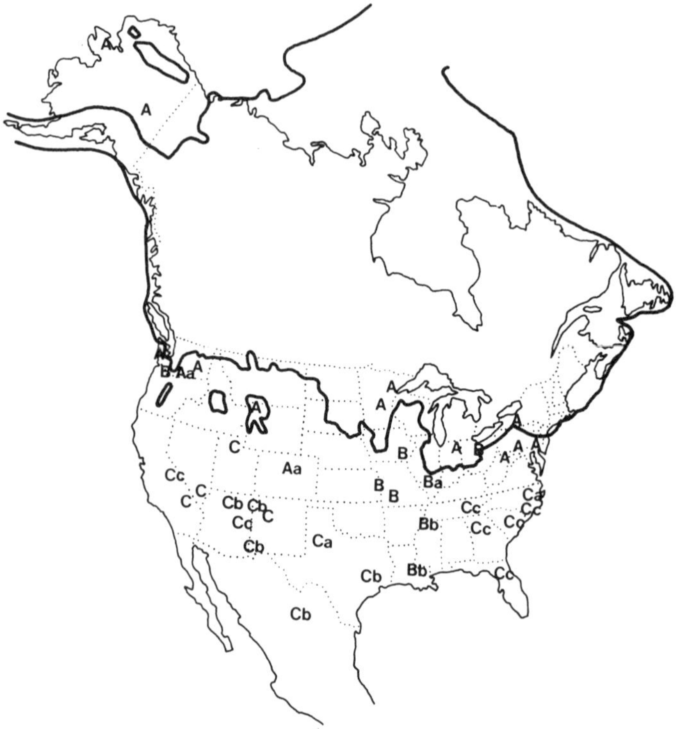

Figure 10.7 Pollen evidence for full-glacial vegetation in North America. The heavy line delimits the margins of the major ice sheets. The positions of some of the more important pollen diagrams extending back to 17–20 ka ago are shown, and their pollen spectra are categorized as follows:
A = tree pollen sparse (<20%);
  Aa = pine or spruce pollen present in significant amounts (>10%);
B = spruce pollen exceeding 20–30%;
  Ba = oak pollen present in significant amounts;
  Bb = pollen of mesic hardwoods present;
C = pine pollen exceeding 20–30%;
  Ca = spruce pollen present in significant amounts;
  Cb = pollen of xeric hardwoods present;
  Cc = pollen of mesic hardwoods/conifers present.

of the ice sheets, and plant populations were not hindered in their migration southwards by high-altitude barriers [52]. Continuous permafrost was limited to a narrow strip, often less than 200 km wide, in the northern United States [53], and a relatively high-altitude treeline was maintained to within a few hundred kilometres of the ice margins. Thus a full-glacial treeline at about 2400 m altitude was reconstructed by Waddington and Wright [54] for Yellowstone Park in the central Rocky Mountains, at 44°N – though in the Appalachian Mountains at 39° 34′N (and within 200 km of the ice margins) the treeline may only have been at 800 m altitude [55].

However, the prevalence of tree pollen in full-glacial pollen spectra across much of the United States should not be interpreted as evidence of unbroken forest cover. A major component of nearly all the pollen spectra is a group of taxa now characteristic of the Great Plains and the Southwestern deserts: grasses (Gramineae), composites (particularly *Artemisia* and *Ambrosia*) and chenopods (almost certainly in part *Atriplex* and *Gayia*). The abundant presence of these pollen types indicates an open woodland or parkland vegetation at full glacial over much of the subcontinent, with scattered groves of pine or spruce set in a matrix of grassland and xeric scrub (not dissimilar to the Eurasian forest–steppe). More continuous forest was probably rather restricted in extent. Thus pollen of mesic hardwoods such as *Fraxinus*, *Juglans* and *Ulmus* occurs consistently only in spectra from the southeastern states, and then in amounts that suggest scattered relict stands rather than continuous forest. Mesic conifers such as redwood (*Sequoia*), douglas fir (*Pseudotsuga*) and white fir (*Abies concolor*) were probably rather more widespread in the western United States – though still restricted in distribution (see below). A similar picture of the full-glacial vegetation is indicated by the faunal evidence. A curious mixture of open-ground herbivores and boreal and even deciduous forest dwellers, suggestive of open-forest or parkland environments, has been recorded consistently at a large number of full-glacial sites throughout the United States. In the former category were bison (*Bison*), ground sloth (*Glossotherium*), mammoth (*Mammuthus*) and peccary (*Platygonus*); in the latter category were mastodon (*Mammuth*), white-tailed deer (*Odocoileus*), chipmunk (*Tamias*) and prairie vole (*Microtus*). Additional open-ground grazers such as camel (*Camelops*) and horse (*Equus*) in the Great Plains region suggest a prevalence of grassland conditions there. Remains of tundra mammals (lemming, *Dicrostonyx*; reindeer, *Rangifer*; and musk-ox, *Ovibos*) have been found only at sites close to the ice-sheet margins – but again in company with boreal forest mammals such as meadow vole, *Microtus* [56,57].

10.4.2 The western United States (except California)

The most detailed vegetation reconstructions at full glacial are those provided by the plant remains preserved in urine-impregnated debris in the dwellings of packrats (*Neotoma* spp.). Radiocarbon-dated records of plant macrofossils from such 'packrat middens' are now available for several hundred sites in the western United States, ranging in latitude from 26° to 42°N, in longitude from 103° to 119°W, and in altitude from 240 to

2680 m [58–60]. These records, often identified to species level, are predominantly from arid and semi-arid habitats, where conventional pollen analysis is not possible [61,62]. They indicate that over a vast and diversified area more than five times the size of Britain only three vegetation types were widespread at full glacial [47]: (1) alpine tundra, at altitudes above 2400–2650 m [60,63]; (2) subalpine parkland, extending downwards to altitudes of about 1500 m, of open stands of spruce (*Picea engelmannii*), pine (*Pinus longaeva, P. flexilis, P. albicaulis*) and fir (*Abies lasiocarpa*), intermixed with typical present-day desert plants such as sagebrush (*Artemisia*) and saltbush (*Atriplex*); (3) a lower woodland zone, of pigmy conifers – notably pinyon pine (*Pinus edulis, P. monophylla, P. remota*) and juniper (particularly *Juniperus osteosperma*) – and xerophytic shrubs and succulents such as mountain mahogany (*Cercocarpus intricatus*) and prickly pear cactus (*Opuntia* spp.). Largely because of topographic limitations, this last zone did not occur north of latitude 37°N, whilst south of latitude 34°N evergreen oaks (*Quercus chrysolepis, Q. pungens, Q. turbinella*) were important constituents of the woodland. All tree taxa were less abundant at low altitudes (often juniper alone persisted), and at a few sites below 900 m altitude, particularly in the Lower Colorado River Valley (Figure 10.9), the midden remains document a treeless desert-scrub vegetation [59,61,64,65].

Within an altitudinal belt from 1350 to 2210 m, and between latitudes 32° and 40°N, there occur a number of sites where the packrat middens contain evident remains of mesic conifers (*Abies concolor, Juniperus scopulorum* and *Pseudotsuga menziesii*) [47,58–60,66]. These records indicate local stands of similar character to the modern montane conifer forests of the Rocky Mountains, but the full-glacial stands were probably confined to sheltered valleys and south-facing slopes, and even there might have formed a mosaic with more xeric subalpine parkland (at higher altitudes) and pinyon–juniper woodland (at lower altitudes). Within the lower-altitude woodlands a similar vegetation mosaic probably existed, of desert scrub and woodland [65].

Fuller details of the packrat midden evidence for the American Southwest are given by Van Devender *et al.* [61]. Tentative summaries of that evidence are given in Figure 10.8, where the reconstructed full-glacial zonation is compared with the modern zonation of vegetation on mountain ranges between 33° and 39°N latitude [63,67,68], and in Figure 10.9, which attempts to map the full-glacial plant cover of the western United States. The 600-m, 1500-m and 3000-m contours have been used to locate the approximate positions of significant vegetation boundaries in Figure 10.9, on the basis of the summary chart in Figure 10.8. It is probable that all high ground above 3000 m altitude was covered by alpine tundra or by permanent ice 17–20 ka ago [69], so that this contour defines their

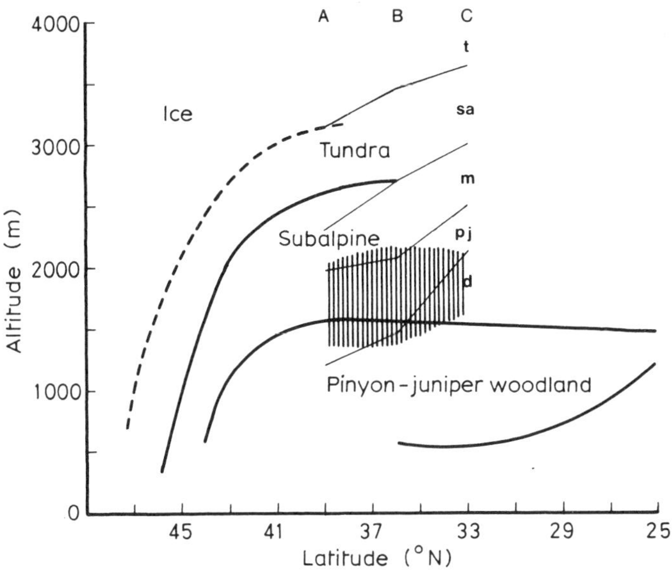

Figure 10.8 Schematic representation of the distribution of full-glacial vegetation types in the western United States, as documented by packrat midden analyses. The shaded area includes sites where mesic conifers (see text) have been recorded. The present-day vegetation zonation on the Great Basin mountains, Utah (A), the Grand Canyon mountains, Colorado (B), and the Sacramento Mountains, New Mexico (C) is also shown (t = tundra, sa = subalpine conifer forest, m = montane conifer forest, pj = pinyon–juniper woodland, d = desert scrub).

minimum extent; the geological evidence for continuous permafrost over large areas of Montana and Wyoming [53] suggests that tundra may also have been widespread at lower altitudes north of latitude 42°N. The packrat midden data identify a pinyon – juniper woodland zone occupying the altitudinal belt between 600 m and 1500 m over much of the region from latitudes 25° to 37°N, and this zone is shaded accordingly in Figure 10.9. The whole area in Figure 10.9 between the 1500-m and 3000-m contours must have been dominated by subalpine or montane vegetation types at full glacial, but whether montane conifer forest was really as localized as the midden records on the map indicate is uncertain.

10.4.3 California

No reconstruction is attempted in Figure 10.9 for the mountain ranges and lowlands to the west of the Great Basin, and for the eastern slopes of the Rocky Mountains. Parts at least of those areas must have been occupied at full glacial either by forest communities (in the west) or by woodland

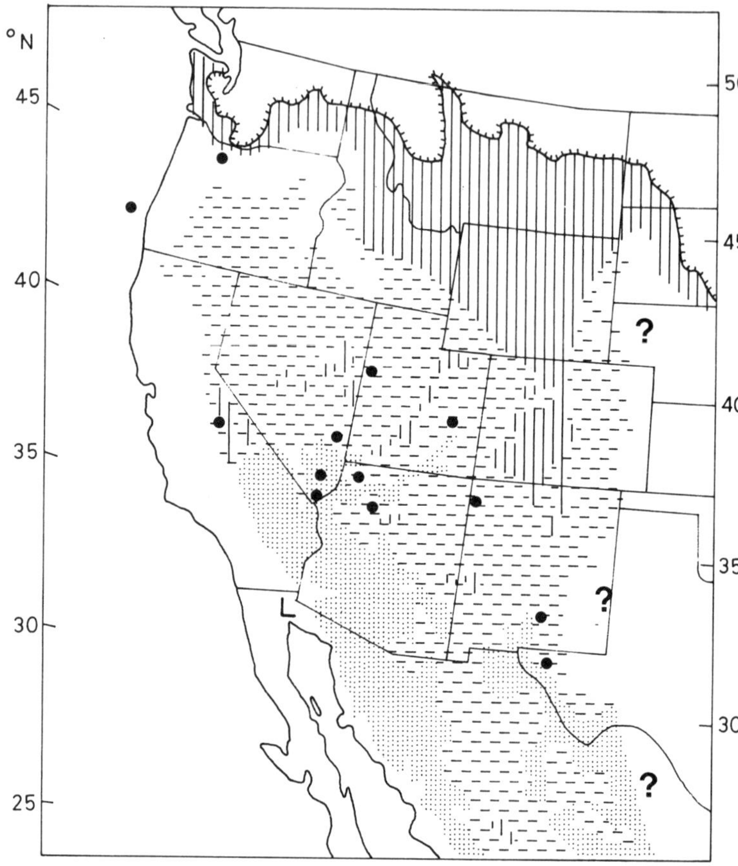

Figure 10.9 Partial reconstruction of conditions in the western United States 17–20 ka ago. ⊥⊥⊥⊥⊥ southern limit of continental ice sheet; | | | | | | | tundra and permanent snow; -‐-‐- subalpine parkland; ⁚⁚⁚⁚ pinyon–juniper woodland; ● sites with mesic conifers present; L, Lower Colorado River Valley. A possible reconstruction of Californian full-glacial vegetation is given in Figure 10.10.

communities (in the east), but the nature and extent of these communities remains speculative. The uncertainty stems both from the scarcity of investigated sites (few sites in California have organic deposits extending back beyond 10 ka ago), and from the need to rely on a pollen record with major ambiguities in identification. Thus in California two important pollen components of full-glacial spectra are *Pinus* and 'TCT' (Taxodiaceae–Cupressaceae–Taxaceae); included within each of these components are woody taxa of diverse stature and behaviour with closely similar pollen grains. *Pinus* pollen may derive, for example, from *P. radiata* (coast forest), *P. lambertiana* and *P. ponderosa* (mesic montane forests), or *P. contorta* and

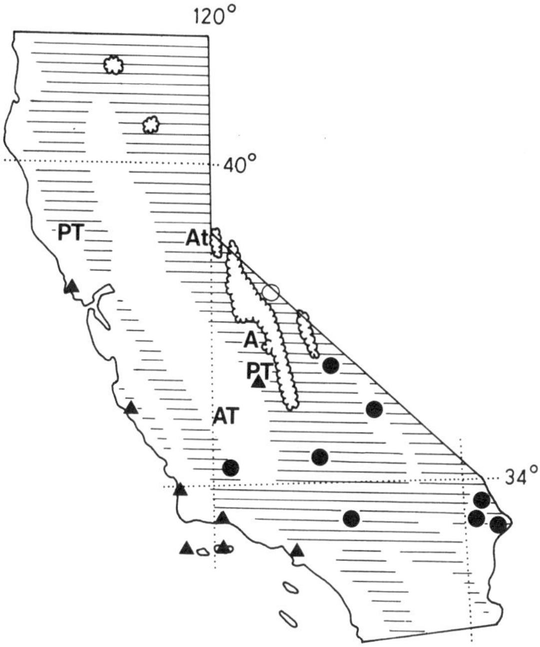

Figure 10.10 Evidence of full-glacial vegetation types in California (higher ground is shaded). ● pinyon–juniper woodland (packrat midden sites); ▲ coastal or mesic conifers present (Table 10.1); ○ Mona Lake; A, *Artemisia* pollen abundant; P, *Pinus* pollen abundant; T/t, 'TCT' pollen (Taxodiaceae + Cupressaceae + Taxaceae) abundant/present; permanent ice and snow.

P. monophylla (xeric montane forests), and TCT pollen may derive from *Cupressus* (coast forest), *Calocedrus* (mesic montane forests), or *Juniperus* (varied forest and woodland types). Midden records are few.

Vegetation reconstructions for the California region are based on the principle, securely established for the Southwestern deserts from the midden evidence, that the plant cover changed little in composition between about 42 and 11 ka ago [61]. Consequently, midden records from any period within that long timespan can be used to reconstruct the full-glacial vegetation. Supporting evidence for such long-term stability is provided by the levels of lakes such as Mona Lake in the west of the Great Basin (Figure 10.10), which were higher than at present for long periods during the last cold stage [70], but fell rapidly between 13 and 12 ka ago [71]. Even if that procedure is adopted for areas to the west of the Great Basin, however – the Sierra Nevada, the Coast Ranges and the California lowlands – the published data still yield only a fragmentary picture (Figure 10.10).

It is clear that vegetation zones on the Sierra Nevada and Coast Ranges were displaced to lower altitudes at full glacial, as in the Southwestern desert region, but the magnitude of the displacement is uncertain. As glacial ice was widespread on the Sierra Nevada above 2000 m altitude, and descended locally on the western side to 1300 m or less [72], full-glacial treelines on the western slopes must have been displaced below 2000 m altitude, as compared with 3000–3500 m at the present day and 2800 –2900 m in the central Rockies at full glacial [61]. Lower altitudes today on the western Sierran slopes are characterized by oak woodland and chaparral, with significant amounts of *Quercus* pollen in the modern pollen spectra up to about 1400 m altitude [73]. However, *Quercus* pollen is scarce or absent in 'full-glacial' sediments from sites ranging in altitude from about 900 m to over 2000 m. The widespread packrat midden evidence of pinyon–juniper woodland at altitudes of 500–1400 m immediately to the south of the main Sierran range [61] (Figure 10.10), and of macrofossils of *Pinus monophylla* and *Juniperus occidentalis* at King's Canyon (920–1270 m altitude) farther to the north [58], suggests that much of the western hillslope region of the Sierra Nevada was covered by pinyon – juniper woodland at full glacial. Records of the more mesic montane conifers, *Abies magnifica, Calocedrus, Pinus ponderosa, Sequoiadendron* and *Torreya* at King's Canyon indicate local survival of montane coniferous stands in protected situations. Even on the floor of the Central California Valley, there is little evidence of the full-glacial presence of oak woodland and chaparral – rather, a desert scrub vegetation similar to that found in the Great Basin today, with *Artemisia, Sarcobatus,* and probably *Juniperus* [74]. Thus it is possible that the full-glacial plant cover that has been reconstructed for the Southwestern deserts extended also to the western side of the Sierra Nevada – and perhaps even to the west slopes of the Coast Ranges, as evidence for Clear Lake at 400 m altitude there (with high values of *Pinus* and TCT pollen, together with some *Artemisia*), indicates some sort of xeric woodland [75,76].

However, macrofossil records ranging in age from 14 ka to more than 45 ka old at a number of coastal and offshore sites in California (Figure 10.10) document the survival of a wide range of taxa characteristic of modern coastal conifer forests and chaparral (Table 10.1). These communities appear to have been restricted in distribution at full glacial to a narrow strip extending down the west coast, with the ranges of individual taxa displaced southwards by up to 160 km [77]. Supporting evidence for that reconstruction comes from the presence of macrofossil remains of douglas fir and sitka spruce (*Picea sitchensis*), dated to 17–20 ka ago, at the south end of the Puget Lowlands, Washington [49], and of pollen of *Alnus, Picea, Pinus, Pseudotsuga* and *Quercus* in offshore marine sediments at 43°N [78]. The likely displacement of westerly storm tracks southwards at full glacial

Table 10.1 Macrofossil records from full-glacial sediments at coastal Californian sites [77]

Coastal conifer forest taxa:
Cupressus forbesii, C. goveniana, C. macrocarpa, Picea sitchensis, Pinus muricata, P. radiata, Pseudotsuga menziesii, Sequoia sempervirens, Torreya californica
Oak woodland and chaparral taxa:
Adenostoma, Arbutus, Arctostaphylos, Artemisia, Ceanothus, Garrya, Pinus sabiniana, Quercus agrifolia, Q. dumosa, Q. lobata, Umbellularia

would have resulted in increased precipitation, particularly during the summer months [79], so that mediterranean climates suitable for the occurrence of chaparral communities might have been poorly developed then. A vegetation mosaic – of forest, woodland and chaparral – is again indicated.

10.4.4 Central and eastern United States

Considerable uncertainty exists also over vegetation reconstructions for the central and eastern United States, as there are fewer than twenty well-dated pollen diagrams (and no midden records) that extend back to 18 ka ago [80,81]. Four major vegetation types are indicated (Figure 10.11):

1. tundra, forming a narrow lowland belt to the south of the ice sheets, but broadening westwards in the Dakotas and Wyoming, and extending discontinuously down the Appalachians at higher altitudes, perhaps as far as the Great Smoky Mountains in Tennessee [82];
2. spruce parkland north of about latitude 36°N, and perhaps extending south down the Mississippi Valley to the Gulf Coast [81];
3. open pine forest in the southeast, but thinning out in Florida, where there may have been substantial treeless areas [83];
4. oak–hickory forest in the central Gulf states [80].

Delcourt and Delcourt [84] see the spruce parkland zone as occupied quite extensively by closed-canopy boreal forest, and suggest that it was bounded to the south by an additional narrow zone of mixed conifer–hardwood forest, coinciding in distribution with the full-glacial position of the polar frontal zone (cf. Figure 3.4), at latitudes 33–34°N. Oak pollen is present at most sites where spruce or pine pollen is well represented, as far north as southern Illinois [45]. Hardier species of *Quercus* could accordingly have been widely distributed at full glacial [84]. Mesic hardwoods, on the other hand, appear to have been more localized, occurring as small populations on steep cliffs and gorges in the southern Appalachians and along the Mississippi Valley [44,85], and at certain sites in Florida [86].

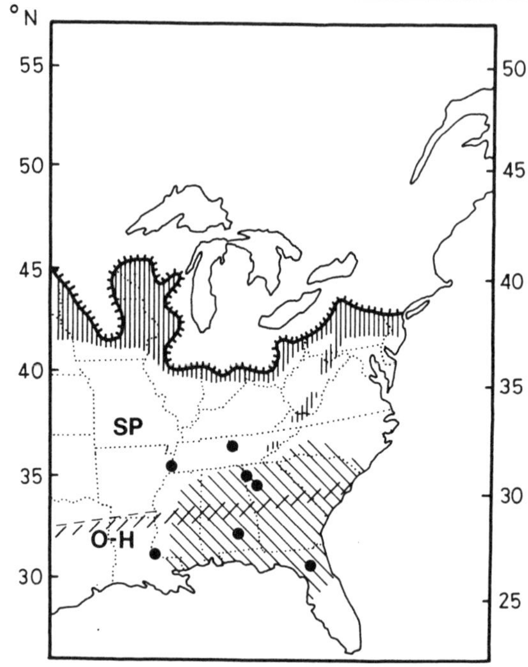

Figure 10.11 Full-glacial conditions in eastern and central North America. Much of the unshaded area, SP, between the tundra |||||||||| and the open pine forests \\\\ was probably spruce parkland, with pine widespread in the northeast of the region; O–H, oak–hickory woodland; ⁄ ⁄ ⁄, possible zone of mixed conifer–hardwood forest; •, sites where pollen of mesic hardwoods has been recorded (Table 10.2); ꟷꟷꟷ, southern limit of Laurentide Ice Sheet. Based on maps in Wright [81], Delcourt and Delcourt [84] and Jacobson *et al.* [80].

Pollen of a variety of tree taxa has been recorded in small amounts at such sites (Table 10.2).

10.4.5 Displacements of range

Most plant taxa in North America had ranges at full glacial that were very different from their current ones. Some species common today in the west were decidedly uncommon then – for example, yellow pine (*Pinus ponderosa*) and creosote bush (*Larrea divaricata*). Other species with wide full-glacial ranges became considerably restricted subsequently – e.g. bristlecone pine (*Pinus longaeva*) and Texan pinyon pine (*P. remota*). The lower temperatures right across the subcontinent resulted in both latitudinal and altitudinal displacements of plant ranges. Aquatic plants common today in the Great Lakes region (*Lobelia dortmanna, Myriophyllum* spp., *Naias* spp.) occurred more than 1000 km to the south in Florida at full glacial [87]. Latitudinal displacement was probably greatest in the east,

Table 10.2 Pollen records of mesic hardwoods at full-glacial sites in the eastern United States*

	1	2	3	4	5	6
Picea		B	C	A	C	
Pinus	A	A	A	C	A	A
Quercus	B	C	C	B	A	A
Abies		C		C		
Acer	C	D		C	D	
Alnus			D		C	
Betula		D	D	C	C	D
Carya	C	D		C	B	D
Castanea					C	
Corylus				C		
Fagus	C	D		C	C	D
Fraxinus	C	C		C	D	D
Ilex	D				D	D
Juglans		D	D	C	D	
Larix				C		
Liquidambar					D	B
Liriodendron				D		
Nyssa					D	D
Ostrya/Carpinus	C	C	C	C	C	D
Platanus					D	
Populus				C		
Salix		D		C		D
Tilia	C					D
Tsuga				D	D	
Ulmus	C	D	D	C	C	D

* 1 = Sheelar Lake, Florida [86]; 2 = Anderson Pond, Tennessee [44]; 3 = Quicksand Pond, Georgia [87]; 4 = Nonconnah Creek, Tennessee [88]; 5 = Pigeon Marsh, Georgia [89]; 6 = Goshen Springs, Alabama [85]. Pollen abundance is indicated as follows: A, >20% TLP; B, 5–10%; C, 1–5%; D, trace amounts.

where the southern limit of *Picea glauca* 17–20 ka ago may have been up to 1500 km south of its current one. Estimates of southward displacements of range in the west for common tree taxa vary from 300 km (*Pinus monophylla, Sequoia*) to 500–640 km (*Abies concolor, Pseudotsuga*) [51,60,77]. Common components of the Great Basin and Mojave Desert floras today occurred at full glacial 400–500 km to the south, in the Grand Canyon region and the Lower Colorado River Valley, respectively [64,90]. However, at lower latitudes in Mexico and in the southwestern United States, many plant ranges were little changed [51,91]. Most common tree taxa in the west were displaced to lower altitudes on the mountains (by amounts ranging from 400 m to 1150 m: Table 10.3), so that, for example, montane conifers occupied at full glacial the altitudinal zone that is today characterized by desert scrub or pinyon–juniper woodland. The zone that today is

Table 10.3 Altitudinal displacements of certain woody taxa in the western United States at full glacial [47,90]

	Lower than present	Higher than present
Abies concolor	720 m	
Pinus flexilis	1070 m	
Pinus longaeva	580 m	
Pinus monophylla	400 m	
Pseudotsuga menziesii	800 m	
Fallugia paradoxa	200 m	
Juniperus osteosperma	1200 m	
Ribes cf. *velutinum*	1150 m	
Symphoricarpos cf. *longiflorus*	670 m	
Atriplex confertifolia		320 m

occupied by montane conifer forest must then have presented conditions of much reduced competition, and certain hardy shrubs that are common components of the present-day lower-altitude desert floras (e.g., *Atriplex confertifolia*) were able to expand upwards to play a major role in the subalpine parkland [47]. On the other side of the subcontinent, too, vegetation zones were similarly lowered; spruce (*Picea glauca*), for example, may have occurred at altitudes of up to 1200 m lower in the southern Appalachian Mountains [87].

These range displacements have been used as a basis for reconstructing climatic conditions at full glacial in North America. The reconstructions are in broad conformity with independent estimates based on geological and ocean-core data [92] and on modelling [93]. Both postulate more extreme changes in the east than in the west, and least change at low latitudes; winters in particular were markedly cooler in the east than at present (perhaps as much as 17–18°C lower [83]), but might conversely have been milder than at present in northern Mexico [47,94]. The lowered temperatures alone would have been sufficient to have made the full-glacial climate wetter (by reducing evaporation), but over much of the United States, except immediately to the south of the ice sheets [93] and in the south-east, actual precipitation was probably in fact higher (with estimates varying from 25 to 100% greater than at present [47]). Therein lay the reasons for the vegetation contrasts at full glacial between North America and Eurasia.

10.5 'FULL-GLACIAL' VEGETATION IN THE TROPICS

The pollen evidence for tropical regions indicates that climatic conditions most different from those pertaining at the present day were developed between about 14 and 21 ka ago [95,96]. These climatic extremes were therefore contemporary with the full-glacial regimes of higher latitudes, and occurred at a time when temperatures are known to have been lowered worldwide. At this time also there is a variety of geological evidence for increased aridity at low latitudes (section 5.2.1). What may be called the 'proxy full glacial' or **hypothermal** period in the tropics was thus characterized by generally cold dry conditions.

Good accounts of the rather scanty pollen evidence for the hypothermal in the tropics are given by Van der Hammen [97], Flenley [7] and Hamilton [95]. Very few of the pollen diagrams, unfortunately, extend back to the 14–21-ka time interval. Those that do extend back that far show pollen spectra from that interval that are very different from contemporary spectra – particularly ones from present-day rainforest areas. These longer-duration pollen diagrams are of four types.

1. In pollen diagrams from the high mountains of South America, Africa and southeast Asia there is an increased representation 14–21 ka ago of pollen taxa indicative of higher-altitude vegetation than that found growing around the sites today. For sites currently in the montane forest belt of the northern Andes, for example, such 'exotic' taxa include *Acaena/Polylepis*, Compositae and *Myriophyllum* supposedly emanating from the alpine paramo [96]; in East Africa, 'exotic' taxa from the alpine scrub found at lower-altitude sites include *Alchemilla* and *Artemisia* [95], and in New Guinea, an 'exotic' component from the subalpine grassland zone includes *Astelia, Cyathea, Ranunculus* and *Styphelia* [98]. Increased levels of grass pollen occurred at all sites 14–21 ka ago.

   A similar pattern is also shown at some lower-altitude sites in the tropics [99], and at some subtropical sites [100].
2. Other lower-altitude sites from the tropics show an increased representation of drier-forest pollen taxa 14–21 ka ago. In northern Queensland, for example, the current wet rainforest taxa such as *Freycinetia* and *Rapanea* were replaced during the hypothermal by *Casuarina* and *Eucalyptus*, which dominate in dry sclerophyll woodland in parts of Queensland today [101].
3. Yet other pollen diagrams from low-altitude sites in or close to present-day rainforest areas show an almost complete absence of tree taxa 14–21 ka ago, and a predominance of grasses instead. The presence of small amounts of *Acacia* pollen at East African sites [102], and of *Bursera, Spondias* and *Acacia* at South American sites [103], has been regarded as indicative of widespread savanna then.
4. However, there is one pollen diagram – from dry woodland at latitude 11°S on the Zambian plateau – which shows almost no change over the

last 22 ka [104]. It is to be expected that other lower-altitude sites will be found, from present-day rainforest, tropical woodland and savanna areas, where the vegetation similarly persisted with little modification through the hypothermal period.

The first three patterns of change represent situations involving contraction of the current range of tropical rainforest – by depression of the upper altitudinal limits and by retraction from the more marginal lower-altitude sites. There is evidence from several areas that 14–21 ka ago the treeline was 1000–1500 m lower than it is at present, with tree growth in the mountains terminating at 2000–2500 m altitude. Such a depression suggests that mean temperatures might have been 6–10°C lower than at present [96,97,105–108]. However, vegetation boundaries at lower altitudes in the mountains were probably depressed by smaller amounts. Thus the lower limit of montane rainforest may have been depressed by only 700 m on Mount Kenya during the hypothermal period [95], 800 m in Sumatra [109] and 700–900 m in the Ecuadorian Andes [99]. Moreover, oxygen-isotope analyses of ocean cores show that temperatures at the water surface in the tropics 18 ka ago were only 2–3°C lower than they are at present. Land surfaces in the tropical lowlands might accordingly have been cooled to a lesser extent than adjacent mountain areas during the hypothermal, resulting in a compression of the montane forest belt [97,108–110]. Pollen evidence from New Guinea [7] suggests that the upper forest zones might have been almost completely 'squeezed out' in the process (Figure 8.9).

There has been some discussion as to whether the pollen changes observed 14–21 ka ago in pollen diagrams from high tropical mountains represent responses of the vegetation to increasing cold or to increasing aridity [111]. Where lake sediments have been used to study the pollen record, then it is clear that lake levels were substantially lower at that time, both at low and at high altitudes [96,102,103,112]. As there is good evidence also of increased cold during the hypothermal period, the climate then must have been both drier and colder (a conclusion which is corroborated by the geological evidence). Estimates of the extent to which precipitation was reduced compared to present-day levels range from 30% [108] to more than 50% [97,101].

Despite the increased stringency of climatic conditions 14–21 ka ago in the tropics, we must nevertheless assume that rainforest was not totally eliminated, and that rainforest trees survived in limited or more extensive enclaves. However, direct pollen evidence for this supposition, in the form of tree pollen continuity through the hypothermal at even one site, is as yet lacking, and accordingly we cannot reconstruct either the location or the composition of these forest enclaves in the same way as for Europe and

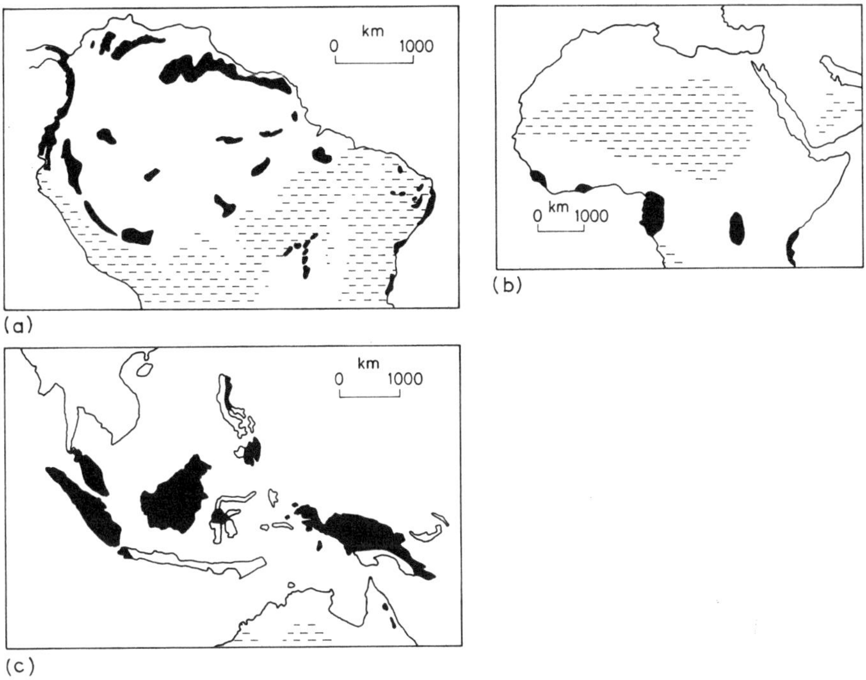

Figure 10.12 Reconstruction of vegetational patterns 20 ka ago in (a) South America; (b) Africa; (c) Southeast Asia. Areas of persistent humid rainforest are shown black; arid and semi-arid areas are hatched. Re-drawn from the maps in Haffer [110], Hamilton [95] and Van Steenis [115], with additional information from Sarnthein [116], Thomas and Goudie [117] and Webb and Tracey [118].

North America. On the other hand, predictions can be made on the basis of the existing biological, geological and climatological data. Direct evidence is provided by the pollen testimony for the likely extent of range displacements during the hypothermal, and by the geological testimony for the former extent of desert and savanna soils [113]. Indirect evidence comes from the present-day distribution patterns of rainforest plant and animal taxa (which can be used to locate possible forest refugia during the hypothermal – section 8.5), and the present-day distribution of rainfall (which can be used to locate areas most likely to become critical for rainforest survival if precipitation levels were to be reduced). The reconstructions presented in Figure 10.12 incorporate these various lines of evidence to differing extents. The reconstruction for South America is based primarily on the geological and climatological evidence, and largely ignores speculations about the locations of forest refugia based on modern distribution patterns; attempts to incorporate the refugia data have produced rather similar pictures – though differing in detail [114]. The reconstruction for Africa utilizes all four types of evidence, but relies heavily on modern distribution patterns of plants and animals to locate the hypothermal forest refugia. The picture for southeast Asia is the least detailed of the three, and is based largely on inferences from present-day rainfall levels. All three reconstructions postulate a spread of drier vegetation types at the expense of rainforest. In both South America and Africa the maps show humid rainforest reduced to a rather small number of separated 'core areas' occupying favourable orographic and topographic situations; in southeast Asia such drastic fragmentation of range is not supposed to have occurred.

The differences between the three tropical regions are only of degree, however. It is unlikely that savanna developed in all the areas shown vacated by rainforest. Recent mapping of the present-day vegetation of Brazilian Amazonia using satellite imagery has demonstrated that about a quarter of the lowlands is covered today not by humid rainforest but by various drier forest types [114]. It is these drier forest types ('transition forest') as well as savanna that probably spread during the hypothermal, so the extent of forest was almost certainly greater then than the maps in Figure 10.12 show. Moreover, wetter 'gallery forest' probably persisted along the main river courses, providing forested corridors linking some at least of the core areas. Savanna was most widespread in the African tropics 14–21 ka ago, but even there it is likely that smaller enclaves of rainforest persisted outside the main core areas [95].

Plant distribution patterns must have been greatly affected by these vegetation changes during the hypothermal. Possibilities of spread were considerably enhanced for many taxa, as lowered sea levels (perhaps as much as 180 m below present levels [108]), lowered tree-lines on the mountains, and the development of seasonally dry or even arid climates, resulted in the linkage of areas that had previously been isolated. In southeast Asia a nearly continuous land corridor with seasonally dry climates, stretching from Taiwan through the Philippines and Sulawesi to northern Australia, may have allowed Asian monsoon-forest species to spread southwards into areas previously occupied by humid rainforest [115]. In Africa, the savanna belts in the northern- and southern-hemisphere subtropics that today are separated by nearly continuous lowland and montane forest were in direct contact over much of the continent 14–21 ka ago [119,120]. Non-forest vegetation expanded at the highest altitudes, linking many mountain peaks that today are isolated by the investing forest [121]. The montane forest, too, must have achieved greater continuity, as it was depressed to lower altitudes where the terrain was less dissected. Rainforest taxa, by contrast, were increasingly confined to the core areas of humid forest.

Despite the contraction of the areal extent of rainforest during the hypothermal period, however, there is no reason to believe that it was accompanied by widespread extinction of plant taxa, as occurred in Eurasia. Nevertheless, the present-day disjunct distributions of many rainforest trees [8] are most easily explained by past fragmentation of formerly continuous ranges, so some degree of extinction must have occurred. The pollen evidence is too scanty to provide any help. Nor does it indicate whether novel assemblages of plant species were formed during the hypothermal, as in mid-latitudes in Eurasia and North America. If we accept the tenet that plant taxa respond individualistically to climatic change (as the evidence reviewed in Chapter 8 suggests), then we must believe that such assemblages were indeed formed. However, the species-richness of many modern tropical plant communities, with several species often capable of occupying the same ecological niche, would probably make it difficult to recognize any fossil assemblages that were different from them. The presence today in the rainforests of central America [122] of certain tree taxa that are more characteristic of temperate forests (e.g., *Cerifera, Myrica, Pinus* and *Quercus*), and of other tree taxa that are apparently xeromorphic (e.g. *Erythroxylum bequaertii, Guettarda gaumeri*), could nonetheless be regarded as a persistent 'memory' of the novel forest assemblages that might have been formed under the drier cooler conditions of 14–21 ka ago (and also earlier in the cold stage, too [105]).

10.6 THE SOUTHERN CONTINENTS AT FULL GLACIAL

Much of Australia, southern Africa and southern South America lies within the subtropical high-pressure belt. Consequently, large areas are currently covered by arid and semi-arid vegetation types, where preservation of long sequences of fossil pollen is rare. Because of this scarcity of biological evidence, the Late Pleistocene vegetation history of the southern land masses is rather poorly known, and often has to be fleshed out by extrapolations from non-biological data. Only in regions affected by glaciation (New Zealand, Tasmania and Chile) is a picture of vegetation change in any way comparable to that for northern-hemisphere land masses beginning to emerge.

Because of the prevalence of drier climates, moisture shifts rather than temperature shifts appear to have dominated the Late Pleistocene climatic record for the southern land masses. Indeed, lowered temperatures, because of the accompanying reduced evaporation, were often registered primarily as a shift to more humid conditions. Moisture shifts occurred with a frequency at least as great as that for stadials and interstadials in the northern hemisphere; for southern South Africa the frequency may have been even greater [123]. Vegetation instability is accordingly more apparent, and the harshness of 'full-glacial' conditions less apparent. 'Unlike in

Europe or North America, there is no representative mode with which to characterize glacial or non-glacial climates' [123]. These moisture shifts were associated particularly with regions close to the margins of the high-pressure cells of the subtropical arid zone, however, and more stable arid conditions may have been maintained in regions lying centrally within the cells. Whether the moisture shifts resulted from an actual migration of the zonal pressure belts [124,125], or from variations in the magnitude and areal extent of seasonal rainfall conditions [126], is still uncertain, though climatic modelling is increasingly emphasizing the latter process at lower latitudes. For certain regions there is evidence also of temperature trends superimposed upon the moisture shifts, which culminated around 19 ka ago in a 'full-glacial' minimum 5–8°C lower than at present [125,127–132]. The following account accordingly concentrates on the vegetation cover of the southern land masses for the period around 19 ka ago.

Although it is recognized that entirely novel plant assemblages might have formed in this 'full-glacial' period [131–133], the present state of our knowledge allows for reconstructions only in terms of modern vegetation

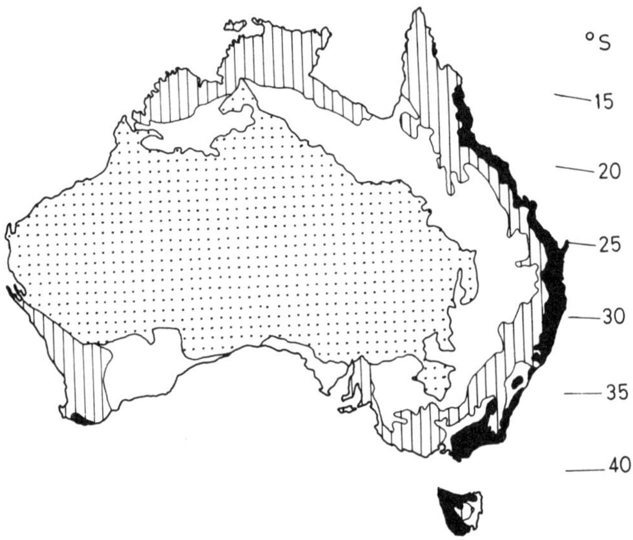

Figure 10.13 The modern plant cover of Australia; forested areas are shown black, savanna vegetation hatched, arid regions stippled, and heathland and mallee unshaded. Based on maps in Specht [134]. Forested areas outside the tropics are dominated by *Nothofagus cunninghamii* and by tall *Eucalyptus* spp.; smaller eucalypts are also prominent in savanna, together with *Acacia, Casuarina* and *Melaleuca* spp., and abundant grasses and composites. Members of the families Epacridaceae, Myrtaceae and Proteaceae are widespread in heathland and mallee (open sclerophyll woodland). Arid-zone communities comprise *Acacia aneura* shrubland, hummock grassland dominated by *Triodia* or *Zygochloa*, and chenopod shrubland (with *Atriplex* and *Maureana*).

types. Even then, surface-sample data by which to characterize these vegetation types palynologically are scanty [132,133]. It is unrealistic as yet to attempt to produce full-glacial vegetation maps for the southern land masses; those that have been produced (e.g., Ab'Sáber [113] and Figure 10.12) should be viewed with some scepticism.

There is most information for the Australian continent, but even that is woefully inadequate. The backcloth to Pleistocene change is provided by the modern plant cover, which can be grouped into four major categories [134] – forest, savanna, heathland and mallee, and arid zone communities (Figure 10.13). Long pollen diagrams through the Australian Pleistocene at some sites [135] show repeated oscillations between rainforest assemblages (with pollen of e.g., Araucariaceae, *Dacrydium*, *Microcachrys*, *Nothofagus*, *Phyllocladus* and *Podocarpus*) and sclerophyll (i.e., savanna and woodland) assemblages (with pollen of *Acacia*, *Casuarina*, Chenopodiaceae, Compositae, Gramineae, Myrtaceae and Proteaceae). These oscillations were primarily in response to moisture shifts, and resulted in local or regional extinctions of certain species (for example: *Drimys*, *Nothofagus cunninghamii* and *Phyllocladus* [136]). Drier conditions prevailed almost everywhere 16–20 ka ago, even in the south-east which should in theory have been affected by rain-bearing westerly winds [125,137]. Pollen spectra then in southern Australia were dominated by grasses and composites, with alpine herbs present at higher-altitude sites and chenopods in the west [133]. The presence of appreciable amounts of pollen of *Eucalyptus* and *Casuarina* at some sites has been taken to indicate a mixture of heathland, woodland and savanna communities in the vicinity [138]. The pollen data, however, are almost entirely confined to the modern savanna and forest regions lying peripherally to the central desert core [136], and these may have been particularly susceptible to moisture shifts in the past. During drier periods we might simplistically envisage a radial spread of drier communities towards the coast in all parts of Australia, and in wetter periods a corresponding inward contraction (Figure 10.13). In drier periods the east coast forests in particular would have been increasingly constricted, and almost certainly fragmented into refugia similar to those reconstructed for the tropical forests further north in Queensland [118].

There are even fewer long pollen sequences for southern Africa than for Australia, and some confusion as to the interpretation of the non-biological evidence for climatic change. Thus although it is clear that the polar front and associated depressions were able to extend further north in southern Africa 16–20 ka ago, bringing winter rains to currently arid and semi-arid areas [139] (Figure 10.14), nevertheless the warm Agulhas Current was less effective as a source of moisture for the southern and eastern coasts, which were accordingly somewhat drier, as well as cooler, at full glacial [123]. Three important sequences of deposits illustrate these complexities.

At Wonderkrater spring, in the dry bushveld of the Transvaal at latitude 24° 26′S, the pollen evidence indicates an open grassland vegetation with many shrubby composites and few trees 25–15 ka ago; the closest modern analogues of this vegetation occur at approximately 1000-m altitude higher, in somewhat moister climates than at Wonderkrater today [132]. At sites along the Gaap Escarpment, at the southeastern margin of the Kalahari Desert at latitudes 27–29°S, the detailed sequence of tufa deposits indicates wetter conditions there 21–14 ka ago, though vegetation reconstructions have not been attempted [140]. At Boomplaas Cave, in the scrubland of the southern Cape, on the other hand, several lines of evidence all point to colder but drier conditions 22–18 ka ago. The large mammal remains, of animals hunted by the human populations living in the cave, are predominantly of larger antelopes and equids (e.g., hartebeest and zebra) characteristic of open grassland environments today, while the small-mammal bones, charcoal remains from wood burnt on the camp fires, and preserved pollen grains blown into the cave by the wind, all show minimum diversity in the full-glacial deposits [128]. The modern vegetation framework against which to view these changes is provided in part by Scott's summary diagram for the Transvaal [132], which indicates the major vegetation shifts to be expected along moisture and temperature gradients (Figure 10.14). However, relevant fossil data for comparison are few, and we must regretfully conclude that the reconstruction of the Pleistocene vegetation cover of southern Africa is still in its infancy.

A similar conclusion applies to southern South America east of the Andes. Ab'Sáber's reconstruction [113], based almost entirely on nonbiological evidence, depicts rather little change in the plant cover from present-day conditions. It is difficult to evaluate the validity of this reconstruction.

Extensive glaciation at higher latitudes in the southern hemisphere (Antarctica excepted) was restricted to Tasmania, the South Island of New Zealand, and the Andes. Maximum glaciation occurred at 18–20 ka ago, when snowlines were lowered by about 800 m and temperatures by about 5°C [130,141–143]. The glaciers reached to sea-level along the west coasts

Figure 10.14 Southern Africa. (a) Present-day seasonal rainfall patterns; humid and subhumid regions are shaded. ▲, sites mentioned in the text (W = Wonderkrater spring, G = Gaap Escarpment, B = Boomplaas Cave). The broken line is the boundary of the Transvaal. (Re-drawn from Butzer [123].) (b) Schematic representation of the relationship between modern vegetation types of the Transvaal and altitudinal and moisture gradients. A = bushveld, B = grassland, C = upland woodland, D = coastal tropical forest, E = montane forest, F = afroalpine vegetation. (Modified from Scott [132].) (c) Reconstructed rainfall patterns 18 ka ago; shaded areas are those receiving summer rainfall. ▲, sites mentioned in the text. (Re-drawn from Van Zinderen Bakker [126].)

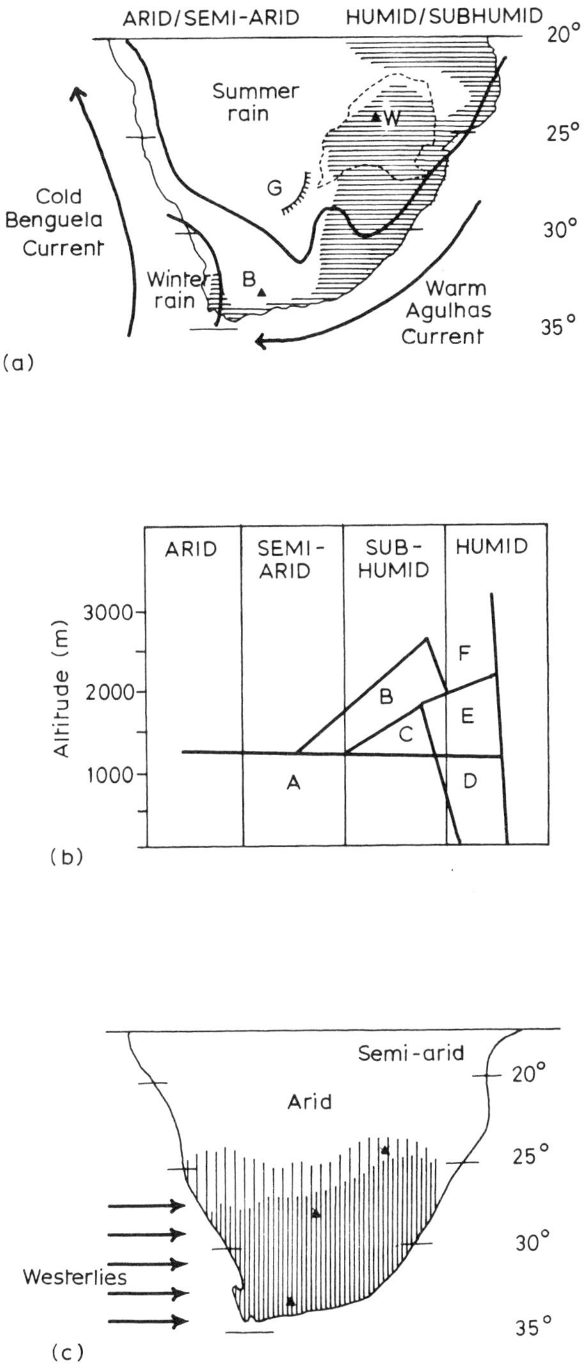

ARID/SEMI-ARID
HUMID/SUBHUMID
20°
25°
30°
35°
Summer rain
W
G
Cold Benguela Current
Winter rain
B
Warm Agulhas Current
(a)
ARID
SEMI-ARID
SUB-HUMID
HUMID
Altitude (m)
3000
2000
1000
A
B
C
D
E
F
(b)
Semi-arid
Arid
20°
25°
30°
35°
Westerlies
(c)

of Tasmania, South Island, and Chile south of latitude 43°S, but ice-free and more arid conditions prevailed generally in the east. In both Tasmania and New Zealand the full-glacial pollen spectra indicate widespread grassland and shrubland, with predominance of pollen of the families Gramineae, Compositae, Epacridaceae and Restionaceae. However, variable and often appreciable amounts of tree pollen occur at almost all investigated sites, and wide-ranging forest survival in a large number of small refugia (particularly in deep ravines and on sheltered north-facing slopes) is envisaged; nevertheless, in South Island all trees except *Nothofagus* were probably extremely restricted in distribution [142]. More extensive forested areas were probably present at full glacial in the North Island of New Zealand, though the podocarp–hardwood forests characteristic of the lowlands of North Island today [144] must have been extremely fragmented 18–20 ka ago [130,131,141,142,145,146].

McGlone and Topping [130] and McGlone [142] make the point that a lowering of temperature by 5°C would not in itself have been sufficient to have brought about such a drastic reduction in the extent of forest in New Zealand at full glacial, and argue that other factors must have been involved as well. They suggest that higher overall wind speeds, increased influence of cold-air masses (leading to more severe frosting), increased incidence of fire, and accentuated drought periods could all have been important.

The most extensive temperate forest in the southern hemisphere today is in Chile, where rainforests dominated by *Nothofagus* spp. extend for about 2000 km on the western side of the Andes between latitudes 35° and 55°S. Several distinct forest types can be recognized within this region, forming defined latitudinal and altitudinal zones (Figure 10.15): lowland deciduous forest, Valdivian rainforest, North Patagonian rainforest, and subantarctic deciduous forest [147]. In northern Chile and Peru the forest grades northwards into mediterranean-type matorral vegetation and then into arid shrubland. Modern pollen spectra from the rainforests are dominated by *Nothofagus* pollen; those from the matorral have little tree pollen, and are dominated instead by Chenopodiaceae and Amaranthaceae [148].

The modern Chilean rainforests occur in a cool wet climate determined by the incidence of rain-bearing westerly winds. At full glacial this high-rainfall region must have been displaced northwards by at least 5° of latitude [149]*, as a site at 34° 30′S, currently in the matorral zone, shows high levels of *Nothofagus* and *Podocarpus* pollen 14–21 ka ago [148]. Further south, in the present-day Valdivian rainforest region, sites at 42°S show reduced amounts of *Nothofagus* pollen 18–19 ka ago, and may have been

* Markgraf [152], however, argues for a corresponding displacement polewards.

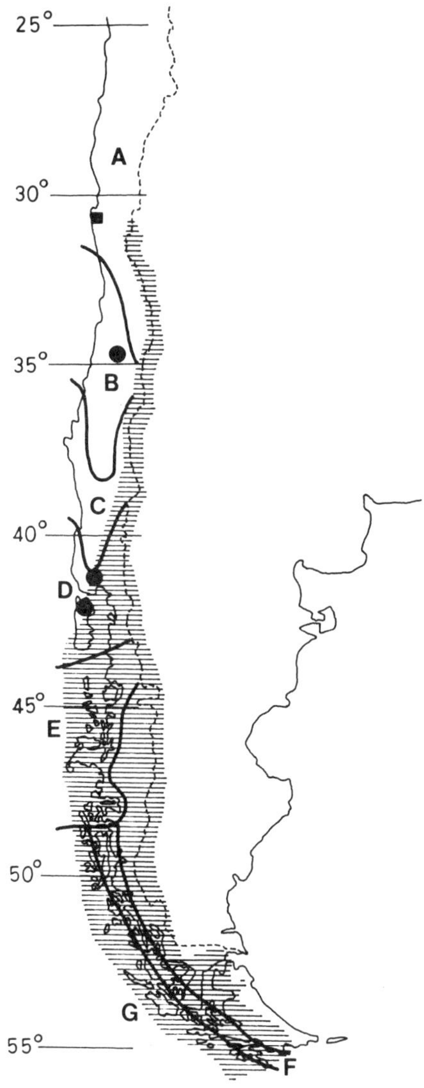

Figure 10.15 Past and present forests of southern South America. A–G modern vegetation zones; ≡ extent of glaciation 18–20 ka ago; ● pollen-analytical sites mentioned in the text; ■ relict lowland deciduous forest stands. A = arid shrubland; B = matorral (broad sclerophyll); C = lowland deciduous forest (*Nothofagus obliqua, Aextoxicon, Laurelia, Persea*); D = Valdivian rainforest (*Nothofagus dombeyi, Eucryphia*); E = North Patagonian rainforest (*Laurelia, Nothofagus dombeyi, Podocarpus, Saxegothea, Weinmannia*); F = subantarctic deciduous forest (*Nothofagus antarctica, N. betuloides, N. pumilio*); G = Magellanic moorland. Based on Heusser [149].

covered instead by highly oceanic moorland and bog communities that are found today in cold wet coastal regions south of latitude 48°S [147,150].

Present-day 'relict' stands of lowland deciduous forest at 30° 30′S [151] may record the northernmost limit of this displacement (Figure 10.15). These relict stands occur on the steep upper slopes of the coastal mountain range, in a locally cool and foggy climate similar to that where *Sequoia sempervirens* grows today in California. Vuilleumier [153] has pointed out that the phyletic structure of the modern bird faunas of the Chilean rainforests (many monotypic and endemic genera) indicates a long period of evolution in isolation within a continuous forest habitat. It is difficult to see how monotypic genera, ranging widely in the present-day forests, could have been maintained if the rainforest had been repeatedly fragmented during the Pleistocene. If continuous forest persisted west of the Andes in northern Chile through the glacial periods, then it is likely that a similar situation pertained also in the equivalent climatic zone of western North America – albeit with a forest of very different composition.

REFERENCES

1. Tanai, T. (1972) Tertiary history of vegetation in Japan, in *Floristics and Palaeofloristics of Asia and Eastern North America* (ed. A. Graham), Elsevier, Amsterdam, pp. 235–55.
2. Watts, W. A. (1988) Europe, in *Vegetation History* (eds B. Huntley and T. Webb III), Kluwer Academic Publishers, Dordrecht, pp. 155–92.
3. Van der Hammen, T., Wijmstra, T. A. and Zagwijn, W. H. (1971) The floral record of the Late Cenozoic of Europe, in *The Late Cenozoic Glacial Ages* (ed. K. K. Turekian), Yale University Press, New Haven, CT, pp. 391–422.
4. Straka, H. (1975) *Pollen- und Sporenkunde. Eine Einführung in die Palynologie*, Fischer, Stuttgart.
5. Ronai, A. (1970) Lower and Middle Pleistocene flora in the Carpathian Basin. *Palaeogeogr. Palaeoclimatol. Palaeoecol.*, **8**, 265–85.
6. Tumajanov, I. I. (1971) Changes of the Great Caucasus forest vegetation during the Pleistocene and Holocene, in *Plant Life of South-West Asia* (eds P. H. Davis, P. C. Harper and I. C. Hedge), Botanical Society of Edinburgh, Edinburgh, pp. 73–87.
7. Flenley, J. R. (1979) *The Equatorial Rain Forest: a Geological History*, Butterworth, London.
8. Prance, G. T. (ed.) (1982) *Biological Diversification in the Tropics*, Columbia University Press, New York.
9. Iversen, J. (1958) The bearing of glacial and interglacial epochs on the formation and extinction of plant taxa. *Uppsala Univ. Årssk.*, **6**, 210–5.
10. Frenzel, B. (1973) *Climatic Fluctuations of the Ice Age* (transl. by A. E. M. Nairn), Case Western Reserve University Press, Cleveland.
11. Birks, H. J. B. (1986) Late-Quaternary biotic changes in terrestrial and

lacustrine environments, with particular reference to northwest Europe, in *Handbook of Holocene Palaeoecology and Palaeohydrology* (ed. B. E. Berglund), Wiley, Chichester, pp. 3–65.
12. Andersen, S. Th. (1966) Interglacial vegetation succession and lake development in Denmark. *Palaeobotanist,* **15**, 117–27.
13. Turner, C. and West, R. G. (1968) The subdivision and zonation of interglacial periods. *Eiszeitalter Gegenw.*, **19**, 93–101.
14. West, R. G. (1977) *Pleistocene Geology and Biology*, 2nd edn, Longman, London.
15. Deacon, J. (1974) The location of refugia of *Corylus avellana* L. during the Weichselian glaciation. *New Phytol.*, **73**, 1055–63.
16. Woillard, G. M. (1978) Grande Pile Bog: a continuous pollen record for the last 140,000 years. *Quat. Res.*, **9**, 1–21.
17. Giterman, R. E., Sher, A. V. and Matthews, J. V. Jr (1982) Comparison of the development of tundra–steppe environments in West and East Beringia: pollen and macrofossil evidence from key sections, in *Paleoecology of Beringia* (eds D. M. Hopkins, J. V. Matthews, Jr, C. E. Schweger and S. B. Young), Academic Press, New York, pp. 43–73.
18. Wijmstra, T. A. (1969) Palynology of the first 30 metres of a 120 m deep section in northern Greece. *Acta bot. neerl.*, **18**, 511–27.
19. Niklewski, J. and Van Zeist, W. (1970) A Late Quaternary pollen diagram from northwestern Syria. *Acta bot. neerl.*, **19**, 737–54.
20. Birks, H. J. B. and Birks, H. H. (1980) *Quaternary Palaeoecology*, Arnold, London.
21. Frenzel, B. (1968) The Pleistocene vegetation of northern Eurasia. *Science*, **161**, 637–49.
22. Grichuk, V. P. (1984) Late Pleistocene vegetation history, in *Late Quaternary Environments of the Soviet Union* (ed. A. A. Velichko), Longman, London, pp. 155–78.
23. Bonatti, E. (1966) North Mediterranean climate during the last Würm glaciation. *Nature*, **209**, 984–5.
24. Thiede, J. (1978) A glacial Mediterranean. *Nature*, **276**, 680–3.
25. Grichuk, V. P. (1971) Aspects of the structure of the Pleistocene vegetational cover across the USSR. *Pollen Spores*, **13**, 101–16.
26. Florschütz, F., Amor, J. M. and Wijmstra, T. A. (1971) Palynology of a thick Quaternary succession in southern Spain. *Palaeogeogr. Palaeoclimatol. Palaeoecol.*, **10**, 233–64.
27. Huntley, B. and Birks, H. J. B. (1983) *An Atlas of Past and Present Pollen Maps for Europe: 0–13,000 Years Ago*, Cambridge University Press, Cambridge.
28. Van Campo, M. (1984) Relations entre la végétation de l'Europe et les températures de surface océaniques après le dernier maximum glaciaire. *Pollen Spores*, **26**, 497–518.
29. Van Campo, M., Aymonin, G., Guinet, P. and Rognon, P. (1964) Contribution à l'étude du peuplement végétale quaternaire des montagnes sahariennes: l'Atakor. *Pollen Spores*, **6**, 169–194.
30. Rognon, P. (1987) Late Quaternary climatic reconstruction for the Maghreb (North Africa). *Palaeogeogr. Palaeoclimatol. Palaeoecol.*, **58**, 11–34.
31. Ben Tiba, B. and Reille, M. (1982) Recherches pollenanalytiques dans les

montagnes de Kroumirie (Tunisie septentrionale): premiers résultats. *Ecol. Mediterr.*, **8**, 75–86.

32. Wright, H. E. Jr (1976) The environmental setting for plant domestication in the Near East. *Science*, **194**, 385–9.
33. Matthews, J. V., Jr (1982) East Beringia during Late Wisconsin time: a review of the biotic evidence, in *Paleoecology of Beringia* (eds D. M. Hopkins, J. V. Matthews Jr, C. E. Schweger and S. B. Young), Academic Press, New York, pp. 127–52.
34. Godwin, H. (1975) *History of the British Flora*, 2nd edn, Cambridge University Press, Cambridge.
35. Hopkins, D. M., Matthews, J. V., Jr, Schweger, C. E. and Young, S. B. (eds) (1982) *Paleoecology of Beringia*, Academic Press, New York.
36. Yurtsev, B. A. (1982) Relics of the xerophytic vegetation of Beringia in northeastern Asia, in *Paleoecology of Beringia* (eds D. M. Hopkins, J. V. Matthews Jr, C. E. Schweger and S. B. Young), Academic Press, New York, pp. 157–77.
37. Schweger, C. E. (1982) Late Pleistocene vegetation of eastern Beringia: pollen analysis of dated alluvium, in *Paleoecology of Beringia* (eds D. M. Hopkins, J. V. Matthews Jr, C. E. Schweger and S. B. Young), Academic Press, New York, pp. 95–112.
38. Vereshchagin, N. K. and Baryshnikov, G. F. (1982) Paleoecology of the mammoth fauna in the Eurasian arctic, in *Paleoecology of Beringia* (eds D. M. Hopkins, J. V. Matthews Jr, C. E. Schweger and S. B. Young), Academic Press, New York, pp. 267–79.
39. Bliss, L. C. and Richards, J. H. (1982) Present-day arctic vegetation and ecosystems as a predictive tool for the arctic-steppe mammoth biome, in *Paleoecology of Beringia* (eds D. M. Hopkins, J. V. Matthews Jr, C. E. Schweger and S. B. Young), Academic Press, New York, pp. 241–57.
40. Guthrie, R. D. (1982) Mammals of the mammoth steppe as palaeoenvironmental indicators, in *Paleoecology of Beringia* (eds D. M. Hopkins, J. V. Matthews Jr, C. E. Schweger and S. B. Young), Academic Press, New York, pp. 307–26.
41. Porter, S. C. (ed.) (1983) *Late-Quaternary Environments of the United States*, Vol. 1, *The Late Pleistocene*, Longman, London.
42. Phillips, F. M., Peeters, L. A., Tansey, M. K. and Davis, S. N. (1986) Paleoclimatic inferences from an isotopic investigation of groundwater in central San Juan Basin, New Mexico. *Quat. Res.*, **26**, 179–93.
43. Smith, G. I. and Street-Perrott, A. (1983) Pluvial lakes of the western United States, in *Late-Quaternary Environments of the United States*, Vol. 1, *The Late Pleistocene* (ed. S. C. Porter), Longman, London, pp. 190–212.
44. Delcourt, H. R. (1979) Late Quaternary vegetation history of the eastern Highland Rim and adjacent Cumberland Plateau of Tennessee. *Ecol. Monogr.*, **49**, 255–80.
45. Watts, W. A. (1983) Vegetational history of the eastern United States 25,000 to 10,000 years ago, in *Late-Quaternary Environments of the United States*, Vol. 1, *The Late Pleistocene* (ed. S. C. Porter), Longman, London, pp. 294–310.

46. Whitehead, D. R. (1973) Late-Wisconsin vegetational changes in unglaciated eastern North America. *Quat. Res.*, **3**, 621–31.
47. Spaulding, W. G., Leopold, E. B. and Van Devender, T. R. (1983). Late Wisconsin paleoecology of the American Southwest, in *Late-Quaternary Environments of the United States*, Vol. 1, *The Late Pleistocene* (ed. S. C. Porter), Longman, London, pp. 259–93.
48. Grüger, J. (1973) Studies on the late Quaternary vegetation history of northeastern Kansas. *Geol. Soc. Am. Bull.*, **84**, 239–50.
49. Barnosky, C. W. (1985) Late Quaternary vegetation near Battle Ground Lake, southern Puget Trough, Washington. *Geol. Soc. Am. Bull.*, **96**, 263–71.
50. Givens, C. R. and Givens, F. M. (1987) Age and significance of fossil white spruce (*Picea glauca*), Tunica Hills, Louisiana–Mississippi. *Quat. Res.*, **27**, 283–96.
51. Lanner, R. M. and Van Devender, T. R. (1981) Late Pleistocene piñon pines in the Chihuahuan Desert. *Quat. Res.*, **15**, 278–90.
52. Wright, H. E. Jr (1971) Late Quaternary vegetational history of North America, in *The Late Cenozoic Glacial Ages* (ed. K. K. Turekian), Yale University Press, New Haven, CT, pp. 421–64.
53. Péwé, T. L. (1983) The periglacial environment in North America during Wisconsin time, in *Late-Quaternary Environments of the United States*, Vol. 1, *The Late Pleistocene* (ed. S. C. Porter), Longman, London, pp. 157–89.
54. Waddington, J. C. B. and Wright, H. E. Jr (1974) Late Quaternary vegetational changes on the east side of Yellowstone Park, Wyoming. *Quat. Res.*, **4**, 175–84.
55. Maxwell, J. A. and Davis, M. B. (1972) Pollen evidence of Pleistocene and Holocene vegetation on the Allegheny Plateau, Maryland. *Quat. Res.*, **2**, 506–30.
56. Graham, R. W. and Mead, J. T. (1987) Environmental fluctuations and evolution of mammalian faunas during the last glaciation in North America, in *North America and Adjacent Oceans during the Last Deglaciation* (eds W. F. Ruddiman and H. E. Wright Jr), Geological Society of America, Boulder, CO, pp. 353–402.
57. Lundelius, E. L. Jr, Graham, R. W., Anderson, E. *et al.* (1983) Terrestrial vertebrate faunas, in *Late-Quaternary Environments of the United States*, Vol. 1, *The Late Pleistocene* (ed. S. C. Porter), Longman, London, pp. 311–53.
58. Cole, K. (1983) Late Pleistocene vegetation of King's Canyon, Sierra Nevada, California. *Quat. Res.*, **19**, 117–29.
59. Van Devender, T. R. and Spaulding, W. G. (1979) Development of vegetation and climate in the southwestern United States. *Science*, **204**, 701–10.
60. Wells, P. V. (1983) Paleobiogeography of montane islands in the Great Basin since the last glaciopluvial. *Ecol. Monogr.*, **53**, 341–82.
61. Van Devender, T. R., Thompson, R. S. and Betancourt, J. L. (1987) Vegetational history of the deserts of southwestern North America: the nature and timing of the Late Wisconsin–Holocene transition, in *North America and Adjacent Oceans during the Last Deglaciation* (eds W. F. Ruddiman and H. E. Wright Jr), Geological Society of America, Boulder, CO, pp. 323–52.
62. Wells, P. V. (1976) Macrofossil analysis of wood rat (*Neotoma*) middens as a

key to the Quaternary vegetational history of arid America. *Quat. Res.*, **6**, 223–48.

63. Cole, K. (1985) Past rates of change, species richness, and a model of vegetational inertia in the Grand Canyon, Arizona. *Am. Nat.*, **125**, 289–303.
64. Cole, K. L. (1986) The Lower Colorado River Valley: a Pleistocene desert. *Quat. Res.*, **25**, 392–400.
65. Spaulding, W. G. (1983) Late Wisconsin macrofossil records of desert vegetation in the American Southwest. *Quat. Res.*, **19**, 256–64.
66. Thompson, R. S. and Mead, J. I. (1982) Late Quaternary environments and biogeography in the Great Basin. *Quat. Res.*, **17**, 39–55.
67. Van Devender, T. R., Betancourt, J. L. and Wimberly, M. (1984) Biogeographic implications of a packrat midden sequence from the Sacramento Mountains, south-central New Mexico. *Quat. Res.*, **22**, 344–60.
68. Woodbury, A. M. (1947) Distribution of pigmy conifers in Utah and northeastern Arizona. *Ecology*, **28**, 113–26.
69. Porter, S. C., Pierce, K. L. and Hamilton, T. D. (1983) Late Wisconsin mountain glaciation in the western United States, in *Late-Quaternary Environments of the United States*, Vol. 1, *The Late Pleistocene* (ed. S. C. Porter), Longman, London, pp. 71–111.
70. Forester, R. M. (1987) Late Quaternary paleoclimate records from lacustrine ostracods, in *North America and Adjacent Oceans during the Last Deglaciation* (eds W. F. Ruddiman and H. E. Wright Jr), Geological Society of America, Boulder, CO, pp. 261–76.
71. Benson, L. and Thompson, R. S. (1987) The physical record of lakes in the Great Basin, in *North America and Adjacent Oceans during the Last Deglaciation* (eds W. F. Ruddiman and H. E. Wright Jr), Geological Society of America, Boulder, CO, pp. 241–60.
72. Wahrhaftig, C. and Birman, J. (1965) The Quaternary of the Pacific mountain system in California, in *The Quaternary of the United States* (eds H. E. Wright Jr and D. G. Frey), Princeton University Press, Princeton, pp. 299–340.
73. Davis, O. K., Anderson, R. S., Fall, P. L. *et al.* (1985) Palynological evidence for early Holocene aridity in the southern Sierra Nevada, California. *Quat. Res.*, **24**, 322–32.
74. Atwater, B. F., Adam, D. P., Bradbury, J. P. *et al.* (1986) A fan dam for Tulare Lake, California, and implications for the Wisconsin glacial history of the Sierra Nevada. *Geol. Soc. Am. Bull.*, **97**, 97–109.
75. Adam, D. P. and West, G. J. (1983) Temperature and precipitation estimates through the last glacial cycle from Clear Lake, California, pollen data. *Science*, **219**, 168–70.
76. Baker, R. G. (1984) Holocene vegetational history of the western United States, in *Late-Quaternary Environments of the United States*, Vol. 2, *The Holocene* (ed. H. E. Wright Jr), Longman, London, pp. 109–27.
77. Johnson, D. L. (1977) The Late-Quaternary climate of coastal California: evidence for an ice-age refugium. *Quat. Res.*, **8**, 154–79.
78. Heusser, L. E. and Shackleton, N. J. (1979) Direct marine–continental correlation: 150,000-year oxygen isotope–pollen record from the North Pacific. *Science*, **204**, 837–9.

79. Barnosky, C. W., Anderson, P. M. and Bartlein, P. J. (1987) The northwestern US during deglaciation: vegetational history and paleoclimatic implications, in *North America and Adjacent Oceans during the Last Deglaciation* (eds W. F. Ruddiman and H. E. Wright Jr), Geological Society of America, Boulder, CO, pp. 289–321.
80. Jacobson, G. L. Jr, Webb, T. III and Grimm, E. C. (1987) Patterns and rates of vegetation change during the deglaciation of eastern North America, in *North America and Adjacent Oceans during the Last Deglaciation* (eds W. F. Ruddiman and H. E. Wright Jr), Geological Society of America, Boulder, CO, pp. 277–88.
81. Wright, H. E. Jr (1981) Vegetation east of the Rocky Mountains 18,000 years ago. *Quat. Res.*, **15**, 113–25.
82. Wright, H. E. Jr (1987) Synthesis; the land south of the ice sheets, in *North America and Adjacent Oceans during the Last Deglaciation* (eds W. F. Ruddiman and H. E. Wright Jr), Geological Society of America, Boulder, CO, pp. 479–88.
83. Watts, W. A. (1980) Late-Quaternary vegetation history at White Pond on the inner coastal plain of South Carolina. *Quat. Res.*, **13**, 187–99.
84. Delcourt, P. A. and Delcourt, H. R. (1987) *Long Term Forest Dynamics of the Temperate Zone*, Springer Verlag, New York.
85. Delcourt, P. A. (1980) Goshen Springs: late-Quaternary vegetation record for southern Alabama. *Ecology*, **61**, 371–86.
86. Watts, W. A. and Stuiver, M. (1980) Late Wisconsin climate of northern Florida and the origin of species-rich deciduous forest. *Science*, **210**, 325–7.
87. Watts, W. A. (1970) The full-glacial vegetation of northwestern Georgia. *Ecology*, **51**, 15–33.
88. Delcourt, P. A., Delcourt, H. R., Brister, R. C. and Lackey, L. E. (1980) Quaternary vegetation history of the Mississippi Embayment. *Quat. Res.*, **13**, 111–32.
89. Watts, W. A. (1975) Vegetation record for the last 20,000 years from a small marsh on Lookout Mountain, northwestern Georgia. *Geol. Soc. Am. Bull.*, **86**, 287–91.
90. Cole, K. (1982) Late Quaternary zonation of vegetation in the eastern Grand Canyon. *Science*, **217**, 1142–5.
91. Meyers, E. R. (1973) Late-Quaternary palaeoecology of the Cuatro Ciénegas Basin, Coahuila, Mexico. *Ecology*, **54**, 982–95.
92. Imbrie, J., McIntyre, A. and Moore, T. C. Jr (1983) The ocean around North America at the last glacial maximum, in *Late-Quaternary Environments of the United States*, Vol. 1, *The Late Pleistocene* (ed. S. C. Porter), Longman, London, pp. 230–6.
93. Kutzbach, J. E. (1987) Model simulations of the climate patterns during the deglaciation of North America, in *North America and Adjacent Oceans during the Last Deglaciation* (eds W. F. Ruddiman and H. E. Wright Jr), Geological Society of America, Boulder, CO, pp. 425–46.
94. Van Devender, T. R. and Burgess, T. L. (1985) Late Pleistocene woodlands in the Bolson de Mapimi: a refugium for the Chihuahuan Desert biota? *Quat. Res.*, **24**, 346–53.
95. Hamilton, A. C. (1982) *Environmental History of East Africa. A Study of the Quaternary*, Academic Press, London.
96. Van Geel, B. and Van der Hammen, T. (1973) Upper Quaternary vegetational

and climatic sequences in the Fuquene area. *Palaeogeogr. Palaeoclimatol. Palaeoecol.*, **14**, 9–92.
97. Van der Hammen, T. (1974) The Pleistocene changes of vegetation and climate in tropical South America. *J. Biogeogr.*, **1**, 3–26.
98. Hope, G. S. (1976) The vegetational history of Mt Wilhelm, Papua New Guinea. *J. Ecol.*, **64**, 627–63.
99. Liu, K. and Colinvaux, P. A. (1985) Forest changes in the Amazon basin during the last glacial maximum. *Nature*, **318**, 556–9.
100. Tsukada, M. (1967) Vegetation in subtropical Formosa during the Pleistocene glaciations and the Holocene. *Palaeogeogr. Palaeoclimatol. Palaeoecol.*, **3**, 49–64.
101. Kershaw, A. P. (1974) A long continuous pollen sequence from northeastern Australia. *Nature*, **251**, 222–3.
102. Kendall, R. L. (1969) An ecological history of the Lake Victoria basin. *Ecol. Monogr.*, **39**, 121–76.
103. Salgado-Labouriau, M. L. (1982) Climatic change at the Pleistocene–Holocene boundary, in *Biological Diversification in the Tropics* (ed. G. T. Prance), Columbia University Press, New York, pp. 74–7.
104. Livingstone, D. A. (1971) A 22,000-year pollen record from the plateau of Zambia. *Limnol. Oceanogr.*, **16**, 349–56.
105. Bonnefille, R. and Riollet, G. (1988) The Kashiru pollen sequence (Burundi) palaeoclimatic implications for the last 40,000 yr BP in tropical Africa. *Quat. Res.*, **30**, 19–35.
106. Rind, D. and Peteet, D. (1985) Terrestrial conditions at the last glacial maximum and CLIMAP sea-surface temperature estimates: are they consistent? *Quat. Res.*, **24**, 1–22.
107. Van der Hammen, T., Barelds, J., de Jong, H. and de Veer, A. A. (1981) Glacial sequence and environmental history in the Sierra Nevada del Coquy (Colombia). *Palaeogeogr. Palaeoclimatol. Palaeoecol.*, **32**, 247–340.
108. Whitmore, T. C. (ed.) (1981) *Wallace's Line and Plate Tectonics*, Clarendon Press, Oxford.
109. Newsome, J. and Flenley, J. R. (1988) Late Quaternary vegetational history of the Central Highlands of Sumatra. II. Palaeopalynology and vegetational history. *J. Biogeogr.*, **15**, 555–78.
110. Haffer, J. (1987) Quaternary history of tropical America, in *Biogeography and Quaternary History in Tropical Latin America* (eds T. C. Whitmore and G. T. Prance), Oxford University Press, New York, pp. 1–18.
111. Coetzee, J. A. (1967) Pollen analytical studies in East and Southern Africa. *Palaeoecol. Africa*, **3**, 1–146.
112. Binford, M. W. (1982) Ecological history of Lake Valencia, Venezuela: interpretation of animal microfossils and some chemical, physical, and geological features. *Ecol. Monogr.*, **52**, 307–33.
113. Ab'Sáber, A. N. (1982) The paleoclimate and paleoecology of Brazilian Amazonia, in *Biological Diversification in the Tropics* (ed. G. T. Prance), Columbia University Press, New York, pp. 41–59.
114. Brown, K. S. Jr (1987) Soils and vegetation, in *Biogeography and Quaternary History in Tropical Latin America* (eds T. C. Whitmore and G. T. Prance), Oxford University Press, New York, pp. 19–45.

115. Van Steenis, C. G. G. T. (1979) Plant-geography of east Malesia. *Bot. J. Linn. Soc.*, **79**, 97–178.
116. Sarnthein, M. (1978) Sand deserts during glacial maximum and climatic optimum. *Nature*, **272**, 43–6.
117. Thomas, D. S. G. and Goudie, A. S. (1984) Ancient ergs of the Southern Hemisphere, in *Late Cainozoic Palaeoclimates of the Southern Hemisphere* (ed. J. C. Vogel), Balkema, Rotterdam, pp. 407–18.
118. Webb, J. L. and Tracey, J. G. (1981) Australian rainforests: patterns and change, in *Ecological Biogeography of Australia*, Vol. 1 (ed. A. Keast), Junk, The Hague, pp. 605–94.
119. Van Zinderen Bakker, E. M. (1982) African palaeoenvironments 18,000 yrs BP. *Palaeoecol. Africa*, **15**, 77–99.
120. Van Zinderen Bakker, E. M. and Mercer, J. H. (1986) Major Late Cainozoic climatic events and palaeoenvironmental changes in Africa viewed in a world wide context. *Palaeogeogr. Palaeoclimatol. Palaeoecol.*, **56**, 217–35.
121. Simpson, B. B. (1975) Pleistocene changes in the flora of the high tropical Andes. *Palaeobiol*, **1**, 273–94.
122. Toledo, V. M. (1982) Pleistocene changes of vegetation in tropical Mexico, in *Biological Diversification in the Tropics* (ed. G. T. Prance), Columbia University Press, New York, pp. 93–111.
123. Butzer, K. W. (1984) Late Quaternary environments in South Africa, in *Late Cainozoic Palaeoclimates of the Southern Hemisphere* (ed. J. C. Vogel), Balkema, Rotterdam, pp. 235–64.
124. Flohn, H. (1984) Climate evolution in the Southern Hemisphere and the equatorial region during the Late Cenozoic, in *Late Cainozoic Palaeoclimates of the Southern Hemisphere* (ed. J. C. Vogel), Balkema, Rotterdam, pp. 5–20.
125. Webster, P. J. and Streten, N. A. (1978) Late Quaternary ice age climates of tropical Australasia: interpretations and reconstructions. *Quat. Res.*, **10**, 279–309.
126. Van Zinderen Bakker, E. M. (1980) Comparison of late-Quaternary climatic conditions in the Sahara and the Namib–Kalahari region. *Palaeoecol. Africa*, **12**, 381–94.
127. Bowler, J. M., Hope, G. S., Jennings, J. N. *et al.* (1976) Late Quaternary climates of Australia and New Guinea. *Quat. Res.*, **6**, 359–94.
128. Deacon, J., Lancaster, N. and Scott, L. (1984) Evidence for Late Quaternary climatic change in southern Africa, in *Late Cainozoic Palaeoclimates of the Southern Hemisphere* (ed. J. C. Vogel), Balkema, Rotterdam, pp. 391–404.
129. Heusser, C. J. (1983) Vegetational history of the northwestern United States including Alaska, in *Late-Quaternary Environments of the United States*, Vol. 1, *The Late Pleistocene* (ed. S. C. Porter), Longman, London, pp. 239–58.
130. McGlone, M. S. and Topping, W. W. (1983) Late Quaternary vegetation, Tongariro region, central North Island, New Zealand. *N. Z. J. Bot.*, **21**, 53–76.
131. MacPhail, M. K. (1979) Vegetation and climates in southern Tasmania since the last glaciation. *Quat. Res.*, **11**, 306–41.
132. Scott, L. (1982) A Late Quaternary pollen record from the Transvaal bushveld, South Africa. *Quat. Res.*, **17**, 339–70.
133. Hope, G. S. (1978) The Late Pleistocene and Holocene vegetational history of

Hunter Island, northwestern Tasmania. *Aust. J. Bot.*, **26**, 493–514.

134. Specht, R. L. (1981) Major vegetation formations in Australia, in *Ecological Biogeography of Australia*, Vol. 1 (ed. A. Keast), Junk, The Hague, pp. 163–297.
135. Singh, G. (1982) Environmental upheaval: vegetation of Australasia during the Quaternary, in *A History of Australasian Vegetation* (ed. J. M. B. Smith), McGraw-Hill, Sydney, pp. 90–108.
136. Kershaw, A. P. (1981) Quaternary vegetation and environments, in *Ecological Biogeography of Australia*, Vol. 1 (ed. A. Keast), Junk, The Hague, pp. 83–101.
137. Rognon, P. and Williams, M. A. J. (1977) Late Quaternary climatic changes in Australia and North Africa: a preliminary interpretation. *Palaeogeogr. Palaeoclimatol. Palaeoecol.*, **21**, 285–327.
138. Dodson, J. (1977) Late Quaternary palaeoecology of Wyrie Swamp, southeastern South Australia. *Quat. Res.*, **8**, 97–114.
139. Van Zinderen Bakker, E. M. and Coetzee, J. A. (1988) A review of Late Quaternary pollen studies in East, Central and Southern Africa. *Rev. Palaeobot, Palynol.*, **55**, 155–74.
140. Butzer, K. W., Stuckenrath, R., Bruzewicz, A. J. and Helgren, D. M. (1978) Late Cenozoic paleoclimates of the Gaap Escarpment, Kalahari margin, South Africa. *Quat. Res.*, **10**, 310–39.
141. Colhoun, E. A. (1985) Pre-last glaciation maximum vegetation history at Henty Bridge, western Tasmania. *New Phytol.*, **100**, 681–90.
142. McGlone, M. S. (1988) New Zealand, in *Vegetation History* (eds B. Huntley and T. Webb III), Kluwer Academic Publishers, Dordrecht, pp. 557–99.
143. Porter, S. C. (1981) Pleistocene glaciation in the southern Lake District of Chile. *Quat. Res.*, **16**, 263–92.
144. Wardle, P. (1964) Facets of the distribution of forest vegetation in New Zealand. *N. Z. J. Bot.*, **2**, 352–66.
145. Dodson, J. R., Enright, N. J. and McLean, R. F. (1988) A Late Quaternary vegetation history for far northern New Zealand. *J. Biogeogr.*, **15**, 647–56.
146. McGlone, M. S. (1985) Plant biogeography and the Late Cenozoic history of New Zealand. *N. Z. J. Bot.*, **23**, 723–49.
147. Heusser, C. J. (1981) Palynology of the last interglacial–glacial cycle in midlatitudes of southern Chile. *Quat. Res.*, **16**, 293–321.
148. Heusser, C. J. (1983) Quaternary pollen record from Laguna de Tagua Tagua, Chile. *Science*, **219**, 1429–32.
149. Heusser, C. J. (1984) Late-glacial–Holocene climate of the Lake District of Chile. *Quat. Res.*, **22**, 77–90.
150. Villagrán, C. (1988) Expansion of Megallanic Moorland during the Late Pleistocene: palynological evidence from northern Isla de Chiloé, Chile. *Quat. Res.*, **30**, 304–14.
151. Rundel, P. W. (1981) The matorral zone of central Chile, in *Ecosystems of the World 11. Mediterranean-type Shrublands* (eds F. di Castri, D. W. Goodall, and R. L. Specht), Elsevier, Amsterdam, pp. 175–201.
152. Markgraf, V. (1989) Reply to C. J. Heusser's 'Southern Westerlies during the Last Glacial Maximum'. *Quat. Res.*, **31**, 426–32.
153. Vuilleumier, F. (1967) Phyletic evolution in modern birds of the Patagonian forests. *Nature*, **215**, 247–8.

CHAPTER ELEVEN

The resurgence of forest

11.1 INTRODUCTION

The evidence presented in the previous chapter shows clearly that many of the major plant assemblages of the last Pleistocene cold stage differed markedly from those of the modern world, and occupied different geographical ranges. The transformation of this Ice-Age plant cover into the contemporary one is the subject of the present chapter and the three that follow. The four chapters deal successively with the spread of forest communities in the Early Holocene, their subsequent recession as man's influence expanded, and the history of dryland biomes (desert, steppe and savanna) and coldland biomes (tundra, boreal forests and mires). The relevant literature is voluminous, particularly for temperate forests and their derivatives. Consequently, the treatment in these four chapters is highly selective – even more so than in previous ones. The emphasis is on the wider-ranging patterns and processes of change that generated large-scale vegetational heterogeneity through the Holocene; and the history of particular sites (with which much of the literature is concerned) is largely ignored. So too are large areas of *terra incognita*, much of it in the southern hemisphere, for which relevant information is scanty or not easily available. Whether the general patterns of change deduced for the better-documented regions are applicable also to these underworked areas is not yet clear – though it is to be hoped that some at least of the processes of change that have been detected in temperate regions operated globally through the Holocene.

11.2 PATTERNS OF FOREST EXPANSION

The glacial maximum of 18–20 ka ago reduced forest communities to their lowest extent globally for almost any time during the last 50 Ma. Slow and intermittent deglaciation over the succeeding 3–4 millennia was accompanied by generally cold (and often dry) conditions, which continued to hold forest communities in check. Not until 15–13 ka ago was there any major amelioration, when temperature estimates from ocean cores [1,2], ice cores [3,4] and preserved beetle remains [5] all show a rapid rise in temperature heralding the start of what may be called the **Protothermal period**. Modelling based on Milankovitch theory (section 6.4) shows clearly that the rise in temperature was largely confined to the warm-season months, as seasonal extremes intensified during the period from 18 to 9 ka ago (Figure 6.4(b)). Consequently, rainfall patterns also changed in the protothermal period in certain regions, with increased monsoonal (summer) precipitation at lower latitudes and in parts of the American Southwest (particularly after 11 ka ago), higher precipitation overall in the Pacific Northwest of the United States after 15 ka ago and reduced precipitation further south [6–9].

The increasing temperatures resulted in a breakdown of the rather stable plant communities of the full glacial, and a rapid re-assembly of new types, as plant ranges expanded or contracted in response [10]. Among the kaleidoscope of change three broad sequences can be recognized that led to an expansion of the total range of forest communities: an invasion of tundra by dwarf shrubs and then by trees; a direct invasion of tundra–steppe by trees (usually pine initially); and a gradual conversion of savanna to tropical forest. These sequences represent the onset of the **protocratic phase** of the glacial/interglacial cycle (section 10.2.2). In periglacial regions adjacent to the melting ice sheets, as well as in recently deglaciated terrain, the protocratic record is of the progressive stabilization and stratification of the raw unsorted and unleached mineral soils of the preceding cryocratic phase. Shrub-tundra may have been a necessary precursor of forest in such higher-latitude situations, promoting the build-up of humus and nitrogen in the developing soils [10,11]. In arid periglacial regions further south, where base-rich and even saline soils had formed below the cryocratic tundra–steppe and steppe communities, a prolonged period of leaching under higher rainfall conditions was almost certainly necessary before invasion of trees was possible. At lower latitudes too, a gradual thickening of the tree cover in the lowland savannas probably occurred, also in response to a wetter climate [12].

Because of the extreme restriction of trees in many regions during the full-glacial, and the known or suspected slow migration rates of certain taxa, a patchwork of tree-dominated communities developed initially in the protothermal period, with separated forest nuclei forming from expanding refugial populations and from long-distance colonists. These nuclei gradually fused. Thus all but one of the eight postulated refugial populations of *Cryptomeria japonica* in Japan were linked into continuous populations along the eastern and western sides of the island of Honshu by 10 ka ago [13,14]. Expansion of trees was temporarily checked in many parts of the world by a return to cold conditions 11–10 ka ago,* but in the **Early Holocene** (10–8 ka ago) forests spread rapidly over large areas of the world's surface, aided by prevailing low sea levels which left extensive dryland bridges connecting land masses now separated by sea barriers. The forest advance was spearheaded by the fast-moving, cold-tolerant trees that had begun to expand their ranges in the early protothermal period: birches, larches, pines, poplars, spruces and willows in North America and Eurasia, for example. Behind them came slower-moving taxa that were to dominate in the succeeding **mesocratic phase**: alder, ash, elm, fir, hazel, hickory and oak, for example. As the forest cover spread to higher latitudes and higher altitudes, so differentiation into distinct zones became increasingly apparent. In the northern hemisphere, forest communities of differing composition came to occupy positions along a latitudinal gradient broadly equivalent to those occupied today by boreal forest, temperate forest, tropical seasonal forest and evergreen rainforest, and at middle and lower latitudes to occupy higher-altitude positions broadly equivalent to modern montane and subalpine forests. Statistical analysis of pollen spectra from a large number of sites in Europe suggests that a considerably greater range of forest types, many of them localized in distribution, was present in the Early Holocene than occurs today [15].

By 7–8 ka ago the advance was slowing down, or had halted altogether, in many parts of the world, at a position of near-stability representing the **forest maximum** of the Holocene 'interglacial'. Temperatures then were slightly (2–3°C) higher than they are today, a least in middle latitudes†, so that forest ranged nearer to the Poles and to higher altitudes. Because the 'climatic optimum' of the mid-Holocene was also moist over much of Europe, forest communities occurred then in some areas that are now too dry to support trees: the steppes of eastern Europe, for example [16]. The corresponding **Altithermal period** in North America, by contrast, was drier than today in middle latitudes, so that forest was rather more restricted in extent. These altithermal forests of the mesocratic phase differed substantially in their floristic make-up from present-day ones. Modernization was achieved only slowly in the **Neoglacial** (the last 5 ka approximately – see section 6.3), in response to a variety of biotic, climatic and edaphic factors. Man's influence, diagnostic of the Holocene, became increasingly pervasive during this period, and is considered in Chapter 12.

* In those parts of the world the period from about 14 to 10 ka ago may be referred to as the **lateglacial period**.

† Though summer temperatures were probably slightly lower then in southern Europe [15].

11.3 THE PROTOCRATIC PHASE

11.3.1 The early Protothermal period

The effects of the marked rise in global temperatures 15–13 ka ago that heralded the start of the Protothermal period are clearly seen in Europe, where a latitudinal variation in the response of the plant cover is discernible. In areas recently uncovered by melting ice (southern Sweden, for example), biogenic sediments – indicative of a well-developed, but still herbaceous, vegetation cover – first began to accumulate 13.5 – 13.0 ka ago [17]. In parts of northern and western Britain juniper pollen increased dramatically at the same time [18]; at other sites tree birches expanded then [19], though in Britain their expansion was delayed until 12.5–12.0 ka ago [11,17]. In central Europe pine increased between 13.5 and 13.0 ka ago, and at the same time oak spread into pine woodland in southern Spain [17]. In the eastern Mediterranean and the Near East, however, perpetuation of the arid conditions of the full glacial right through the lateglacial effectively prevented the spread of oak woodland until 10.7–10.1 ka ago [17].

In North America rather earlier plant responses are apparent, perhaps because of the steeper temperature gradient to the south of the ice sheet than in Europe, and the moister conditions at full glacial there that allowed greater survival of trees. Thus broadleaved forest communities expanded in the southeastern United States around 16.3 ka ago [20], and even in areas such as Florida that had been dry at full glacial, mesic forest was established by 13.5–14.0 ka ago [21]. In coastal regions of the northwest, cold-tolerant conifers currently found in subalpine forest (*Picea engelmannii, Pinus contorta, Tsuga mertensiana*) were spreading into lowland tundra by 15–17 ka ago [22], probably from nearby refugial populations. Immediately to the south of the ice sheets in Minnesota, shrub-tundra expanded 14.7 ka ago, and was then replaced by open conifer forest 13.6 ka ago [23].

In the southern hemisphere, too, temperate forest communities expanded early: about 14.5 ka ago in the North Island of New Zealand and in central Chile, for example [24,25]. In tropical regions, by contrast, there is less evidence of early forest expansion. In the mountains of Sumatra and New Guinea, it is true, forest expansion was apparent by 13–14 ka ago [26,27], but long-continuing aridity in other parts apparently checked the spread of trees until 10 or 11 ka ago [28–30]. At higher altitudes in East Africa, forest expansion may have started earlier (perhaps by 12.5 ka ago) on the wetter mountains in the west of the region (e.g. Ruwenzori) than on the drier mountains in the east – where it could have been delayed until 9 ka ago [31].

Another cause of delayed spread of trees was proximity to slow-melting continental ice sheets. In southern Sweden and Finland, where Fennoscandian ice persisted until nearly 10 ka ago, birch woodland did not expand until 9.8 ka ago [32]. Further north, near the Arctic Circle in Finnish Lappland, woodland became established only about 8.5 ka ago [33]. Effects were even more pronounced in North America, where residues of the Laurentide Ice Sheet persisted long into the Holocene. In the centre of the continent, open spruce forest did not reach the latitude of southern Sweden (57°N) until 6.5 ka ago [34], and in the northeast, in Labrador, not until 4.4 ka ago [35].

In the northeastern United States, Laurentide ice did not finally melt from the higher ground of the White Mountains until about 11 ka ago. Scattered cold-hardy trees in the uplands (*Abies, Betula, Picea* and *Populus*) then quickly spread, to form a continuous forest cover by 10 ka ago. In the adjacent New England lowlands, by contrast, coniferous forest had been expanding northwards since 13 ka ago [36]. A similar retardatory effect may have operated in high-mountain regions at lower latitudes too, where long-persistent snow and ice could have delayed the upward spread of trees. In western North America, major increases in tree pollen (notably *Pinus*) in the pollen spectra from higher-altitude sites are not apparent before 13 ka ago [37] – at least 1 ka later than equivalent changes at some low-altitude sites.

There is limited evidence, based on the statistical analysis of lateglacial pollen patterns, that groups of taxa may in fact have moved together during this initial phase of sorting and re-assembly. In northeast Scotland the synchronous expansion of *Betula* (cf. *nana*), Cyperaceae, *Empetrum* and *Juniperus* about 11.6 ka ago [38] presumably documents the spread of shrub-tundra; in central Minnesota, *Artemisia, Betula, Fraxinus, Larix, Picea* and *Pinus* (presumed open coniferous forest) expanded together 13.6 ka ago [39], and in the New Guinea highlands the present-day montane forest taxa *Ascarina, Macaranga, Nothofagus* and *Phyllocladus* apparently all moved upslope together about 14 ka ago [27]. Although these data form an insufficient basis for drawing firm conclusions, they nevertheless indicate that interdependent species-assemblages could have been formed early in the protocratic phase.

11.3.2 Forest taxa in the Early Holocene

(a) Europe and eastern North America

The main phase of forest expansion in many parts of the world occurred in the Early Holocene, between about 10 and 8 ka ago. During this period

Table 11.1 Pollen records of woody taxa 8–10 ka ago in Europe and North America

	Site											
	1	2	3	4	5	6	7	8	9	10	11	12
Juniperus		**	+		+	+					+	–
Salix	**	**	+	+	+	+		+	+			–
Betula	**	**	**	**	+	+	**	**	+	**	+	+
Pinus	**	**	**	**	**	**	**	**	**	**	**	**
Corylus	+	**	**	**	**	**		+	+	+	+	
Ulmus	+	+	**	**	**	**	**	**	**	+	+	+
Quercus	+	+	**	**	**	**	**	**	**	**	**	**
Alnus	+	–	–	+	+	+	+	+	+	+	+	–
Fraxinus		–	–	+	+	+	–	+	**	+	+	**
Tilia			–	+	+	+			+			+
Picea				+		+	**	**	+	+	+	+
Fagus			–	–	–	+		–		+	+	+
Acer			–	–		+	–	+	**	+	+	+
Carpinus/Ostrya			–	–		+	+	+	**	**	**	**
Abies						**	+	**	+			+
Phillyrea					**							
Pistacia					+							
Larix							+	+	+	–		
Carya							+		**	+	–	–
Tsuga							–	+		**	+	+
Juglans									+	+	+	+
Platanus									+	+	+	+
Celtis									+			
Liquidambar										+	**	+
Magnolia											+	+
Nyssa											+	+
Castanea										–		+
Liriodendron												–
Aesculus												–

1 = S. Finland, 61°N [32]; 2 = N.W. Scotland, 58°N [40]; 3 = East Anglia, 52°N [41]; 4 = S. Poland, 50°N [42]; 5 = S.W. France, 45°N [43]; 6 = Dalmatia, 43°N [44,45]; 7 = N.W. Michigan, 46°N [46]; 8=N. Maine, 46°N [47]; 9 = N. Illinois, 42°N [48]; 10 = Pennsylvania, 41°N [49]; 11 = N. and S. Carolina, 34–36°N [50,51]; 12=Tennessee, 36°N [20].
** Major pollen type; +, minor pollen type; –, present today only.

forest communities spread over 8–10° of latitude and by up to 1000 m of altitude (Figure 8.9). Documentation of this spread requires a close network of adequately dated pollen diagrams (or packrat midden records), and is most complete for Europe and eastern North America. Accordingly it is this evidence that we consider first, before examining the far less complete data for other parts of the world.

Illustrative of the network of sites in Europe and eastern North America are those listed in Table 11.1. The twelve sites (listed for each region in latitudinal sequence from north to south) cover much of the potential forested area in the two regions, and are located in a range of modern forest types (Figure 11.1). Table 11.1 shows the woody pollen taxa recorded at these sites between 8 and 10 ka ago, classified as major and minor components of the pollen diagrams (continuously present and typically exceeding 5% total pollen, and sporadic low presence, respectively). The data give a broad picture of the floristic changes occurring during that time-interval. Immediately apparent is the much greater number of woody taxa in eastern North America than in Europe, a consequence of the progressive extinction of trees through the Pleistocene in Europe (Figure 10.1). In both land areas, however, nearly all the woody taxa that occur in the modern forests were present 8 ka ago – though many (e.g., *Fagus*) were then more restricted geographically.

The overall dynamics of the Early Holocene forest expansion in Europe are relatively straightforward. Three groups of taxa with different behaviours are apparent in Table 11.1: widespread taxa that were commoner at the more northerly sites 8 ka ago; widespread taxa that were less common at the more northerly sites; and taxa found only at the more southerly sites 8 ka ago. As the first group includes cold-tolerant trees and shrubs characteristic of modern boreal forests (*Betula, Juniperus, Pinus* and *Salix*), and the second group includes taxa of modern temperate broad-leaved forest (*Corylus, Quercus* and *Ulmus*), it is tempting to invoke a basic climatic control to these migration patterns. However, in the third group are included trees that are climatically as hardy as those in the second group (*Alnus, Fraxinus* and *Picea*), as well as trees that today dominate middle-altitude forests in the mountains of central Europe, above the lowland oak forests (*Abies, Fagus*). Group 3 trees must thus be seen as taxa that expanded their ranges rather slowly in the Early Holocene, not so much because of climatic restrictions as because of slow migration rates.

The three groupings probably do not represent species-assemblages responding as interdependent units to the changing environmental conditions of the Early Holocene. In detail, the picture is much more complicated, as the maps of migration patterns compiled by Huntley and Birks [16] reveal. These maps show largely individualistic responses by the

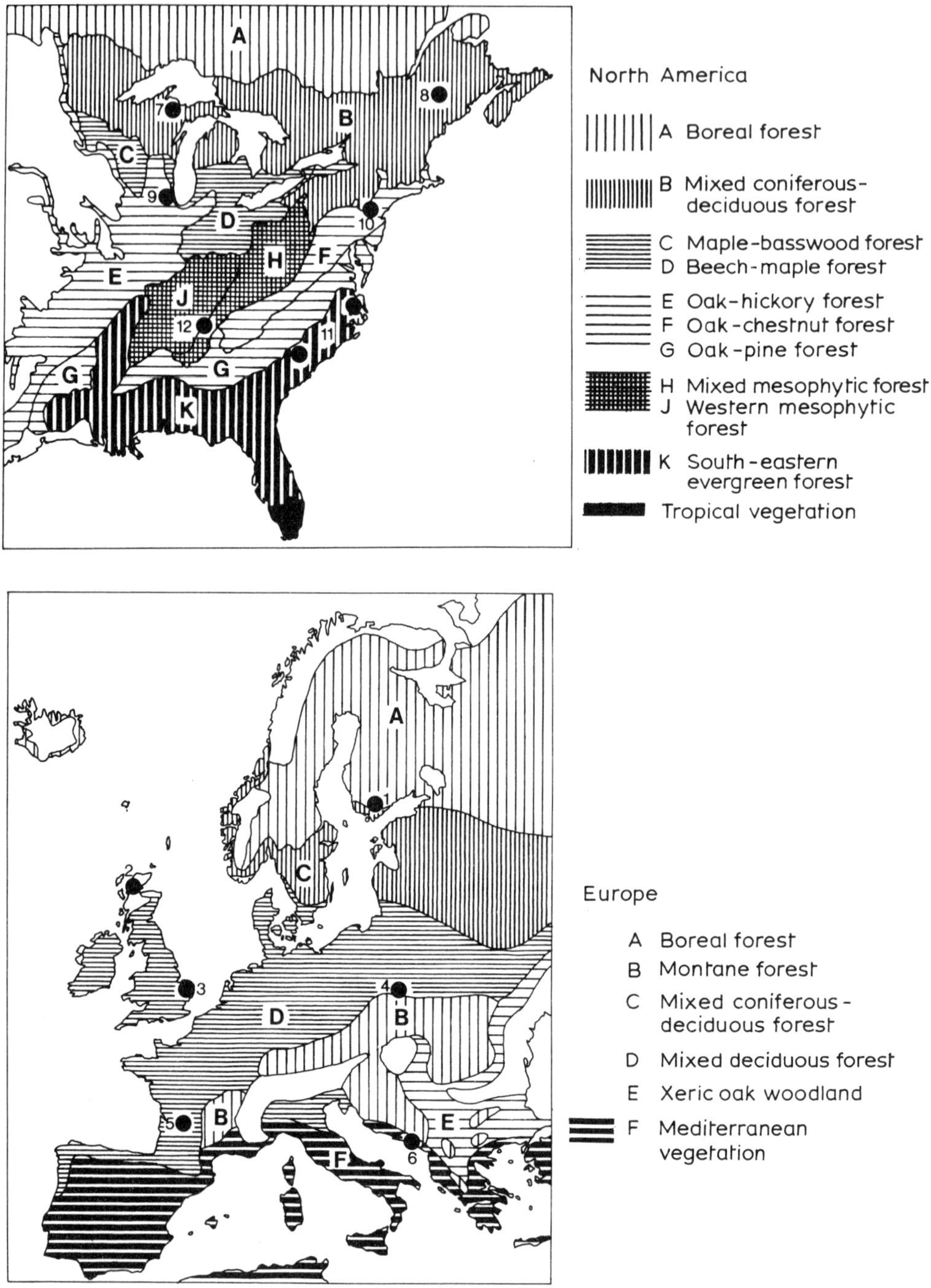

Figure 11.1 Location of sites in Table 11.1 in relation to modern forest types. Re-drawn from Huntley and Birks [16] and Watts [21].

various tree taxa, which can be explained in part by varying location of full-glacial refugia in relation to topographical barriers, and by varying dispersal efficiencies. Figure 11.2 shows simplified versions of some of Huntley and Birks' maps. The range-limits achieved by many trees 8 ka ago demonstrate how the massive east–west mountain systems in south-central Europe formed a major climatic and topographic barrier that penned in trees that had survived in full-glacial refugia in the Balkans and Italy. *Albies, Carpinus* and *Fagus* had not yet surmounted this obstacle by 8 ka ago; and *Acer* and *Fraxinus* were only just beginning to spread into the lowlands to the north at that time. However, *Tilia* and *Ulmus*, with apparently comparable refugia in the Balkans and Italy, spread rapidly to the north and west beyond the mountain barrier between 10 and 8 ka ago. It is difficult to relate such patterns of expansion to the presumed climatic tolerances of the seven tree taxa: *Fraxinus*, for example, ranges further north and west in Europe at the present-day than *Tilia*.

The dominant tree taxa in the Early Holocene forests of Europe were those with widely scattered full-glacial refugia (*Betula, Juniperus, Pinus* and *Salix*), or with western and eastern refugia adjacent to the lowland plains of central Europe. For these taxa the Pyrenees, Alps and Carpathians presented no barrier. *Corylus*, and to a lesser extent *Quercus*, expanded northwards most rapidly in the west, perhaps from refugia in the Biscay area. *Alnus* and *Picea*, by contrast, spread westwards from refugia somewhere in the east, and were late-comers to much of Europe. The maps in Figure 11.2 show that clear-cut regional differentiation in the floristic composition of European forest communities had been achieved by 8 ka ago, but although the total extent of forest was approaching stability, there was nevertheless within it considerable ongoing flux.

Comparable groupings of tree taxa to the European ones can be recognized in the lists in Table 11.1 for the six eastern North American sites. Taxa characteristic of the modern Canadian boreal forests (*Abies, Larix* and *Picea*) and of the mixed conifer–hardwood forests of the Lake States (*Betula* and *Ulmus*) were commonest at the more northerly sites 8 ka ago; *Quercus* and *Pinus* spp. were common at all sites; and a range of taxa were commoner in the south and east of the United States (*Acer, Carpinus/Ostrya, Fagus, Fraxinus, Tilia, Tsuga*, and the North American 'exclusives' in Table 11.1). Maps of migration pathways [50,52] (Figure 11.3) show, as for Europe, a diversity of behaviour by the migrating trees. Between 10 and 8 ka ago many broadleaved tree taxa spread at differing rates and by different routes throughout the whole of the eastern United States (cf. Figure 8.5), from the 'mixed mesic forests' established in the south-east during the lateglacial (Figure 11.1). Some trees (notably *Fraxinus, Populus* and *Ulmus*) were more abundant during that period than they are today [53]. Although the Appalachian Mountains did not form a major obstacle to plant migration as

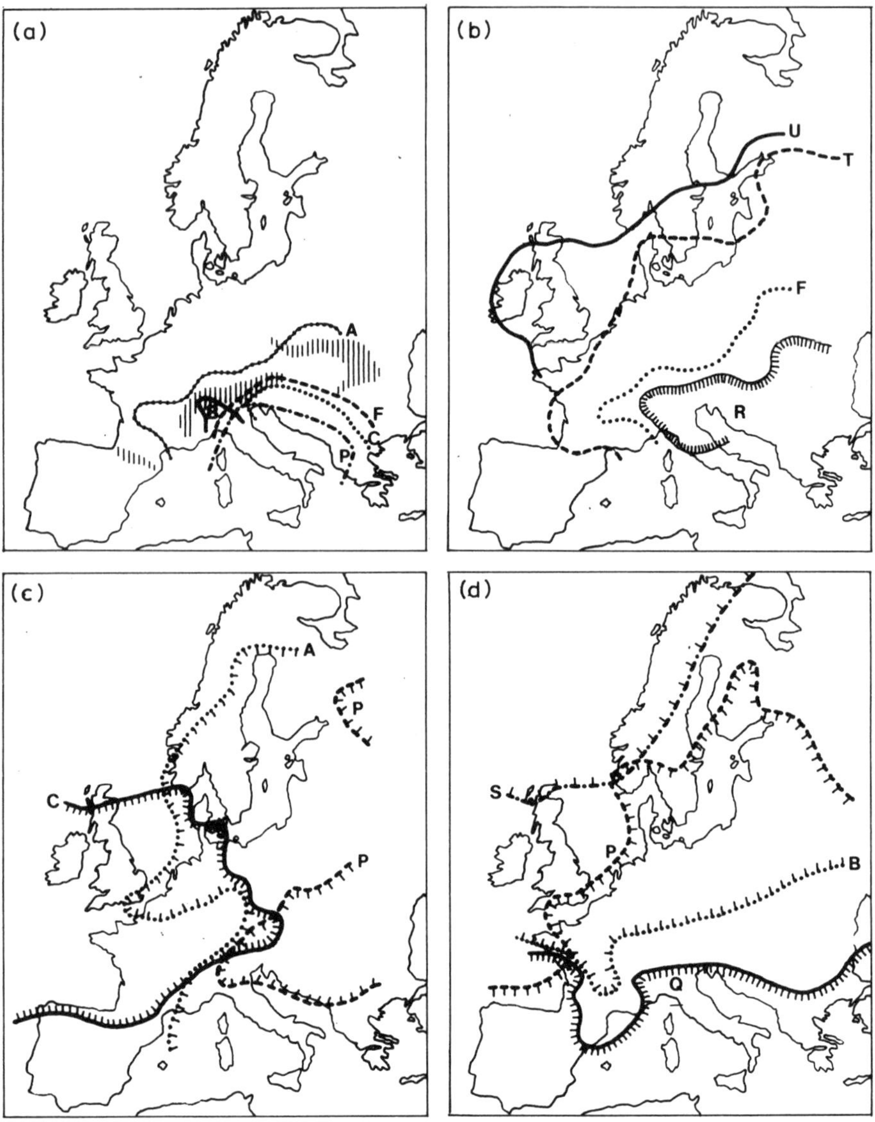

Figure 11.2 (a) Northern range-limits 8 ka ago of *Acer* (A); *Abies* (B); *Carpinus* (C); *Fagus* (F); and *Phillyrea* (P). Shading denotes mountain regions; (b) Northern range-limits 8 ka ago of *Ulmus* (U); *Tilia* (T) and *Fraxinus* (F); R is the range of these taxa 10 ka ago; (c) Eastern range-limits of *Corylus* (C); and western range-limits of *Alnus* (A); and *Picea* (P) 8 ka ago; (d) Southern limits of *Salix* ≥10% (S) and *Betula* ≥25% (B), and northern limits of *Pinus* ≥10% (P) and *Quercus* ≥10% (Q) at 10 ka ago. Re-drawn from Huntley and Birks [16].

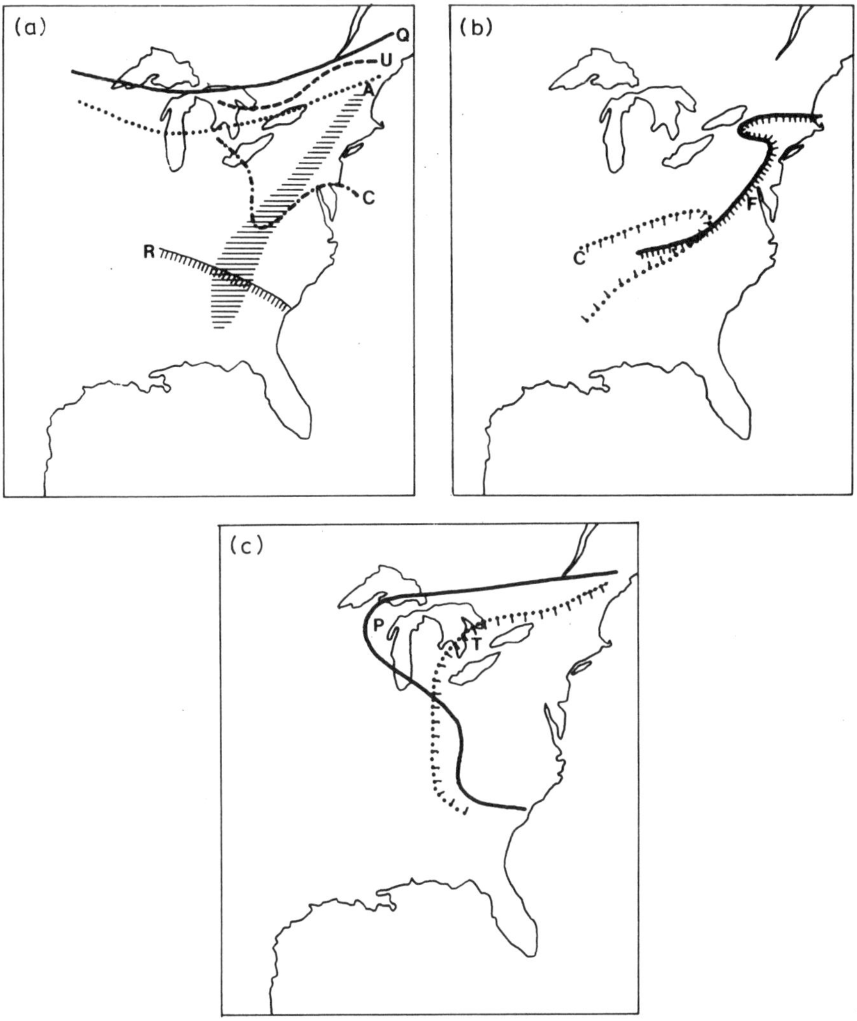

Figure 11.3 Range limits of tree taxa in eastern North America 8 ka ago. (a) Northern Limits of *Quercus* (Q); *Ulmus* (U); *Acer* (A); and *Carya* (C); R is the northern limit of these taxa 13 ka ago. The Appalachian Mts are shaded. (b) Northern and western limits of *Fagus* (F); and northern and eastern limits of *Castanea* (C); (c) Northern and western limits of *Pinus strobus* (P); and *Tsuga canadensis* (T). Re-drawn from Davis [50].

did the Alps and the Carpathians in Europe, nevertheless they did influence the pathways followed by many tree taxa. Thus *Acer*, *Carya*, *Quercus* and *Ulmus* reached the Lake States in the Early Holocene by direct northward migration along the western side of the Appalachians. *Fagus*, by contrast, migrated northwards largely to the east of the Appalachians in the Early Holocene, and reached the Lake States only in the mid-Holocene by turning west across New York State. The northward movement of *Carya* was noticeably slower along the Appalachian chain than to the east and west of it, whereas *Castanea* migrated almost entirely along it in the Early and mid-Holocene.

Other tree taxa spread into the Lake States, like *Fagus*, from the east (*Pinus banksiana*, *P. strobus* and *Tsuga canadensis*), but from refugia that were almost certainly situated offshore of the present-day Coastal Plain of New Jersey and Virginia. Yet other tree taxa (*Liquidambar*, *Liriodendron*, *Magnolia* and *Nyssa*, for example) spread only slowly northwards, if at all, from their southern refugia. By 8 ka ago, therefore, the forests of eastern North America consisted of a series of floristically differentiated regions, achieved by plants responding individualistically to climatic change by migrating at different speeds along different pathways [52] (Figure 8.5). A state of stability, however, had not yet been reached. The picture is strikingly similar to that for the European forests.

(b) Western North America

The picture of Early Holocene changes that has been reconstructed for the coniferous forests and woodlands of the western United States is less complete than that for the eastern forests. Large parts of the region, as mentioned previously (section 10.4.3), either lack a close network of investigated sites or else possess important woody taxa that cannot be identified unambiguously in the pollen record. Nevertheless, the picture revealed so far highlights controlling factors similar to those in the east: proximity of full-glacial refugial populations, efficacy of topographic barriers, and variable rates of migration. We concentrate here on the evidence for the far northwest and the intermontane region (between the Sierra–Cascades and the Rocky Mountains).

The woody taxa that were probably most restricted in range at full glacial in the far northwest were the massive conifers that now dominate the lowland forests on the western flanks of the Coast Ranges and the Sierra–Cascades. The presence of refugial populations of several of these taxa at 46°N in the Puget Lowlands of Washington and Oregon 15–20 ka ago [54] clearly made possible the rapid development after 10 ka ago of coniferous forests of largely modern aspect in northwestern Washington: with dominant *Picea sitchensis*, *Pseudotsuga menziesii* and *Tsuga heterophylla*, accompanied by *Abies* (cf. *grandis*) and *Thuja plicata* [37,54]. Similar forests must undoubtedly have been present further south also. However, the mountain barrier of the Cascades, coupled with increasing climatic dryness inland, apparently prevented any of these taxa from spreading eastwards in the Early Holocene, for sites in northeastern Washington that today are covered by forests of *Pseudotsuga*, *Tsuga* and *Abies* were dominated then, and for several millennia subsequently, by open *Pinus contorta* forests [37].

Elsewhere in the western United States the picture is complicated by altitudinal as well as latitudinal shifts in the ranges of the major tree taxa [55]. Between 12 and 9 ka ago a broad altitudinal belt of forest and woodland, that during the full glacial extended very approximately from 500 to 2500 m altitude (Figure 10.8), moved rapidly upslope, to occupy its present position (about 1000 m higher) by about 9 ka ago [56,57]. The lower altitudes progressively vacated by the full-glacial pinyon–juniper woodland after 11 ka ago were taken over ultimately by desert scrub, as the climate became drier in the Early Holocene (Figure 13.2), but continued to be dominated by juniper alone (*Juniperus osteosperma*) in the Mojave Desert for at least 1 ka after its accompanying pinyon pines (*Pinus monophylla*) had gone; by 7.8 ka ago, however, the replacement was complete [9]. On the western slopes of the Sierra Nevada, xeric oak woodland spread rapidly into the pinyon–juniper zone after 11 ka ago [6,58].

Pinyon–juniper woodland also migrated north into the higher-lying Great Basin. Montane conifers there, markedly restricted in range at full glacial (Figure 10.9), spread latitudinally at varying speeds through the upwardly-moving tree belt, to interpose a new montane forest zone in many regions between the expanding pinyon–juniper woodland and the contracting subalpine forest above it. Limber pine (*Pinus flexilis*) was probably the most successful of these taxa, spreading to nearly all the Great Basin ranges between 12 and 8 ka ago, from full-glacial 'staging posts' around the southern and western margins of the region [57]. *Abies concolor*, *Juniperus scopulorum*, *Pinus ponderosa* and *Pseudotsuga menziesii* migrated northwards more slowly into the Great Basin. In the initial stages of this migration, nearly continuous forests must have linked the various separate mountain peaks and ranges in the central part of the Great Basin to the main Rocky Mountain area to the east, along a chain of high connecting ridges in the north and south [57], but ultimately this forest link was severed as the lower tree limit continued to move upwards. *Pseudotsuga* and *J. scopulorum*, but probably not *Abies concolor* and *Pinus ponderosa*, had penetrated some way into the Great Basin along these migration routes by 8 ka ago. Subalpine conifers that were wide-ranging over the Great Basin at full glacial (*Pinus longaeva*, *Picea engelmannii* and *Juniperus communis*), underwent continuing contraction of range through the Holocene, but the failure of the montane conifers to penetrate into the west of the Basin (beyond longitude 116°W) has left these subalpine conifers even today in undisputed control of an altitudinal interval that in the east is partly occupied by the montane conifers. Remoteness from full-glacial montane-forest refugia, absence of low-altitude habitats, an increasingly arid climate westwards, and the isolation of the forested peaks in an investing 'sea' of desert and semi-desert, have all been contributory factors in protecting the western Great Basin forests from change from their full-glacial state [57]. Even the restricted ranges of certain subalpine conifers in the Great Basin at the present day – *Pinus albicaulis* in the northwest and *Abies lasiocarpa* in the northeast – probably perpetuate full-glacial rather than Holocene migrations into the region.

The Great Basin is almost certainly atypical in the protracted scale of its Holocene vegetation changes. For other mountain regions of the western United States the evidence is much more fragmentary. Nevertheless, patterns of change similar to those in the Great Basin, though over a speeded-up time-interval, probably occurred. The extent to which modernization of the vegetation had proceeded by the mid-Holocene is considered in the next chapter.

(c) Other temperate forests

There are three other regions of the world where temperate forest communities form part of a north/south vegetational gradient that extends over a latitudinal range similar to that in Europe and North America, and where extensive pollen-analytical investigations have been carried out. These regions are Japan, New Zealand and Chile – where glaciation was confined to the higher 'backbone' mountains, and, because of the prevailing oceanic character of the climate of all three regions, refugial tree populations were probably widely distributed at full glacial. For the very important temperate forest communities of China, however, there have been as yet insufficient studies to enable any generalizations to be made (see the series of papers by Professor D. Walker and co-workers in *Journal of Biogeography* **13**, 1986).

Pollen diagrams over much of the Japanese Archipelago (Figure 11.4(a)) show a dramatic replacement of the boreal and temperate conifers that were predominant at full glacial (species of *Pinus, Picea, Abies* and *Tsuga*), by broadleaved 'temperate elements' at the start of the Holocene, about 10.5–9.5 ka ago [14,59]. At this stage the archipelago presented a continuous land surface, with the main island of Honshu linked to the smaller islands of Hokkaido in the north and Kyushu and Shikoku in the south, as a result of lowered sea levels. The replacement of cold-stage conifers was achieved largely by rapid spread of *Fagus* and *Quercus*, with other taxa (*Carpinus, Castanea, Celtis, Cryptomeria, Juglans, Pterocarya, Sciadopitys, Tilia,*

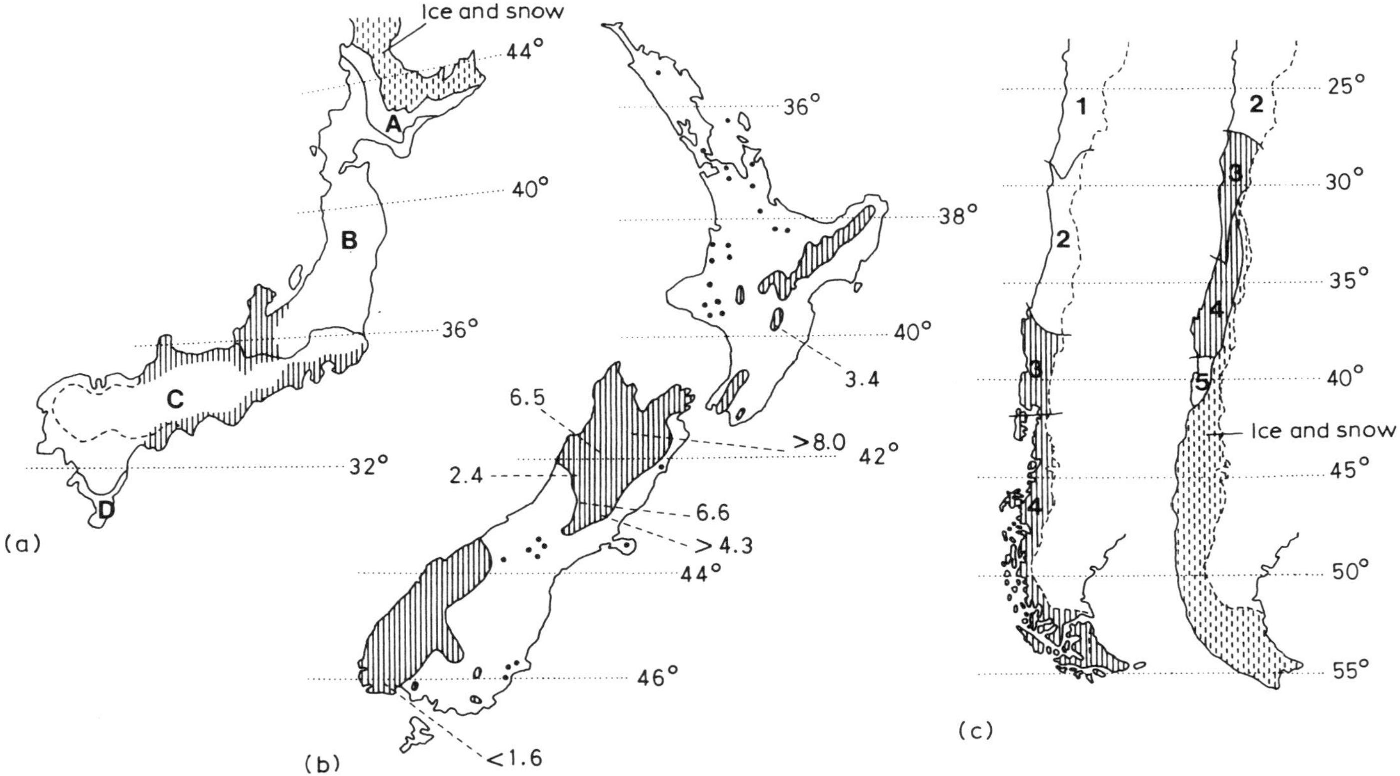

Figure 11.4 Temperate forests in the Early Holocene. (a) Vegetation types in the Japanese Archipelago, and range of *Cryptomeria* (shaded) 8 ka ago. A = tundra and parkland; B = boreal conifer forest; C = temperate conifer forest; D = broadleaf evergreen forest. (b) Present-day *Nothofagus* forests in New Zealand (shaded and dots), and approximate time of origin (ka ago). (c) Climatic regions of Chile at the present-day (left) and at full glacial (right); areas influenced by westerly winds are shaded. 1 = subtropical arid; 2 = semi-arid, summer-dry; 3 = temperate, humid; 4 = cool temperate, humid; 5 = polar. Re-drawn from Tsukada [13], Wardle [60] and Heusser [61].

Ulmus and *Zelkova*) expanding more slowly. All these taxa were probably relatively minor components of the temperate conifer forest at full glacial, with their main populations in the extreme southwest (to the west of the dashed line in Figure 11.4(a)); however, *Fagus* almost certainly survived locally throughout Honshu, as far north as latitude 38°N, and *Quercus* probably to 43°N in Hokkaido [13,62]. No major migrations were thus involved in the rapid expansion of *Fagus* and *Quercus* after 10.5 ka ago. Of the other temperate taxa, *Cryptomeria* at least was much more restricted in range at full glacial (Figure 8.12), and Early Holocene spread was predominantly upwards, to higher altitudes, and not latitudinal. The approximate extent of *Cryptomeria* 8 ka ago is shown by the shaded area in Figure 11.4(a) – when populations occurred up to about 1500 m altitude. Only when such higher-altitude populations were established did rapid latitudinal spread take place (between about 7 and 4 ka ago) – almost entirely at these higher altitudes [14]. The earlier spread of *Fagus* and *Quercus* was probably similarly at higher altitudes, and displacement of the 'full-glacial' temperate conifers from the lower altitudes resulted largely from the slower spread northwards after 8.5 ka ago of warm–temperate evergreen taxa such as *Cyclobalanopsis* from their full-glacial refugia in the extreme south of the archipelago (zone D in Fig. 11.4(a) [13]).

In New Zealand, shrublands and grasslands that predominated at full glacial were replaced during the lateglacial period by podocarp–hardwood forest (principally of *Podocarpus spicatus, Dacrydium cupressinum* and *Weinmannia racemosa*). This replacement occurred before 14 ka ago in central North Island [25], by 13 ka ago in the south of North Island [63], and between 10.4 and 9.4 ka ago in South Island [64,65], and was achieved very rapidly, perhaps in as short a period as 400 years [65,66]. The inference must be that refugial tree populations survived through the full glacial close to every site so far investigated [67]. Moreover, nearly all the tree taxa that are present today were involved in this early forest expansion, and over considerable areas of New Zealand little change occurred subsequently in the podocarp–hardwood forests [66]. In other areas, however, there was a slow rise to prominence of *Nothofagus* which was clearly diachronous, ranging in time from 8.5 to 1.5 ka ago (Figure 11.4(b)). Although the earlier dates come generally from more northerly sites, no obvious latitudinal pattern in the spread of *Nothofagus* can be discerned; in view of the slow rates of spread that have been calculated for *Nothofagus*, it has been suggested that the present-day disjunct distribution of *Nothofagus* forests in New Zealand [60] is a direct consequence of extreme fragmentation of range at full glacial [68].

In Chile, the pollen evidence shows clearly that both climatic zones and vegetation zones were displaced northwards at full glacial (Figure 10.15). Rapidly rising temperatures after about 14 ka ago led to a gradual reversal of this displacement, and modern zone positions were probably assumed by the end of the Early Holocene. Figure 11.4(c) shows the likely extent of these changes, based on the maps of Heusser [61] and Ab'Sáber [69]. It is apparent that sites in central Chile, north of about latitude 39°S, would have become progressively drier through the Early Holocene, but that sites to poleward of latitude 39°S would have become progressively more forested as the zone of rain-bearing westerly winds was displaced southwards. The gradual increase in *Nothofagus* pollen at Puerto del Hambre at latitude 53° 36′S [61], and Isla Grande de Chiloé at latitude 42°S [70], and of macroremains of the evergreen *N. betuloides* at the cave of Ultima Esperanza at latitude 51° 35′S [71], between 11 and 8 ka ago, is evidence of the latter change, while the contemporaneous replacement of *Nothofagus* pollen successively by Myrtaceae, *Aextoxicon* and Graminaeae pollen at Rucañancu at 39° 33′S [72] records the antithetic change from temperate rainforest to deciduous forest and then matorral as the climate became drier. Major latitudinal, and probably also altitudinal, migrations of taxa must have been involved, but the existing information is too scanty to be able to assess the relative importances of these two processes, and whether or not the various tree taxa behaved individualistically, as in Europe and North America.

(d) Tropical forests

From what is known of ice-age (hypothermal) conditions in the tropics (section 10.5) it might be expected that resurgence of forest in these regions in the Early Holocene would have followed two patterns: upward expansion to higher altitudes, and radial expansion from full-glacial centres of survival. The rather limited pollen evidence is adequate to demonstrate that these processes did indeed occur, but is as yet inadequate to fill in the picture with the sorts of detail that have been worked out for temperate regions – though tantalizing glimpses are beginning to emerge of comparable complexity in the patterns of forest spread in both temperate and tropical regions.

The pioneer studies of J. A. Coetzee [73] on Mount Kenya in East Africa remain among the clearest demonstrations of the effects of altitude on the pollen record. The present-day vegetation of the East African mountains is ordered into a series of altitudinal belts (Table 11.2), which can be distinguished to some extent on the basis of their contemporary pollen spectra [28]. A simplistic view of the full-glacial environment is that these vegetational belts were positioned lower down on the mountains, and subsequently moved upwards again as temperatures rose between 13.5 and 8 ka ago. The successive high values over this time-interval in the pollen diagram from Sacred Lake at 2400 m altitude on Mount Kenya

Table 11.2 Altitudinal zonation of vegetation on East African mountains at the present day [28]

Altitudinal range (m)	Vegetation Belt	Vegetation Zone
>3500–3600	Afroalpine Belt	
3000–3700	Ericaceous Belt	
2000–3450	Montane Forest Belt	
	2800–3450 m	Upper Montane Forest Zone (*Hagenia–Hypericum* Zone)
	2450–3200 m	Bamboo Zone ('moist' mountains only)
	2000–3000 m	Lower Montane Forest Zone
<2150	Lowland Forest Belt (savanna vegetation on 'dry' mountains)	

(Figure 11.5), for grasses and ericaceous shrubs, the small tree *Hagenia*, and various evergreen tree taxa, could be interpreted as resulting from the presence around the site successively of non-forest vegetation, treeline vegetation and then montane forest – i.e., a progressive upward movement of the Early Holocene equivalents of the modern Ericaceous Belt, the *Hagenia–Hypericum* Zone and the Montane Forest Zone. At Lake Rutundu, higher up the mountain at 3140 m altitude, tree pollen remained low over the same time-interval, and the pollen diagram is dominated successively

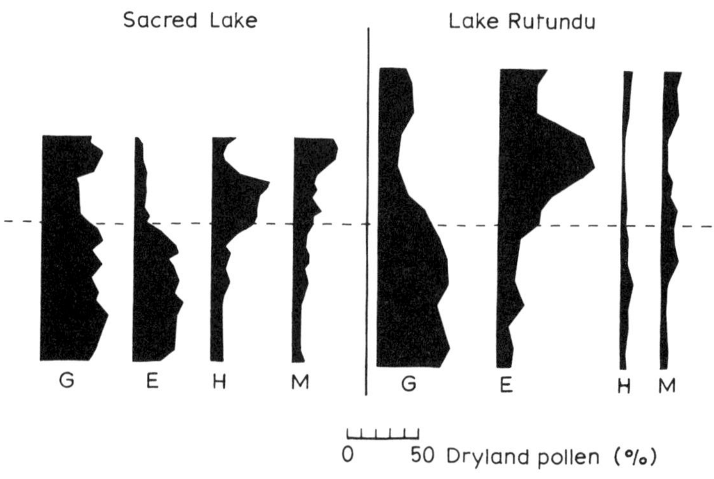

Figure 11.5 Pollen values for Gramineae (G), Ericaceae (E), *Hagenia* (H) and montane forest trees (M) at two sites on Mount Kenya for the time-period 14–7 ka ago approximately. The dashed line is a radiocarbon-dated horizon of 10.6–10.8 ka ago on both diagrams. Re-drawn from Flenley [31].

by grass pollen (high also in modern pollen spectra from afroalpine vegetation) and Ericaceae pollen. About 12 ka ago, therefore, afroalpine vegetation probably occurred around Lake Rutundu and ericaceous scrub around Sacred Lake, whilst by 10 ka ago the ericaceous scrub had moved upwards to the Lake Rutundu area, and Sacred Lake was below the treeline. Forest must thus have spread upwards by several hundred metres during this time-period. The summary diagrams in Flenley [31] show upward movements of the treeline between 14 and 8 ka ago of 700–800 m in East Africa, 1500 m in the Andes, and at least 1500 m in the New Guinea highlands.

Pollen evidence from New Guinea, however, suggests that the full-glacial forest vegetation was less differentiated altitudinally than it is today (Figure 8.9), and that the modern forest zonation developed only in the Early Holocene. It is believed that the modern montane forest species there are a subset that segregated from the full-glacial pool of species in the Early Holocene, in response to the rapid availability then of new habitats for tree growth at progressively higher altitudes. The subset is thus seen as a largely fortuitous assemblage of opportunist, efficiently-dispersed species rather than one specialized to higher-altitude conditions. In support of this view is the fact that few of the taxa characteristic of montane forests generally in the tropics today are restricted to higher altitudes; most occur widely at lower altitudes also, though there they are confined to forest edges and clearings, and to secondary forest [28,31]. Many lowland taxa, on the other hand, occur only at low altitudes, and presumably lack the efficient dispersal mechanisms that would have enabled them to take advantage of the new habitats opened up in the Early Holocene.

For East Africa, similarly, there are differences between pollen diagrams from the various mountain regions in the Early Holocene that suggest that some of the floristic heterogeneity observable today in the montane forests could have originated during this initial period of differential range expansion by the full-glacial pool of species. Thus the higher values of *Podocarpus* pollen at the Cherangani Hills site than at Ruwenzori, which experiences higher rainfall today, and, conversely, the increased prominence of *Celtis*, *Myrsine* and Urticaceae at Ruwenzori, might indicate an early differentiation of 'dry' and 'moist' forest types [31]. Such differentiation would presumably have been in response to climatic gradients which are still a feature of the East African mountains today. Other floristic variation in modern montane forests that is less obviously related to environmental heterogeneity might nevertheless have been generated in the Early Holocene too by the patchy colonization of the new montane habitats by different tree taxa [31].

Modern distributional patterns have also formed one of the lines of evidence used to locate possible full-glacial forest refugia in the tropics (sections 8.5 and 10.5). The assumption is that tree taxa spread at differing rates over the lowlands from these refugia in the Holocene, to give present-day ranges that are either continuous over large areas or fragmented to varying extents. Although the assumption is a perfectly reasonable one on the basis of evidence from temperate forests, there is almost no direct supporting evidence from the tropics. Indications that tree taxa did indeed move at different speeds in the Early Holocene come from two areas only: East Africa and northeastern Australia. In northeastern Australia, rainforest was almost entirely replaced by dry sclerophyllous woodland at full glacial (section 10.5), when the pollen spectra at three sites on the Atherton Tableland, Queensland (within 20 km of each other), were dominated by *Casuarina* and *Eucalyptus*-type. Pollen values of rainforest taxa (Cunoniaceae, *Elaeocarpus, Freycinetia Macaranga/Mallotus* and *Rapanea*) increased in the Early Holocene, but the rise occurred at different times at the three sites: at 9.5–8.4, 8.0–7.5 and 7.0–6.0 ka ago [31]. The implication is of slow spread from a limited number of source areas – a suggestion which is supported by the failure of certain tree taxa (notably *Araucaria* and *Dacrydium*) that were prominent in earlier cold-stage forests on the Atherton Tableland to return after the dry woodland period [12]. In East Africa, Kendall's pollen diagram [74] from the north end of Lake Victoria shows a sequential spread of a range of tree taxa over the period 12.5–8.5 ka ago. Prominent between 12.5 and 10.0 ka ago were certain taxa characteristic of modern semi-deciduous forests in East Africa (e.g., *Diospyros*, Oleaceae, Rhamnaceae and Rutaceae), whilst taxa of evergreen forests increased only after 9.5 ka ago (e.g., *Alchornea, Bosqueia, Mallotus* and *Tetrorchidium*). This sequence of change in the pollen diagram was clearly in response to a progressive increase in climatic wetness between 13.5 and 9.5 ka ago, but varying rates of migration could also have been involved in some of the changes.

(e) Resumé

What is here called the protocratic phase was a time of rapid forest expansion, when, within a span of a few thousand years, forest communities spread to cover nearly 60% of the earth's land surface (cf. Table 1.2). This spread occurred under conditions of major climatic change, when the rate of change could have outstripped the capacity of some plants to respond conformably. As more sites are studied, so it becomes increasingly likely that the major tree taxa migrated independently of each other, and that integrated and interdependent ecosystems did not exist for any length of time. Two opposing views have been propounded about the patterns of change in this Protothermal period [53,75,76]: that rates of plant migration were controlled by rates of climatic change, and accordingly that plant ranges were in equilibrium with the climate at any point in time; or, alternatively, that plant migrations, although triggered by climatic change, proceeded at speeds governed primarily by the intrinsic dispersal capabilities of the plants themselves. In the latter case, plant ranges would not necessarily have been in equilibrium with the climate. Moreover, there is still dispute as to the degree of interdependence shown by the populations of the migrating tree taxa. One view is that the major forest types in eastern North America persisted as recognizable entities through the climatic vicissitudes of the last 20 ka, though their spatial locations and floristic composition changed [52]; the alternative view is that 'ecosystems and plant assemblages are to the biosphere what clouds, fronts, and storms are to the atmosphere . . . They are features that come and go' [53]. The debate is ongoing [92], and further detailed evidence of migration patterns in areas other than Europe and North America is clearly desirable.

It is clear from the preceding pages that our state of knowledge varies enormously from one country to another, and that any conclusions about global patterns of change are premature. Nevertheless, the existence of plant distribution patterns in, for example, tropical regions today that can be explained in terms of processes that are known, or believed, to have operated in temperate regions in the protothermal period is perhaps an indication that common patterns and processes of change could have existed then.

11.4 THE FOREST MAXIMUM

After the environmental and vegetational instability of the Early Holocene, the earth entered a period of relative calm lasting from about 8 to 5 ka ago. The evidence reviewed in Chapter 6 suggests that nearly all the major glaciers were quiescent during this period, before experiencing a surge of renewed activity beginning about 5.3 ka ago. Fluctuations of lake levels similarly became more pronounced after 4.5 ka ago. A variety of lines of evidence, moreover, indicates that northern-hemisphere temperatures were higher 8–5 kg ago than they are today by 2–3°C. Comparison with the reconstructed temperature curves for previous interglacials (Figure 5.3) leads to the conclusion that the **Altithermal period** of 8–5 kg ago represented the **thermal maximum** of the present interglacial (the Holocene), delayed in many regions from its Milankovitch-predicted date of 12–9 ka ago by slow-melting ice residues, and that the renewed environmental instability that developed about 5 ka ago marked its close. The thermal maximum was also a forest maximum. During it, forests were more extensive globally than at any time since the last interglacial, more than 100 ka ago. Evidence of this forest maximum comes from high latitudes and high altitudes, where long-dead (and sometimes burnt) tree remains beyond the present treeline register its former limits. In arctic regions, trees advanced up to 400 km nearer to the North Pole. In Siberia, fossil timber in present-day tundra vegetation dates variously from 9.0 to 2.5 ka ago, but the most northerly remains, 300–400 km beyond the present treeline, are 6.0–4.9 ka old [77]. In central Canada, burnt tree remains 3.4–3.6 ka old occur 250–280 km north of the present treeline – but the trees may in fact have been established much earlier (Figure 8.6 [78,79]). In the far northwest of Canada the fossil outposts (50–60 km north of the modern treeline) are 7.5–4.9 ka old [79]. In mountain regions, evidence for the former extension of trees to higher altitudes comes not only from preserved timber but also from insect remains [56] and pollen grains [80,81]. In the White Mountains of California trees occurred 120–150 m altitude higher 5.8–3.1 ka ago [82]; in the southern Sierra Nevada, 70 m higher 6.5–3.5 ka ago [83]; and in the Cairngorms, Scotland, perhaps 90–130 m higher 7.4–4.6 ka ago [84,85]. Estimates for the European Alps, based on pollen analyses, range from 100 m to 400 m altitude higher [86].

At lower altitudes and latitudes, similar, or even greater, displacements of range undoubtedly occurred during the Altithermal period, but the evidence is rather scanty. Both individual taxa and species-assemblages were affected. The southern border of mixed conifer–hardwood forest in the Great Lakes region, for example, was probably 100–150 km north of its present position 6 ka ago [87]; lowland mixed oak forest occurred up to 700 m altitude higher in the Jura Mountains of France [88]; and the northern boundary of rainforest in Nigeria may have been displaced polewards by 20–120 km [89]. *Corylus* occurred up to 400 km north of its present limits in the forests of Fennoscandia [90]; and *Pinus strobus* and *Tsuga canadensis* ranged 300–400 m altitude higher in the forests of the White Mountains of New England 7–5 ka ago [91].

Any vegetational stability at the forest maximum, however, was only superficial and temporary. Migration of individual taxa continued to occur, the extent of human influence on the forest continued to grow, and environmental instability increased again after 5.3 ka ago. The effects of these processes are considered in the next chapter.

REFERENCES

1. Berger, W. H., Killingley, J. S. and Vincent, E. (1987) Time scale of the Wisconsin/Holocene transition: oxygen isotope records in the western equatorial Pacific. *Quat. Res.*, **28**, 295–306.

2. Ruddiman, W. F., Sancetta, C. D. and McIntyre, A. (1977) Glacial/interglacial response rate of subpolar North Atlantic waters to climatic change: the record in ocean sediments. *Phil. Trans. R. Soc. Lond. B*, **280**, 119–42.
3. Dansgaard, W., Johnsen, S. J., Møller, J. and Langway, C. C. Jr (1969) One thousand centuries of climatic record from Camp Century on the Greenland ice sheet. *Science*, **166**, 377–81.
4. Lorius, C., Merlivat, L., Jouzel, J. and Pourchet, M. (1979) A 30,000-yr isotope climatic record from Antarctic ice. *Nature*, **280**, 644–8.
5. Atkinson, T. C., Briffa, K. R. and Coope, G. R. (1987) Seasonal temperatures in Britain during the last 22,000 years, reconstructed using beetle remains. *Nature*, **325**, 587–92.
6. Davis, O. K. and Moratto, M. J. (1988) Evidence for a warm dry Early Holocene in the western Sierra Nevada of California. *Madroño*, **35**, 132–49.
7. Kutzbach, J. E. and Guetter, P. J. (1986) The influence of changing orbital parameters and surface boundary conditions on climatic simulations for the past 18,000 years. *J. Atmos. Sci.*, **43**, 1726–59.
8. Spaulding, W. G. and Graumlich, L. J. (1986) The last pluvial climatic episodes in the deserts of southwestern North America. *Nature*, **320**, 441–4.
9. Van Devender, T. R., Thompson, R. S. and Betancourt, J. L. (1987) Vegetational history of the deserts of southwestern North America; the nature and timing of the Late Wisconsin–Holocene transition, in *North America and Adjacent Oceans during the Last Deglaciation* (eds W. F. Ruddiman and H. E. Wright Jr), Geological Society of America, Boulder, CO, pp. 323–52.
10. Birks, H. J. B. (1986) Late-Quaternary biotic changes in terrestrial and lacustrine environments, with particular reference to northwest Europe, in *Handbook of Holocene Palaeoecology and Palaeohydrology* (ed. B. E. Berglund), Wiley, Chichester, pp. 3–65.
11. Pennington, W. (1986) Lags in adjustment of vegetation to climate caused by the pace of soil development: evidence from Britain. *Vegetatio*, **67**, 105–18.
12. Kershaw, A. P. (1978) Record of last interglacial/glacial cycle from northeastern Queensland. *Nature*, **272**, 159–61.
13. Tsukada, M. (1982) *Cryptomeria japonica*: glacial refugia and late-glacial and postglacial migration. *Ecology*, **63**, 1091–105.
14. Tsukada, M. (1986) Altitudinal and latitudinal migration of *Cryptomeria japonica* for the past 20,000 years in Japan. *Quat. Res.*, **26**, 135–52.
15. Huntley, B. (1988) Europe, in *Vegetation History* (eds B. Huntley and T. Webb III), Kluwer Academic Publishers, Dordrecht, pp. 341–83.
16. Huntley, B. and Birks, H. J. B. (1983) *An Atlas of Past and Present Pollen Maps for Europe: 0–13,000 Years Ago*, Cambridge University Press, Cambridge.
17. Watts, W. A. (1980) Regional variation in the response of vegetation to Lateglacial climatic events in Europe, in *Studies in the Lateglacial of Northwest Europe* (eds J. J. Lowe, J. M. Gray and J. E. Robinson), Pergamon, Oxford, pp. 1–21.
18. Pennington, W. (1975) The Late Devensian flora and vegetation of Britain. *Phil. Trans. R. Soc. Lond. B*, **280**, 247–71.
19. Jalut, G., Delibrias, G., Dagnac, J. *et al.* (1982) A palaeoecological approach to

the last 21,000 years in the Pyrenees: the peat bog of Freychinède (alt. 1350 m, Ariège, South France). *Palaeogeogr. Palaeoclimatol. Palaeoecol.*, **40**, 321–59.

20. Delcourt, H. R. (1979) Late Quaternary vegetation history of the eastern Highland Rim and adjacent Cumberland Plateau of Tennessee. *Ecol. Monogr.*, **49**, 255–80.
21. Watts, W. A. (1983) Vegetational history of the eastern United States 25,000 to 10,000 years ago, in *Late-Quaternary Environments of the United States*, Vol. 1, *The Late Pleistocene* (ed. S. C. Porter), Longman, London, pp. 294–310.
22. Heusser, C. J. (1983) Vegetational history of the northwestern United States including Alaska, in *Late-Quaternary Environments of the United States*, Vol. 1, *The Late Pleistocene* (ed. S. C. Porter), Longman, London, pp. 239–58.
23. Birks, H. J. B. (1976) Late-Wisconsinan vegetational history at Wolf Creek, central Minnesota. *Ecol. Monogr.*, **46**, 395–429.
24. Heusser, C. J. (1983) Quaternary pollen record from Laguna de Tagua Tagua, Chile. *Science*, **219**, 1429–32.
25. McGlone, M. S. and Topping, W. W. (1977) Aranuian (post-glacial) pollen diagrams from the Tongariro region, North Island, New Zealand. *N. Z. J. Bot.*, **15**, 749–60.
26. Newsome, J. and Flenley, J. R. (1988) Late Quaternary vegetational history of the Central Highlands of Sumatra. II. Palaeopalynology and vegetational history. *J. Biogeogr.*, **15**, 555–78.
27. Walker, D. (1982) Vegetation's fourth dimension. *New Phytol.*, **90**, 419–29.
28. Hamilton, A. C. (1982) *Environmental History of East Africa. A Study of the Quaternary*, Academic Press, London.
29. Hope, G. S. (1976) The vegetational history of Mt Wilhelm, Papua New Guinea. *J. Ecol.*, **64**, 627–63.
30. Salgado-Labouriau, M. L. (1982) Climatic change at the Pleistocene–Holocene boundary, in *Biological Diversification in the Tropics* (ed. G. T. Prance), Columbia University Press, New York, pp. 74–7.
31. Flenley, J. R. (1979) *The Equatorial Rain Forest: a Geological History*, Butterworth, London.
32. Tolonen, K. and Ruuhijärvi, R. (1976) Standard pollen diagrams from the Salpausselkä region of southern Finland. *Ann. Bot. Fenn.*, **13**, 155–96.
33. Reynaud, C. (1974) The pioneer vegetational phase in Central Lapland (Finland) based on new palynological data. *Ann. Bot. Fenn.*, **11**, 197–201.
34. Nichols, H. (1967) The postglacial history of vegetation and climate at Ennadai Lake, Keewatin, and Lynn Lake, Manitoba, *Eiszeitalter Gegenw.*, **18**, 176–97.
35. Short, S. K. and Nichols, H. (1977) Holocene pollen diagrams from subarctic Labrador-Ungava: vegetational history and climatic change. *Arctic Alpine Res.*, **9**, 265–90.
36. Davis, R. B. and Jacobson, G. L. Jr (1985) Late glacial and Early Holocene landscapes in northern New England and adjacent areas of Canada. *Quat. Res.*, **23**, 341–68.
37. Baker R. G. (1984) Holocene vegetational history of the western United States, in *Late-Quaternary Environments of the United States*, Vol. 2, *The Holocene* (ed. H. E. Wright Jr), Longman, London, pp. 109–27.

38. Birks, H. J. B. and Gordon, A. D. (1985) *Numerical Methods in Quaternary Pollen Analysis*, Academic Press, London.
39. Walker, D. and Pittelkow, Y. (1981) Some applications of the independent treatment of taxa in pollen analysis. *J. Biogeogr.*, **8**, 37–51.
40. Birks, H. H. (1972) Studies in the vegetational history of Scotland. III. A radiocarbon-dated pollen diagram from Loch Maree, Ross and Cromarty. *New Phytol.*, **71**, 731–54.
41. Bennett, K. D. (1983) Devensian late-glacial and Flandrian vegetational history at Hockham Mere, Norfolk, England. 1. Pollen percentages and concentrations. *New Phytol.*, **95**, 457–87.
42. Ralska-Jasiewiczowa, M. (1983) Isopollen maps for Poland 0–11,000 years B.P. *New Phytol.*, **94**, 133–75.
43. Donner, J. J. (1969) Holocene pollen diagrams from the Beune Valley, Dordogne. *Pollen Spores*, **11**, 97–116.
44. Beug, H.-J. (1961) Beiträge zur postglazialen Flora- und Vegetationsgeschichte in Süddalmatien: der See 'Malo–Jezero' auf Mljet. *Flora, Jena*, **150**, 600–56.
45. Brande, A. (1973) Untersuchungen zur postglazialen Vegetationsgeschichte im Gebiet der Neretva-Niederungen (Dalmatien, Herzegowina). *Flora, Jena*, **162**, 1–44.
46. Brubaker, L. B. (1975) Postglacial forest patterns associated with till and outwash in northcentral Upper Michigan. *Quat. Res.*, **5**, 499–527.
47. Anderson, R. S., Davis, R. B., Miller, N. G. and Stuckenrath, R. (1986) History of late- and post-glacial vegetation and disturbance around Upper South Brand Pond, northern Maine. *Can. J. Bot.*, **64**, 1977–86.
48. King, J. E. (1981) Late Quaternary vegetational history of Illinois. *Ecol. Monogr.*, **51**, 43–62.
49. Watts, W. A. (1979) Late-Quaternary vegetation of central Appalachia and the New Jersey coastal plain. *Ecol. Monogr.*, **49**, 427–69.
50. Davis, M. B. (1984) Holocene vegetational history of the eastern United States, in *Late-Quaternary Environments of the United States*, Vol. 2, *The Holocene* (ed. H. E. Wright Jr), Longman, London, pp. 166–81.
51. Watts, W. A. (1980) Late-Quaternary vegetation history at White Pond on the inner coastal plain of South Carolina. *Quat. Res.*, **13**, 187–99.
52. Delcourt, P. A. and Delcourt, H. R. (1987) *Long Term Forest Dynamics of the Temperate Zone*, Springer Verlag, New York.
53. Webb, T. III (1988) Eastern North America, in *Vegetation History* (eds B. Huntley and T. Webb III), Kluwer Academic Publishers, Dordrecht, pp. 385–414.
54. Barnosky, C. W. (1985) Late Quaternary vegetation near Battle Ground Lake, southern Puget Trough, Washington. *Geol. Soc. Am. Bull.*, **96**, 263–71.
55. Thompson, R. S. (1988) Western North America, in *Vegetation History* (eds B. Huntley and T. Webb III), Kluwer Academic Publishers, Dordrecht, pp. 415–58.
56. Elias, S. A. (1985) Paleoenvironmental interpretations of Holocene insect fossil assemblages from four high-altitude sites in the Front Range, Colorado, USA. *Arctic Alpine Res.*, **17**, 31–48.
57. Wells, P. V. (1983) Paleobiogeography of montane islands in the Great Basin since the last glaciopluvial. *Ecol. Monogr.*, **53**, 341–82.

58. Davis, O. K., Anderson, R. S., Fall, P. L. *et al.* (1985) Palynological evidence for early Holocene aridity in the southern Sierra Nevada, California. *Quat. Res.*, **24**, 322–32.
59. Tsukada, M. (1967) Vegetation in subtropical Formosa during the Pleistocene glaciations and the Holocene. *Palaeogeogr. Palaeoclimatol. Palaeoecol.*, **3**, 49–64.
60. Wardle, P. (1964) Facets of the distribution of forest vegetation in New Zealand. *N. Z. J. Bot.*, **2**, 352–66.
61. Heusser, C. J. (1984) Late-glacial–Holocene climate of the Lake District of Chile. *Quat. Res.*, **22**, 77–90.
62. Tsukada, M. (1985) Map of vegetation during the last glacial maximum in Japan. *Quat. Res.*, **23**, 369–81.
63. Lees, C. M. (1986) Late Quaternary palynology of the southern Ruahine Range, North Island, New Zealand. *N. Z. J. Bot.*, **24**, 315–29.
64. Dodson, J. R. (1978) A vegetation history from northeast Nelson, New Zealand. *N. Z. J. Bot.*, **16**, 371–8.
65. McGlone, M. S. and Bathgate, J. L. (1983) Vegetation and climatic history of the Longwood Range, South Island, New Zealand, 12,000 B.P. to the present. *N. Z. J. Bot.*, **21**, 293–315.
66. McGlone, M. S. (1988) New Zealand, in *Vegetation History* (eds B. Huntley and T. Webb III), Kluwer Academic Publishers, Dordrecht, pp. 557–99.
67. McGlone, M. S. (1985) Plant biogeography and the Late Cenozoic history of New Zealand. *N. Z. J. Bot.*, **23**, 723–49.
68. Moar, N. J. (1971) Contributions to the Quaternary history of the New Zealand flora. 6. Aranuian pollen diagrams from Canterbury, Nelson, and North Westland, South Island. *N. Z. J. Bot.*, **9**, 80–145.
69. Ab'Sáber, A. N. (1982) The paleoclimate and paleoecology of Brazilian Amazonia, in *Biological Diversification in the Tropics* (ed. G. T. Prance), Columbia University Press, New York, pp. 41–59.
70. Villagrán, C. (1988) Expansion of Magellanic Moorland during the Late Pleistocene: palynological evidence from northern Isla de Chiloé, Chile. *Quat. Res.*, **30**, 304–14.
71. Moore, D. M. (1978) Post-glacial vegetation in the South Patagonian territory of the giant ground sloth *Mylodon*. *Bot. J. Linn. Soc.*, **77**, 177–202.
72. Heusser, C. J. (1984) Late Quaternary climates of Chile, in *Late Cainozoic Palaeoclimates of the Southern Hemisphere* (ed. J. C. Vogel), Balkema, Rotterdam, pp. 59–83.
73. Coetzee, J. A. (1967) Pollen analytical studies in East and Southern Africa. *Palaeoecol. Africa*, **3**, 1–146.
74. Kendall, R. L. (1969) An ecological history of the Lake Victoria basin. *Ecol. Monogr.*, **39**, 121–76.
75. Davis, M. B., Woods, K. D., Webb, S. L. and Futyma, R. B. (1986) Dispersal versus climate: expansion of *Fagus* and *Tsuga* into the upper Great Lakes region. *Vegetatio*, **67**, 93–103.
76. Webb, T. III (1986) Is vegetation in equilibrium with climate? How to interpret Late-Quaternary pollen data. *Vegetatio*, **67**, 75–91.
77. Khotinskiy, N. A. (1984) Holocene vegetation history, in *Late Quaternary*

Environments of the Soviet Union (ed. A. A. Velichko), Longman, London, pp. 179–200.

78. Nichols, H. (1970) Late Quaternary pollen diagrams from the Canadian Arctic barren grounds at Pelly Lake, Northern Keewatin, N.W.T. *Arctic Alpine Res.*, **2**, 43–61.
79. Ritchie, J. C. (1984) *Past and Present Vegetation of the Far Northwest of Canada*, University of Toronto Press, Toronto.
80. Baker, R. G. (1976) Late Quaternary vegetation history of the Yellowstone Lake Basin, Wyoming. *U.S. Geol. Surv. Prof. Pap.*, **729-E**, 1–48.
81. Petersen, K. L. and Mehringer, P. J. Jr (1976) Postglacial timberline fluctuations, La Plata Mountains, southwestern Colorado. *Arctic Alpine Res.*, **8**, 275–88.
82. Lamarche, V. C. and Mooney, H. A. (1967) Altithermal timberline advance in western United States. *Nature*, **213**, 980–2.
83. Scuderi, L. A. (1987) Late-Holocene upper timberline variation in the southern Sierra Nevada. *Nature*, **325**, 242–4.
84. Dubois, A. D. and Ferguson, D. K. (1985) The climatic history of pine in the Cairngorms based on radiocarbon dates and stable isotope analysis, with an account of the events leading up to its colonization. *Rev. Palaeobot. Palynol.*, **46**, 55–80.
85. Pears, N. V. (1967) Present tree-lines of the Cairngorm Mountains, Scotland, *J. Ecol.*, **55**, 815–30.
86. Beug, H.-J. (1982) Vegetation history and climatic changes in central and southern Europe, in *Climatic Change in Later Prehistory* (ed. A. F. Harding), Edinburgh University Press, Edinburgh, pp. 85–102.
87. Webb, T. III, Cushing, E. J. and Wright, H. E. Jr (1984) Holocene changes in the vegetation of the Midwest, in *Late-Quaternary Environments of the United States*, Vol. 2, *The Holocene* (ed. H. E. Wright Jr), Longman, London, pp. 142–65.
88. Magny, M. (1982) Atlantic and Sub-boreal: dampness and dryness? in *Climatic Change in Later Prehistory* (ed. A. F. Harding), Edinburgh University Press, Edinburgh, pp. 33–43.
89. Sowunmi, M. A. (1981) Aspects of Late Quaternary vegetational changes in West Africa. *J. Biogeogr.*, **8**, 457–74.
90. Tallantire, P. A. (1981) Some reflections on hazel (*Corylus avellana* L.) and its boundary in Fennoscandia during the post-glacial. *Acta Palaeobot.*, **21**, 161–71.
91. Davis, M. B., Spear, R. W. and Shane, L. C. K. (1980) Holocene climate of New England. *Quat. Res.*, **14**, 240–50.
92. Birks, H. J. B. (1989) Holocene isochrome maps and patterns of tree-spreading in the British Isles. *J. Biogeogr.*, **16**, 503–40.

CHAPTER TWELVE

Forests under threat

12.1 INTRODUCTION

The environmental calm of the Altithermal period was probably of sufficient duration for forest communities worldwide to achieve some degree of stability, both in composition and in range, by the end of the period. Stability cannot have been total, as migration of individual tree species continued to occur, but any change was muted compared to the Early Holocene. Nevertheless, a traveller in time to these Altithermal forests would have noticed considerable differences from modern forests – differences which emphasize the magnitude of the changes that must have occurred subsequently in the Neoglacial period. Thus many tree species, in many geographical regions, were less common 6 ka ago (part-way through the Altithermal period) than they are today. The time-traveller to Europe, for example, might be struck by the lack of forests dominated by *Picea abies* over much of Scandinavia, the scarcity of *Carpinus betulus* and *Fagus sylvatica* north of latitude 52°N, and the simplified altitudinal zonation in some European mountain areas (because of the scarcity there of *Abies alba*, *Fagus* and *Picea*). To offset this, *Quercus*, *Tilia* and *Ulmus* spp. were more common. In parts of the Mediterranean he might observe extensive forests dominated by deciduous oaks (notably *Quercus pubescens*), in contrast to the prevalence of evergreen oaks (*Q. ilex*), *Buxus sempervirens*, *Olea europaea* and *Pinus halepensis* in the surviving modern fragments of forest. In the eastern United States, also, the time-traveller to 6 ka ago might observe the scarcity of *Carya* spp. and *Fagus grandifolia* in the forests of northern New England; of *Acer* spp., *Fagus* and *Tsuga canadensis* in the Great Lakes region; of *Pinus* spp. in Florida and the Coastal Plain; and of *Picea rubens* at higher altitudes in the Appalachians. In some of these forests *Pinus strobus*, *Quercus* spp. and *Tsuga canadensis* were more abundant than they are today. He might notice, too, that forests dominated by *Pseudotsuga menziesii*, *Thuja plicata* and *Tsuga heterophylla* were less widespread in the Pacific Northwest six millennia ago. In New Zealand, *Eucalyptus* and *Nothofagus* forests were likewise less widespread, but podocarp forest was more common. In tropical forests by contrast, *Podocarpus* spp. were more restricted in range 6 ka ago than they are today.

If some degree of stability had indeed been reached in these late altithermal forests, then further change could only result from perturbations to the forest ecosystems – perturbations that affected the balance between competing tree species (allowing invading species to displace resident ones), and between the forests and adjoining ecosystems (allowing the spread of derivative non-forest communities). Prominent among these perturbations was Man, who in the 5 ka of the Neoglacial period has nearly halved the extent of the world's forests, to provide land and materials for cultivation and industry. Six ka ago most human societies still relied on hunting of wild animals and gathering of wild plants for their subsistence, and agricultural/pastoral economies were probably being practised only locally, in the Mediterranean countries and the Near East, southeastern Europe, India, eastern Asia, Mexico and Peru. A great upsurge in human activity occurred in the Neoglacial. Nevertheless, to attribute all change in forest communities over the last 5 ka to human activities would clearly be erroneous. The onset of the Neoglacial period was marked by renewed glacial activity and a downturn in the general trend of global climate; prolonged periods of climatic deterioration were a feature of the Neoglacial (Chapter 6). There is evidence also of soil deterioration in temperate latitudes, with continued leaching, acidification and waterlogging leading to irreversible soil changes [1]. Changes in both soil and climate would have affected the competitive balance between tree species, and enhanced the ability of certain species to expand their ranges; and as range-expansion of trees in the Neoglacial almost always involved spread within existing forests, any such expansion would also have been facilitated by local destruction or opening-up of the forest canopy. At the present day, in addition to deliberate clearance, gap-formation in forests can result from freak weather events, accidental fires, tree death from disease, volcanic activity, etc. Such processes probably operated in the past, too, so that a variety of factors, with interactive effects, could have been responsible for forest changes in the Neoglacial (Figure 12.1). None of these processes is exclusive to the Neoglacial. Some, however, increased in intensity and frequency then, to confer a special character to the Neoglacial in many parts of the world.

12.2 PLANT RESPONSES TO NEOGLACIAL CLIMATIC CHANGE

The separation of an Altithermal from a Neoglacial period was recognized many years ago by Lennart von Post [2]*, who demonstrated parallel

* Though Von Post did not use this terminology.

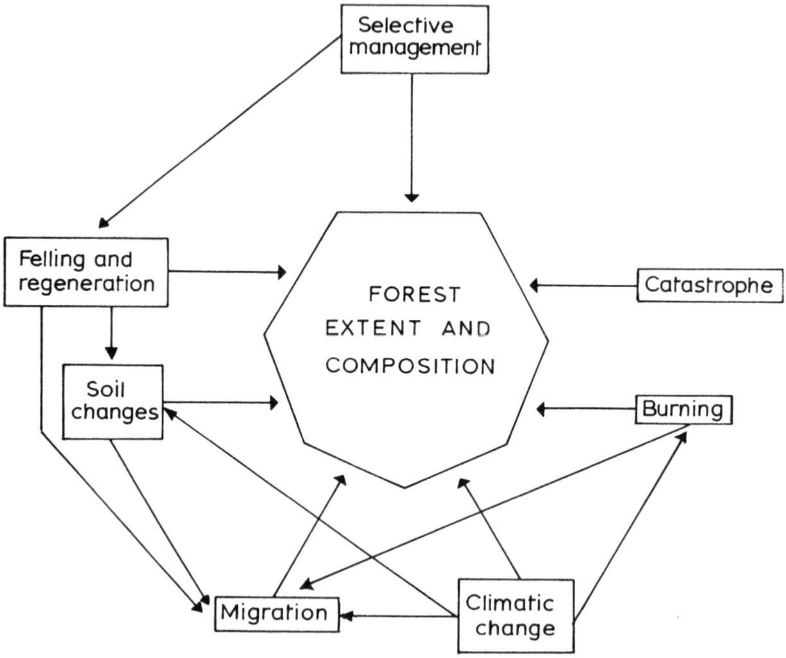

Figure 12.1 Forests under threat. Possible factors involved in forest changes during the Neoglacial period.

changes in pollen diagrams from widely different geographical regions that indicated a **revertence** to less favourable conditions in recent millennia. The evidence was most convincing for Europe, where the tree pollen taxa that increased in abundance in the upper part of the pollen diagrams were either those characteristic of the earlier protocratic phase (*Betula, Pinus*), or those that had expanded in the late temperate (oligocratic) phase of interglacial pollen diagrams (section 10.2.3 – *Abies, Carpinus, Fagus* and *Picea*). Convincing identification of vegetational changes that might have been brought about by the downturn in the climatic cycle that constitutes the Neoglacial period [3] is difficult to find, however, as shifts in floristic composition of vegetation or in geographical ranges of plants through the Neoglacial could also have resulted from long-term migrational movements of slower-moving taxa that were lagging behind climatic change, or from gradual soil deterioration (Figure 12.1). Thus the spread of *Abies, Carpinus, Fagus* and *Picea* in Europe in the Neoglacial can be seen as essentially a continuation of the slow range-expansion that had been ongoing through the earlier Holocene – though Huntley [4] suggests that colder winters in Scandinavia and warmer winters further south during the Neoglacial controlled the later migration patterns of *Picea* and *Fagus*, respectively. Among what appear to be valid examples of true revertence may be mentioned the southward spread of *Picea glauca* into mixed conifer–hardwood forest and prairie in central and eastern North America after 4 ka ago [5–7], and the spread of *Fagus* and subalpine conifers into evergreen oak–laurel forest in south-western Japan, also after 4 ka ago [8]. More ambiguous examples involve the spread of reputedly cold-tolerant taxa, often at the expense of more warmth-demanding ones: the spread of *Picea rubens* at higher altitudes in the Appalachian Mountains of the eastern United States in the last 2 ka [9]; the spread of *Fitzroya* and *Podocarpus* into the Valdivian rainforest of central Chile after 6.5 ka ago [10]; and the partial replacement of *Dacrydium* by *Nothofagus* in the North Island of New Zealand after 5 ka ago [11,12]. Some changes were probably directed more by shifts in moisture status: the spread of pines capable of growing in areas of higher water table in the southeastern United States after 5.5 ka ago [13]; and the partial replacement of the evergreen *Nothofagus betuloides* by the deciduous *N. pumilio* in southern Chile after 5.6 ka ago [14], for example.

The variation in the timing of these floristic shifts (from 6.5 to 2.0 ka ago) emphasizes the protracted nature of the transition from Altithermal to Neoglacial conditions. No single date can be set for the start of the Neoglacial period. Moreover, subsequent climatic change within the Neoglacial did not follow a uniform downward trend. The picture in detail is rather of alternating periods of rapid and slow change, and of upturns and downturns in both temperature and moisture [15–19]. The broad pattern of forest change in the Neoglacial is also similar, with periods of relative stability (either of floristic composition or of geographical range) punctuated by periods of rapid change. Such spasmodic forest change could have resulted from climatic change, or from the periodic accentuation of other processes like forest clearance or epidemic disease. Separation of the latter effects from those of climatic change is not easy, especially in densely-settled regions, and as climatic change may render certain areas of forest more susceptible to perturbation than others by shifting the competitive balance between co-existing taxa over parts of their ranges, the role of climate in punctuated forest change can never be ruled out completely. Nevertheless, perturbations are a special feature of the Holocene, and their significance as a process of forest change needs to be examined independently of any climatic effects.

12.3 PERTURBATION

12.3.1 The significance of perturbation

Perpetuation of forest vegetation at a particular site over many centuries requires repeated recruitment of new individuals of its constituent woody species. These new individuals may already be present as subordinate members of the community, but enter the tree canopy only when space is made available by the death or removal of an existing mature individual; or they may originate from dormant or dispersed seed only when a gap in the canopy is created. Woody species of the former group (shade-bearers, late successional or climax species, dryads) have seedlings capable of establishing themselves in shade, often on a humus-rich substrate, and then of persisting for many years under low light conditions; woody plants of the second group (light-demanders, pioneer or early successional species, nomads) have shade-intolerant seedlings that will often only establish on a mineral substrate [20,21]. Over a long period of time the expectation is that a stable forest ecosystem would come to be dominated by shade-bearers, and that the light-demanders would be gradually eliminated. For light-demanders to persist, or for new (immigrant) species to invade, there would have to be regular creation of sizeable gaps in the canopy – and the forest would then come to develop a **mosaic** structure, with newly-formed gaps where the light-demanders could flourish, partially-stabilized gaps where the shade-bearers could establish, and patches of mature canopy dominated by the shade-bearers. Gaps suitable for the establishment of light-demanding species are typically larger than those resulting from the death of an individual tree ($> 1000\ m^2$), and are almost always associated with some form of perturbation. The important role played by light-demanding species in many forest communities, both modern and ancient, suggests that perturbation has been a continuing process in forest development, and a contributory factor to the maintenance of floristic diversity [21]. Wherever light-demanding trees with shade-intolerant seedlings are dominant in modern forests, regular perturbation must be suspected. Thus the anomalous dominance of *Picea glauca* throughout the boreal forests of Canada can be explained as the result of recurrent perturbation by fire [22]. The data recently published for the mixed conifer–hardwood forests of the Changbai Mountains of northeast China indicate that their floristic richness may be a consequence not only of excellent species-survival through the Tertiary and Pleistocene, but also of periodic perturbation. Three of the main canopy trees in the Changbai forests (*Fraxinus mandshurica, Pinus koraiensis, Quercus mongolica*) at the present day are light-demanding, and show a discontinuous age-structure (Figure 12.2), with

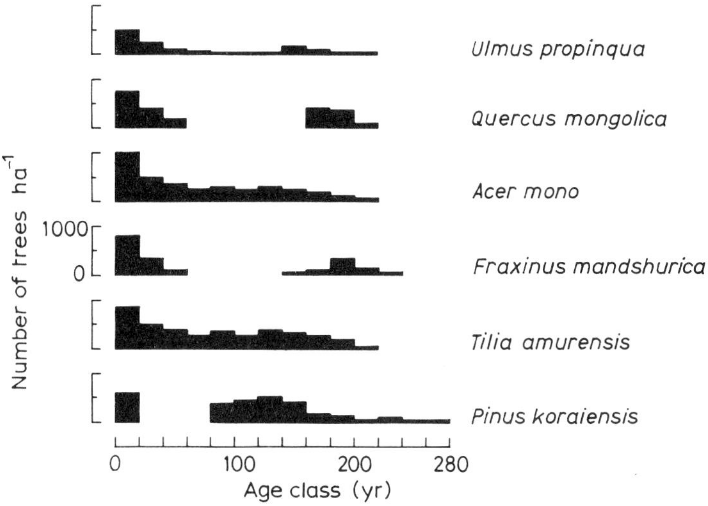

Figure 12.2 Age-structure of populations of the six major tree species of the Changbai forests, northeast China. The histograms show numbers of individuals in successive 20-yr age-classes. Re-drawn from Hanxi and Yegang [23].

the oldest individuals having entered the forest in a wave 200–280 yr ago [23]. The Baiyun volcano, the highest peak of the Changbai range, has a history of eruptions (Table 12.6), and periodic opening-up of the forest by such action may govern the persistence of these tree species in the same way as has been suggested for *Nothofagus* spp. in the Chilean forests [24].

12.3.2 The human factor

It is questionable whether Man was ever by choice a forest-dweller. The early hominids probably evolved in non-forested environments, and were expanding their range as forests contracted before the oncoming Pleistocene Ice Age. Modern man (*Homo sapiens*) spread widely during the last cold stage of the Pleistocene, but then was partially engulfed by forest as the climate warmed again in the Early Holocene. These forest-dwelling societies retained the hunter–gatherer mode of life of their open-ground ancestors, in some cases almost to the present day; the change to a settled farming economy, begun in the Early Holocene, probably took place in areas with a strongly seasonal climate, where woodland was only one element in a landscape mosaic that included substantial and varied open-ground plant communities. Not till later, as populations grew and climates changed, did these early farmers tackle the problem of converting forest to arable and pasture land. There would thus have been an intensifying interaction between man and forest during the Holocene. This interaction must have generated an ambivalence in human attitudes towards the forest. On the one hand, forest was an impediment, hampering movement across country, harbouring dangerous predators, and infesting land apparently suitable for growing crops and pasturing animals. As such, it needed to be destroyed. On the other hand, forest was a provider of wood, for building, heating and equipping dwelling places, and for enclosing and protecting property, as well as providing a continuing supply of wild animal and plant foods and products. As such, forest needed to be conserved. Initially there was no conflict: forest was seemingly an inexhaustible resource, and one that returned readily when cultivation ceased and the land was abandoned. Only later, sometimes too late, when clearance had taken its toll, did human societies recognize that forest was an integral part of their life and landscape, and needed to be managed and conserved in a rational way. This change in attitude occurred at different times in different regions. Consequently, the world's forests today exhibit a diversity in their structure and composition that reflects in part these varying interactions with human societies through the Holocene.

The interactions of human societies with forest over the centuries have taken many forms. There is well-founded archaeological or documentary evidence to substantiate, *inter alia*, the following processes: the burning of limited areas of forest by hunter–gatherer societies to provide open areas where game animals might congregate; the selective clearance of forests from certain soil types for either short-term or long-term cultivation; the selective gathering of leafy branches as winter fodder for stalled cattle, sheep and pigs; the grazing of domesticated animals within forest; the selective removal of timber and underwood, either in an uncontrolled way or as part of a rotational management regime involving coppicing or pollarding of certain tree species; the interplanting of crops among the residual native trees in partially felled forest; and the deliberate planting of saplings of certain tree species. Examples of such processes abound in the literature – see, for example, Bell and Limbrey [25] and Jones [26]. Almost certainly they would have operated selectively, favouring the survival and spread of some tree species and discouraging others; in time they would have profoundly modified the floristic composition of the forest – as well as progressively reducing its extent. Although these modifications can in part be predicted from a knowledge of present-day forest ecology, nevertheless it is preferable wherever possible to utilize the direct evidence of past forest change preserved in the pollen record.

Before the pollen record can be utilized, however, some alternative and independent assessment of human activities in a particular region needs to be made; that assessment can then be set against the pollen evidence for forest change, and cause-and-effect relationships sought. Archaeological or documentary evidence often provides that assessment, and is most valuable when the pollen analyses are undertaken at the same site as that to which the specific historical information relates. Where such specific independent information is not available, intrinsic features of the pollen record may have to be used instead as proxy documentation of human activities. Because much of our knowledge of human influence on forests in the Neoglacial derives from this sort of evidence, it is important to be sure that the pollen features constitute a real proxy record. What these features are, and how they should be interpreted, is considered below.

Forest clearance, by definition, opens up the forest canopy and modifies the local pollen rain. The short-term results of an event similar to local clearance at the present-day can be seen in the pollen analyses from the raw humus layer of a Kentish woodland in which all the mature elm trees were killed by Dutch elm disease between 1978 and 1982 [27]. Three years later the local pollen record showed a marked increase in the values of light-demanding taxa such as cf. *Bidens*, *Campanula*, cf. *Digitalis*, *Epilobium*, grasses, *Hedera*, *Plantago lanceolata*, *Pteridium*, cf. *Rhinanthus* and *Rumex acetosa* (Figure 12.3). Some of these taxa were local woodland plants that flourished in the gaps created by the death of the elm trees; other taxa, however, were not present in the woodland, and the partial removal of the

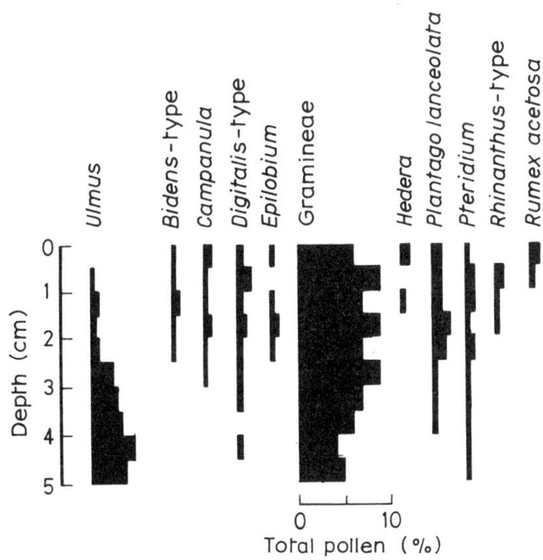

Figure 12.3 Pollen analyses carried out in 1985 through the top 5 cm of mor humus at Scords Wood, Kent. The diagram covers the period of death of the mature elm trees from Dutch elm disease between 1978 and 1982. Only certain pollen taxa are shown. Re-drawn from Perry and Moore [27].

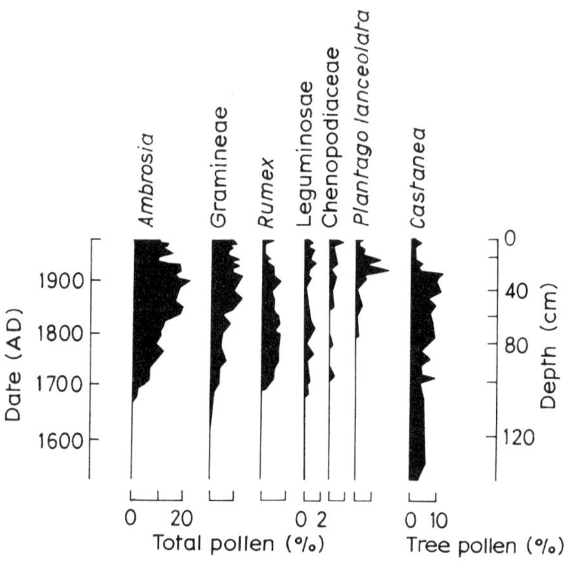

Figure 12.4 Pollen diagram from the upper sediment layers of Linsley Pond, Connecticut. Only certain pollen taxa are shown. Re-drawn from Brugam [28].

filtering effect of the tree canopy had clearly allowed pollen to be blown in from more distant sources. In a densely farmed landscape such as that of southern England it is not surprising that at least two of the 'exotic' pollen taxa were of species that are closely associated with cultivation today (*Plantago lanceolata* and *Rumex acetosa*).

This example suggests that the pollen rain after forest clearance is a sensitive monitor of local vegetation changes, but might be affected also by windblown pollen from farther afield. The linkage between the pollen record and local forest clearance is easy to demonstrate in North America, where widespread forest clearance has occurred only in recent centuries with the coming of European settlers. An elegant example is the work of Brugam [28], who compared the pollen changes in the upper 2 m of sediment at Linsley Pond, Connecticut, with the documented record of settlement around the lake. A time-scale for the sediment was derived from lead-210 and carbon-14 dating of selected levels, together with the position of the abrupt decline in *Castanea* pollen resulting from chestnut blight (which reached Linsley Pond in 1913). The first homestead was built in 1700 AD, 0.75 km from the lake, and at this level in the pollen diagram (Figure 12.4) there are major rises in *Ambrosia*, Chenopodiaceae, Gramineae, Leguminosae and *Rumex*. More intensive livestock rearing by 1790 is mirrored in the appearance of *Plantago lanceolata* in the pollen diagram. The general decline in pollen of open-ground plants above 30 cm depth

Table 12.1 Hemerophilous pollen taxa in Holocene sediments of central and northern Europe. The taxa are scored on a five-point scale (+ to 4) according to their abundance in modern arable, grassland and ruderal communities*

	A	B	C	D	E	F	G
Secale	4	+	+				
Hordeum	2	3	+				
Triticum	1	3	+				
Avena		4	+				
Linum usitatissimum		3	+				
Fagopyrum esculentum		3	+				
Vicia (faba)		2	+				
Cannabis		3	+				
Centaurea cyanus	2	+	+				
Convolvulus	2	1	1				1
Lychnis/Agrostemma	2	+	+	1			
Scleranthus annuus	2	+	+		+		
Spergula arvensis	1	2	1				+
Caryophyllaceae *p.p.*	1	1	1	1	+	+	1
Rumex cf. *acetosella*	3	1	2	+	2	1	1
Polygonum aviculare	2	2	2	1		1	2
Polygonum persicaria-type	1	2	2			+	1
Cruciferae	1	2	2	1	+	1	2
Compositae, Tubuliflorae	1	1	2	2	1	1	2
Compositae, Liguliflorae	1	1	2	2	1	2	2
Gramineae	2	2	3	4	3	3	1
Trifolium (repens)	+	1	1	2	+	2	2
Ranunculaceae	+	+	1	3		1	2
Rumex cf. *acetosa*			1	3		1	+
Plantago lanceolata			1	3	+	+	+
Plantago major + media		+	1	2	1	1	4
Cyperaceae			1	3	1	1	
Umbelliferae			1	2	+	2	1
Succisa				2	1		
Jasione + Campanula			1	+	2	1	
Calluna					4	1	
Juniperus					3	1	
Melampyrum (pratense)					1	3	
Pteridium			+	1	1	3	
Polypodium					1	1	
Chenopodiaceae	+	2	2				3
Urtica		1	1			1	4
Artemisia (vulgaris)			1			1	4

* A, winter cereals; B, summer cereals and root crops; C, fallow land; D, wet meadows and pastures; E, dry pastures and heaths; F, wood pasture; G, footpaths and waysides. Data of Behre [30].

(the 1913 chestnut blight level) is believed to reflect the agricultural recession in New England this century, brought about by competition from farmers on the more fertile soils of the American Midwest.

In the past, forest clearance was often followed by a period of cultivation. In such circumstances we might expect appropriate pollen indicators of cultivation to occur in the contemporary spectra. Such pollen indicators would include not only crop plants, but also associated weedy species. The first demonstration of this effect in a prehistoric context was by Iversen in 1941; he found that the first occurrences of pollen of cereals, *Artemisia* and *Plantago* at Ordrup Mose, Denmark, coincided with a thin layer of charcoal in the sediment and archaeological evidence of a local Early Neolithic settlement [29]. As changes were also visible in the tree pollen spectra at this level (see below), Iversen interpreted his findings as indicative of local forest clearance. Subsequent workers in Europe have refined and extended Iversen's conclusions to encompass a wide range of so-called 'clearance indicators'. Behre [30], for example, lists thirty-eight pollen taxa that occur in twentieth-century arable, grassland and ruderal communities in central and northern Europe, and which have also been recorded in Holocene deposits there (Table 12.1). Such **hemerophilous** (shade-intolerant) plants are dependent upon human settlement patterns at the present day for their survival and spread. In the pollen record they are thus an index of human interaction with the forest.

Attempts to extend this approach beyond Europe and North America have likewise relied on the recognition of 'clearance indicators' in modern vegetation. In tropical regions, where archaeological data are scanty, the occurrence of supposedly hemerophilous plants in the pollen diagrams may constitute the only evidence for forest clearance. Some of the pollen taxa recognized in this context are listed in Table 12.2.

The recognition of hemerophilous plants in pollen diagrams has enabled palaeoecologists to study the impact of human settlement on the world's forests in more detail than is possible from the archaeological record alone. In conjunction, the two lines of evidence show that in some parts of the world human impact has extended back well beyond the Neoglacial, whereas in other parts human impact has been minimal until recent centuries. Some assessment of this impact is provided by Figure 12.5. Forest clearance has been most intense in regions of dense settlement, cross-hatched in Figure 12.5: the temperate forests of Europe, eastern North America and eastern Asia, and the subtropical and tropical forests of India and southeast Asia. In Europe and Asia the impact has been long-continued, initiated by early (Neolithic) farmers spreading progressively from 'centres of primary domestication' in the Near East, the Indus Valley, northern China and Thailand. In both continents clearances probably began 7–8 ka ago. In the American continent, by contrast, farming cultures associated with the Mexican and Peruvian centres of domestication remained far more localized in distribution and influence, whilst the forests of eastern North America persisted in a largely undisturbed state until the coming of European settlers (Spanish, British and French) from the late sixteenth century onwards [31,32]. In all continents, extensive areas of forest, on soils or in climates unsuitable for permanent agriculture, effectively escaped human impact until the twentieth century. Included in this category is much of the boreal coniferous forest, and certain areas of tropical forest – though evidence is accumulating of localized or more widespread forest clearance dating back several thousand years in parts of the tropics [33,34,36].

The long-term pattern of forest clearance in space and time is well-documented only for Europe, where there is a variety of both palynological and archaeological evidence. Human impact on woodland has probably been of longer duration there than anywhere else in the world. The first farming communities originated outside the main European forest region, in semi-arid parts of the Near East, where they probably avoided patches of dense forest initially. Thus in Turkey (Anatolia), early agriculture was

Table 12.2 Pollen taxa that have been used as evidence of forest clearance in tropical pollen diagrams [33–37]*

Cultigens	*Manihot* (1)
	Zea (1)
genera containing cultivated species	*Arenga* (3)
	Casuarina (3)†
	Ipomoea (1)
Non-woody plants of open habitats	*Ambrosia* (1)
	Chenopodiaceae (1,2)
	Compositae (1)
	Gramineae (1,2,3)
	Plantago (2)
	Rumex (2)
Woody plants found in regenerating and secondary forest, but rarely in primary lowland forest	*Acalypha* (3)
	Celtis (2)
	Cecropia (1)
	Dodonaea (1,2)
	Hagenia (2)
	Justicia (2)
	Macaranga (3)
	Trema (1,3)
	Urticaceae (3)

* 1 = Central and South America; 2 = Africa; 3 = Southeast Asia.
† Planted for timber and to improve soil fertility.

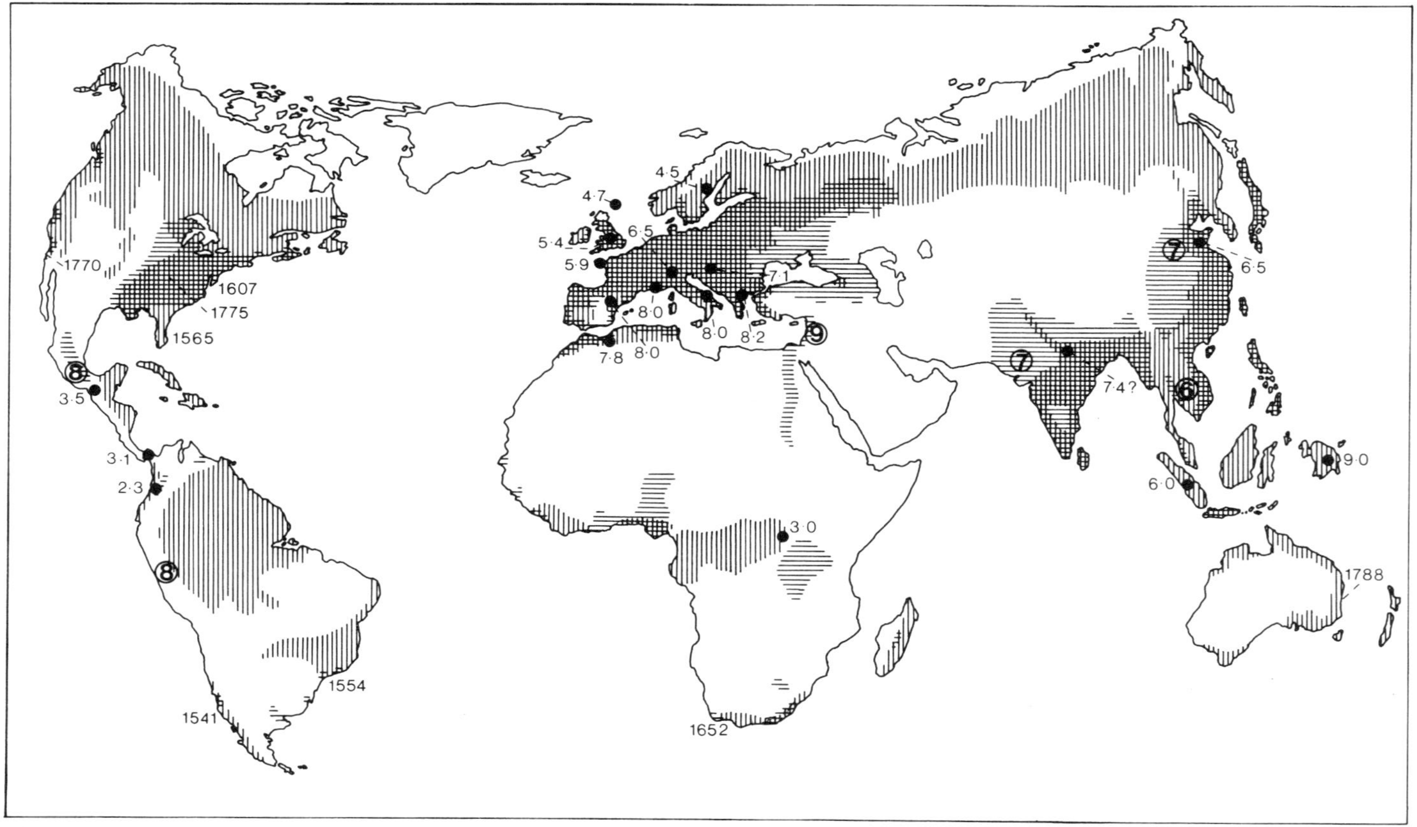

Figure 12.5 Patterns of forest clearance. The map shows (a) the potential extent of forest and woodland communities at the present day |||||||| – a conservative picture based on the map in Whittaker [38]; (b) the approximate extent of regions of dense settlement ≡ – with more than 25 persons per km^2; (c) the location of probable centres of primary domestication ○ – the numeral within the circle is the approximate date of first domestication, in ka ago; and (d) dates of first farming communities within forested areas – either as yr AD (e.g., 1565) or as ka ago (e.g., ● 8.2).

concentrated in the south, in regions of mediterranean climate, and the summer-green forests in the north along the shores of the Black Sea remained unsettled until much later [39]. Expansion from the Near Eastern centre, about 8 ka ago, took place first into the dry woodlands around the Mediterranean Sea [40], and the main penetration of the continent's forests was delayed until about 7 ka ago. By 4.5 ka ago all parts of the European temperate forest region were under attack. The forest cover 7 ka ago must have been highly heterogeneous, with large-scale patterning determined by climatic gradients latitudinally, meridionally and altitudinally in Europe, smaller-scale patterning resulting from the late migration of certain tree species, and fine-scale patterning governed by soil and topography [41]. Consequently the initial Neolithic settlement was highly selective, and restricted to certain elements of the forest mosaic by reason of technological limitations to clearance and tillage, and by climate–soil interactions. Thus Early Neolithic sites over much of central Europe were concentrated on the fertile but easily-worked loessic soils [42], though in the drier climate of Hungary alluvial floodplain situations were favoured [43], and in the wetter north-west the well-drained calcareous soils of southern England were preferentially settled [44]. On the poorest soils some sort of shifting cultivation may even have been practised in the Neolithic [44,45]. Subsequently, settlement spread to other soils, but when this happened previous settlements were sometimes abandoned. Forest clearance in prehistoric Europe was thus not uniformly progressive and destructive. Rather there was a fluctuating balance between clearance and regeneration, with the periods of clearance and settlement at a particular site ranging from a few decades to many centuries. When cultivation ceased the forest returned, spearheaded by opportunistic taxa with efficient dispersal. As the forests retreated and advanced through the Neoglacial, so their composition was gradually changed.

This process is documented in the pollen record. If the pollen frequency of clearance indicators is regarded as a crude quantitative index of human impact, then any changes in tree pollen values that accompany the changes in hemerophilous pollen can perhaps be interpreted as responses of the forest to human impact. Such changes indicate that where areas of cleared forest were cultivated for only a short period and were then abandoned, the regenerating forest normally resumed its original composition fairly quickly, but that more prolonged cultivation usually led to irreversible changes in composition. Examples from northern Europe, where the pollen record of forest clearance has been studied in considerable detail [46], are shown in Figures 12.6 and 12.7: short-duration forest clearance in Denmark, where there are only temporary fluctuations in the pollen values of *Betula*, *Corylus* and *Tilia*, but more lasting changes in *Fraxinus* and *Ulmus*; and long-duration forest clearance in England, where *Tilia*

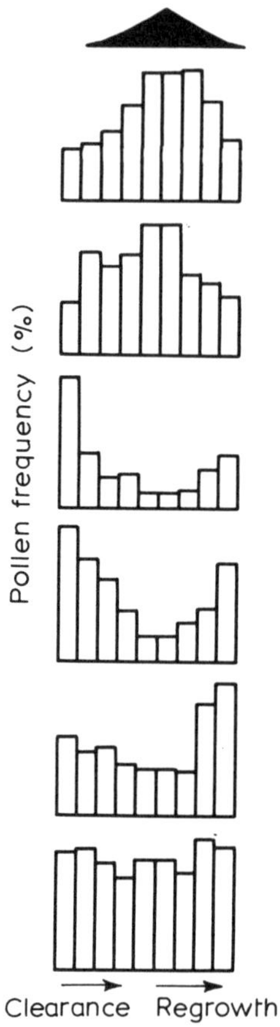

Figure 12.6 Pollen changes during the Early Neolithic period in Denmark. The diagram is a composite one, compiled from several pollen diagrams. *Plantago* values in the diagrams were grouped into eight categories, four of ascending values (to a maximum of 25% total herbaceous pollen) and four of descending values. Pollen values of individual tree taxa were then grouped according to the eight *Plantago* categories, and mean values calculated for each category and for a 'zero *Plantago*' category immediately prior to the first clearances. The histograms show these values expressed as a percentage of the maximum value attained across the nine categories. Re-drawn from Iversen [41].

was permanently eliminated from the forest and *Fraxinus* progressively increased in importance.

These two examples are illustrative of general patterns of change in forest composition following clearance, and do not necessarily define invariable responses by these particular tree taxa. Characteristic of forest clearance in prehistoric Europe, and the later phase of more active conservation, was its heterogeneity. Different areas of forest were subjected to

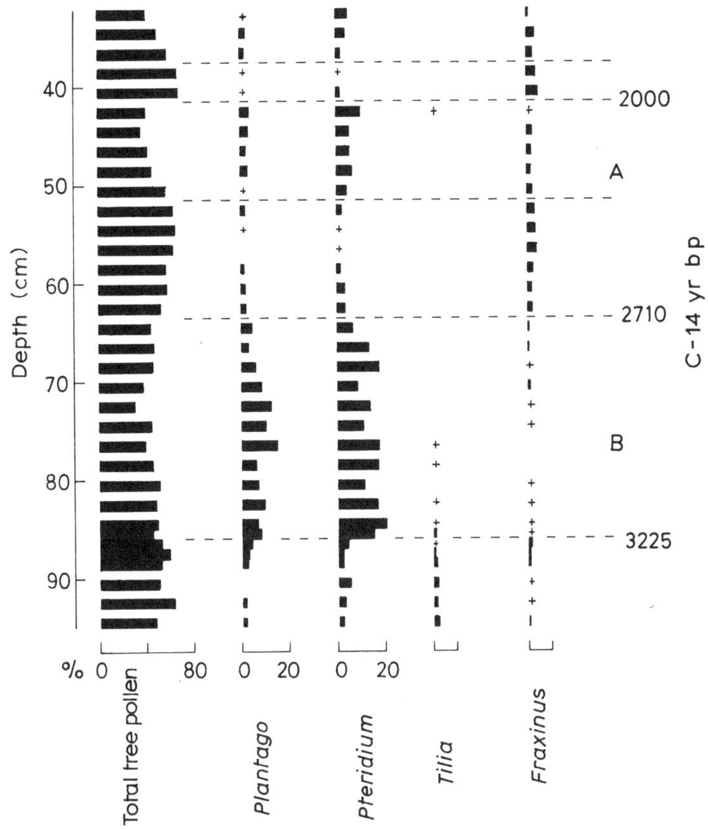

Figure 12.7 Selected pollen components from peat deposits at Whixall Moss, Shropshire, showing two major periods of forest clearance, A and B, between 2000 and 3225 yr bp. Individual pollen taxa are shown as % tree pollen, and tree pollen as % total pollen. Re-drawn from Turner [46].

differing intensities of human impact, often on quite a local scale, so that the present-day composition of the surviving woodland records a multiplicity of effects. Each individual wood in Europe has had its own specific history, and two woods of similar floristic composition today might have achieved this in entirely different ways. Moreover, human settlement patterns in the prehistoric period were inextricably linked to climatic change, and the consequences of these two processes on forest history are often difficult to separate – particularly when long-term migration of tree species was also involved. The complex interactions between vegetation and environment are exemplified by Figure 12.8, which summarizes the changing pollen spectra at a site in northeastern Sweden over the last 3.3 ka [47]. Several questions are raised by this diagram. Why, for example, did

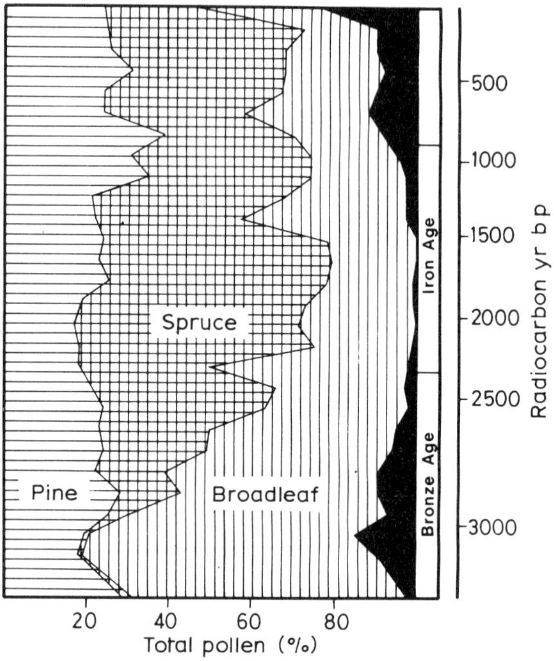

Figure 12.8 Simplified pollen diagram from Prästsjön, Västerbotten, Sweden, showing the approximate composition of the forest over the last 3.3 ka. Broadleaf = birch or alder forest; black = herbaceous pollen. Re-drawn from Engelmark [47].

spruce forest largely replace birch forest by 2 ka ago: was the spread of spruce controlled solely by a slow migration rate, or was it aided by the known change to a cooler wetter climate about 2.5 ka ago? Did Bronze Age settlers, by opening up the birch forest for cultivation and then later abandoning their farms again, facilitate the spread of spruce, or did the settlers abandon their land because of the encroachment of spruce under deteriorating climate? These are questions that cannot yet be answered with any certainty.

12.3.3 Fire

Several major vegetation types, all associated with strongly seasonal climates, are affected by recurrent natural fires at the present day: dry grasslands and savannas, dry sclerophyll woodlands, and boreal conifer forests. Most natural fires are started by lightning, and it is reasonable to assume that fire has been a recurrent hazard for these and similar vegetation types in the past also. For the eucalypt woodlands of southeastern

Table 12.3 Some first dates of major charcoal presence in sediments (ka ago)

Europe	Finland [55]	9.5
	Sweden [55]	10.7
North America	Manitoba [55]	6.0
	Minnesota [56]	10.0
	Massachusetts [57]	9.0
	Nova Scotia [58]	9.7
	Washington [59]	12.0
Southeast Asia	Sumatra [36]	17.8
Australasia	Tasmania [60]	22.0
	New Zealand [61]	6.5

Australia, for example, there is continuous presence of charcoal in the sediments accumulated in Lake George, New South Wales, over the last 120 ka [48].

Fire has also been known and used by humans, probably for at least 500 ka [49,50]. Over the last 20 ka fire has been used increasingly to modify and control the environment: by hunter – gatherers to drive animals to their destruction, and to produce new succulent growth to attract grazers [51,52], and by farmers to effect land clearance and to maintain good grazing land [53]. The fire record for the last 20 ka, preserved as charcoal and carbonized plant material in suitable sediments and at archaeological sites, thus comprises both natural and man-induced events, which in most cases cannot be separated. Fire effects extend back well beyond the Neoglacial (Table 12.3), but the increasing prominence of human activity through the Neoglacial makes it appropriate to consider these effects here.

The frequency of major fires differs in the various regions today, being more frequent in grasslands and savannas than in the fire-susceptible conifer forests. Fire frequency is normally expressed as **fire-return interval**, defined as 'the total area of the region under consideration'

Table 12.4 Fire-return intervals (yr) for certain modern forest and woodland types

Californian chaparral [54,65]	15–40
Sierran conifer forest [50]	35
Pine forest [50,54]	10–60
Spruce forest [54,66]	80–100
Mixed conifer–hardwood forest [54,67]	100–>200
Pacific coast conifer forest [50]	150–>500
Montane conifer forest [68]	300

divided by 'the long-term mean area burnt each year' [54]. The fire-return interval varies considerably between different forest and woodland types at the present-day (Table 12.4).

Long-term fire frequencies in the past have been calculated for a small number of sites, utilizing major peaks in charcoal or associated fire effects in well-dated lake sediments. These associated effects include increased rates of sediment accumulation (resulting from erosion), increased ash content, increased levels of aluminium, vanadium and ferrimagnetic minerals, and birch pollen peaks [55,62,63]. Fire frequencies calculated in this way are probably not directly comparable with modern return-interval estimates, but they have clearly varied considerably at different times in the past (Table 12.5).

The different fire-return intervals found in modern forest and woodland types result from a complex of factors: the quantity of ignitable fuel, the dryness of the fuel (which is a function of local climate), the occurrence of a suitable agent of ignition [64], the floristic composition of the vegetation (conifers being in general more flammable than mesic hardwoods, and pine more so than spruce, for example), the local topography, and the level of human activity [50]. The influence, and the interaction, of these factors in the past also can be discerned in the estimates of past fire frequencies in Table 12.5. Thus the shorter fire rotations over the last 2.0 ka in Finland reflect in large part the increasing human impact on these forests. In North America this is less likely though. The lengthened fire rotation in the Nova Scotian forests between 4.5 and 2.1 ka ago might be in part a reflection of the increasing proportion of hardwoods in the forest, and in part a result of the cooler wetter climatic conditions after 4 ka ago. Ultimately, however, it is suggested, this climatic deterioration led to a resurgence of the conifers, increased flammability, and a shorter fire rotation [69].

Recurrent fires in the past must have affected forest vegetation in two ways: by reducing its extent, and by changing its floristic composition. Most of the relevant studies have been carried out in North America, but it

Table 12.5 Estimated fire frequencies in lake sediments

Time-interval (ka ago)	Location	Fire frequency (yr)
7.5–3.5	Southern Finland [55]	180
6.6–4.5	Nova Scotia [69]	250
4.5–2.1	Nova Scotia [69]	350
3.5–2.0	Southern Finland [55]	135
2.0–0.3	Wisconsin [63]	120
2.0–0	Southern Finland [55]	<100
1.2–0.7	Ontario [62]	80
1.0–0	Minnesota [56]	60–70

must be emphasized that fire was not a uniquely American phenomenon. Similar changes must have occurred in the past in appropriate forest communities throughout the world.

In North America the boundary between prairie and forest is known to have changed its position through the Holocene in response to climatic change (section 13.3), but the documented spread of forest trees in the last 150 years in areas where grassland fires have been controlled since European settlement [70] is clear evidence of the importance of burning also. The presence of woodland trees (notably *Pinus ponderosa* and *Juniperus scopulorum*) throughout the central prairies of the United States from Indiana to Wyoming, in topographic situations such as rocky escarpments where they might be protected from the annual late-summer fires, is regarded by some ecologists as indicative of former more widespread woodland [71]. An extreme view is that the spruce parkland communities of the lateglacial in the American Midwest (section 10.4) were destroyed by fire at or before the start of the Holocene, and then replaced by opportunist grasses and composites [71]. A similar process undoubtedly operated later in the Holocene in the South Island of New Zealand, where abundant woody remains and charcoal of *Dacrydium*, *Metrosideros*, *Nothofagus* and *Podocarpus* (sometimes buried to a depth of 45 cm in the soil) in the now treeless eastern half of the island are clear testimony of a former forest cover destroyed by burning. The majority of the charcoal has been radiocarbon-dated to between 500 and 1700 yr bp [61]. The treeless steppes of eastern Europe may have a similar human origin [72].

Buried charcoal fragments in tundra soils north of the treeline in arctic Canada, some dating back to 3.6 ka ago [73], record the presence of fire as an agent of destruction at this vegetation boundary also. Indeed, the gradual loss of trees in the forest–tundra zone there over the last 3 ka may owe more to fire effects than to climatic change [74]. The two processes are interlinked. Thus in northern Quebec, the charcoal record documents widespread fires 400 years ago that killed most individuals of *Picea mariana* near the treeline; re-establishment may then have been prevented by the colder conditions of the Little Ice Age. A zone of scattered tree 'islands' (clones of vegetatively-reproducing *Picea*) set in tundra north of the present treeline could represent the few individual trees that survived this fire [68].

An opposing effect of the Little Ice Age is suggested by Grimm [75] for the 'bigwoods' region of southern Minnesota. Stratigraphic and pollen evidence shows that these mixed hardwood forests (of *Acer saccharum*, *Ostrya virginiana*, *Tilia americana* and *Ulmus* spp.) occupy land that was covered by prairie 5 ka ago. Oak and aspen gradually replaced the prairie vegetation between 5.0 and 2.4 ka ago, and then about 300 years ago the bigwood trees spread rapidly westwards into this oak woodland for a distance of

60–70 km. Reduced fire frequency under a cooler wetter climate, first in the early Neoglacial, and subsequently in the Little Ice Age, is believed to have been responsible for these changes.

Recurrent fires are also known to have changed the floristic balance between the species in the North American forests in the past. Those pollen analyses that have been carried out at close intervals in conjunction with charcoal assessments typically show striking and repeated fluctuations in the values of many of the major pollen taxa. The fluctuations are geared to the positions of the major charcoal peaks (Figure 12.9). Green [69] examined the pattern of these fluctuations in a pollen diagram from Everitt Lake, Nova Scotia, over the time-interval 4450–2100 yr bp, and was able to show statistically that the pollen peaks of individual tree taxa occurred at characteristic and different time-intervals after each major charcoal (= fire) peak: *Larix, Picea, Pinus, Salix* and *Ulmus* 50 years after; *Fraxinus* and *Quercus* 100–150 years after; and *Abies, Acer, Fagus* and *Ostrya* 250 years after. A similar pattern was found by Swain [63] at Hell's Kitchen, Minnesota: *Betula* soon after a fire; *Pinus strobus* 40–80 years after; and *Tsuga* 80–170 years after. The sequences of the tree taxa in these two pollen diagrams exactly correspond to their modern responses to fire. It is known that certain trees recover rapidly after forest fire; this is because the mature individuals are resistant to the effects of fire, because of rapid regeneration from seed, or because of efficient seed dispersal. Other tree taxa lack adaptations to fire, but are able instead to invade, spread, and ultimately dominate within closed forest [76]. The scheme of recovery after fire that Green [69] drew up on the basis of his pollen analyses at Everitt Lake closely matches present-day forest dynamics in response to fire. Trees characteristic of the early stages of recovery return after each fire; late-stage species return only if the fire rotation is of sufficient duration to allow their entry and establishment. At Hell's Kitchen, for example, where *Betula* peaks were used as proxy indicators of fire, a peak at 110 AD was followed only by a *Pinus* peak, as the next fire occurred sufficiently soon after to preclude the return of *Tsuga* [63]. Green suggests that the change to a longer fire rotation at Everitt Lake after 4450 yr bp (Table 12.5) allowed fire-intolerant species such as *Fagus grandifolia* and *Abies balsamea* to spread within the forest, after having been kept at low frequencies previously by a shorter fire rotation. At an earlier stage in the forest history, infrequent but large firest between 10 and 7 ka ago probably opened up the forest sufficiently to allow the initial entry of a variety of the less fire-tolerant tree taxa that were unable to spread into closed-canopy forest: *Abies, Acer, Fagus, Fraxinus, Ostrya, Quercus* and *Tsuga* [58].

On a shorter time-scale, at Lake of the Clouds, Minnesota, periods of more frequent fires within the last millennium may have allowed *Pinus banksiana* to spread at the expense of fire-susceptible species

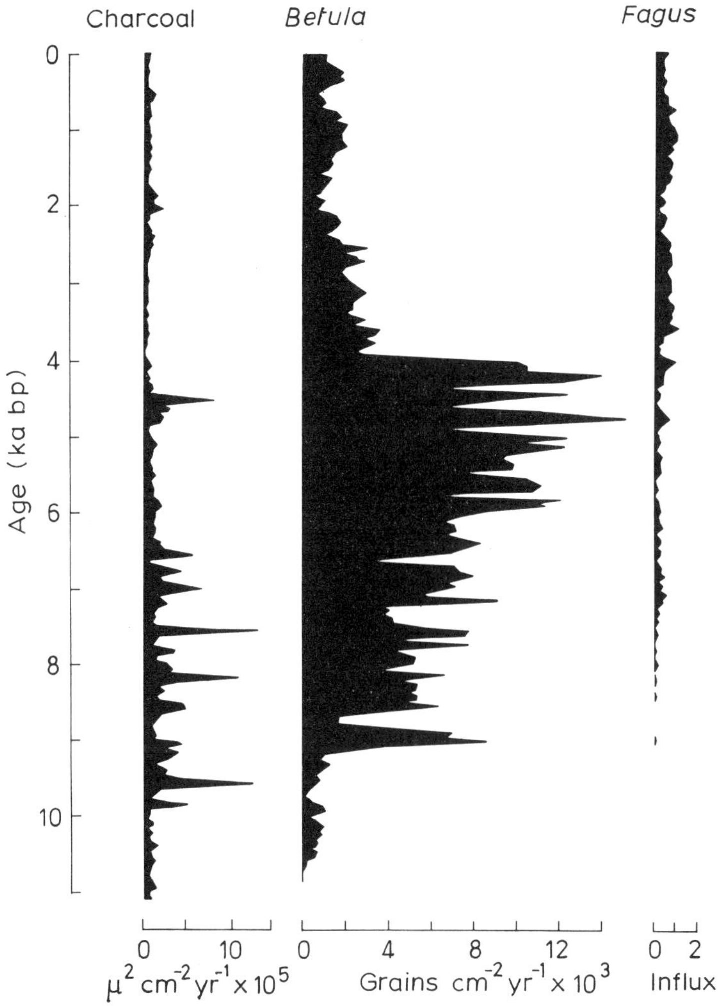

Figure 12.9 Influx of charcoal and pollen grains of *Betula* and *Fagus* to sediments at Everitt Lake, Novia Scotia. Re-drawn from Green [69].

such as *P. strobus*, *Picea glauca*, *Abies balsamea* and *Tsuga occidentalis* [77].

These eastern North American examples demonstrate the very important consequences that recurrent natural fires could have had on the floristic composition of certain forest types in the Holocene. In Australia and New Zealand, too, the prevalence of *Acacia* and *Eucalyptus* spp. in the modern woodlands has long been ascribed to regular burning. However in

Europe, even in the fire-sensitive Mediterranean region, the role of fire, apart from its use by Man as an agent of forest clearance, has largely been ignored. Clearly, however, fire should be regarded as a potentially major source of perturbation to many of the world's forests.

12.3.4 Pathogens

Plant and animal pathogens occur in probably all natural ecosystems, but are normally held in check by a system of regulatory mechanisms and by some degree of immunity by the host organisms. When these controls fail, serious pest or disease outbreaks may result. Two such outbreaks in indigenous forest communities this century, that will lead almost certainly to long-term changes in floristic composition and are well documented, have led to speculations that similar outbreaks might also have occurred in the past.

Chestnut blight, caused by the fungus *Endothia parasitica*, devastated populations of *Castanea dentata* in eastern North America between 1904 (when it was first recorded in New York City) and 1950 (by which time *Castanea* was virtually exterminated as a forest tree). In that period the disease spread over a distance of about 1500 km (Figure 12.10). Up to 85% of the canopy trees in the oak–chestnut forests of the Appalachian Mountains were killed [78], and in the long term might be expected to be replaced by *Quercus* and *Carya* spp. [79]. Lake sediments laid down over the last 100 years record the demise of *Castanea* by a dramatic decline in pollen values within a depth-interval of 1–5 cm (Figure 12.4).

In Europe, Dutch elm disease, caused by the fungus *Ceratocystis ulmi* and vectored by the bark-beetle *Scolytus*, spread from France to western Russia (a distance of about 1500 km) between 1918 and 1936, and resulted in considerable mortality of native and planted *Ulmus* populations (Figure 12.10). In Britain, where the disease was worst from 1927 to 1937, 10–20% of the mature elm trees were killed [80]. Dutch elm disease was recorded in the United States in 1930, and reached epidemic proportions in some states from the 1950s onwards. Up to 100 million trees may have been killed by the late 1970s [81]. The disease erupted again in Britain in 1965, probably as the result of the accidental importation of a new virulent strain of the fungus from Canada, and subsequently spread back to mainland Europe. As with chestnut blight, mortality of canopy trees has been almost total in some countries.

The pollen record of the chestnut blight outbreak appears as a sudden and nearly synchronous decline in pollen values over a large geographical area. Given sufficient analyses from suitable sediments (Figure 12.3), we might expect a similar picture for the Dutch elm disease outbreak. In the

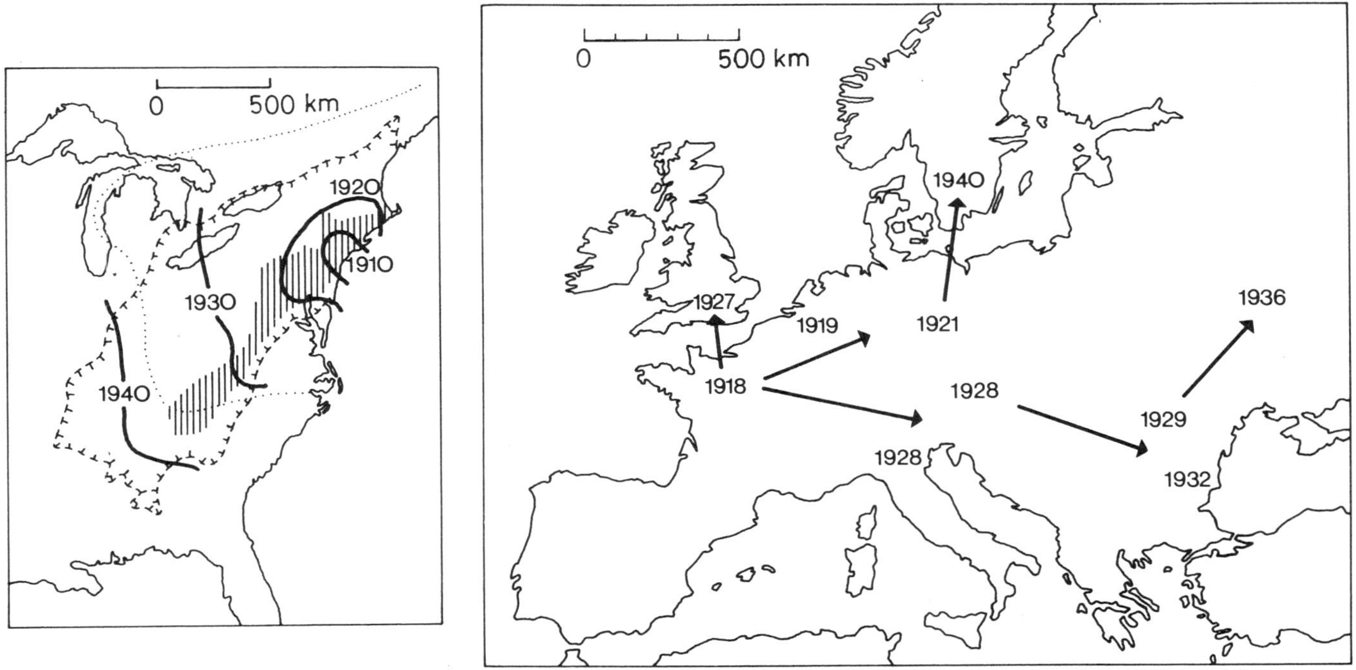

Figure 12.10 Twentieth-century disease epidemics in forests. (a) The spread of chestnut blight in eastern North America (re-drawn from Anderson [78]). |||||||, approximate extent of oak–chestnut forests; ⊥⊥⊥, range-limit of *Castanea dentata*; (........, approximate range-limits of *Tsuga canadensis* 5 ka ago). (b) The first spread of Dutch elm disease in Europe.

fossil pollen record of the Holocene there are two further events which show resemblances to the chestnut decline: the hemlock decline in North America and the elm decline in northwestern Europe.

Pollen of hemlock (*Tsuga canadensis*) underwent a drastic decline in frequency about 4.85 ka ago over an area of the eastern United States and southern Canada nearly as large as that ravaged by chestnut blight (Figure 12.10). Detailed examination of laminated sediments at one site in New Hampshire suggested that the decline occupied a time-period of only 7–8 yr, and that hemlock pollen remained at low frequency thereafter for about 500 yr [82]. The place of hemlock in the forests was taken over first by birch and then by other longer-lived canopy trees: maple and beech in northern New Hampshire, white pine (*Pinus strobus*) and oak in southern New Hampshire, and white pine and other conifers in Michigan, for example. Only after nearly 2000 years did hemlock recover and resume its position as a major canopy tree [82].

The likely cause of the hemlock decline is not clear. Margaret Davis [83] suggests that a pathogen outbreak could have been responsible, perhaps of caterpillars of the looper moth (*Lamdina fiscellaria*). A parallel may be drawn with the high mortality of *Abies fraseri* that has occurred in the southern Appalachians since 1960 as a result of severe infestation by a woolly aphid, *Adelges piceae* [84], and of *Fagus grandifolia* in New England this century as a result of beech-bark disease, caused jointly by a fungus and a scale insect [77].

The elm decline in northwestern Europe occurred within a few centuries either side of 5.0 ka ago, over an area rather larger than that affected by the recent chestnut blight outbreak. After nearly 50 years of study, palaeoecologists are still uncertain of its causes. A number of different environmental processes have been identified as operating over the period of the elm decline in Europe, but no single process can satisfactorily explain the diversity of effects recorded then in the pollen diagrams. There is increasing acceptance in recent years of the view that Dutch elm disease, or a similar infestation, was the underlying cause, and that other processes either aided its spread or weakened the vigour of the host trees [85,86].

In northwestern Europe *Ulmus glabra* was probably the only elm species affected [87], but in southeastern Europe, where the decline took place rather earlier (Figure 12.11), other species were involved too. To seek a unifying climatic effect that would disadvantage all these different species progressively along a SE/NW gradient of 1800 km and over a time-period of 2.0 ka is perhaps unrealistic, though a probable shift to a more continental climate in parts of northwestern Europe about 5.0 ka ago [41,88] could have lowered summer water tables sufficiently to affect *U. glabra* adversely [87,89]. There are clear signs too that soil deterioration had reached critical levels at certain sites by the start of the Neoglacial [90,91], and as all elms

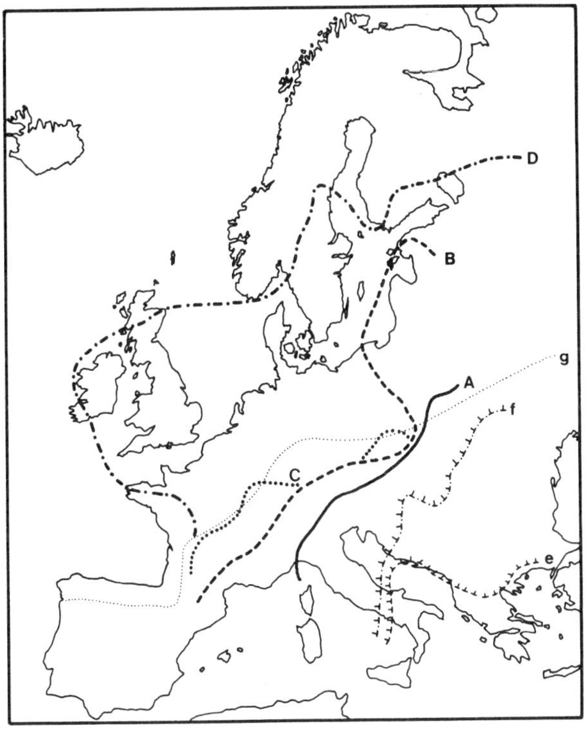

Figure 12.11 The elm decline in Europe. The lines A, B, C and D mark the approximate northwestern limits of sites where elm pollen values declined markedly over the periods 6.5–7.0 ka ago (A), 6.0–6.5 ka ago (B), 5.5–6.0 ka ago (C), and 5.0–5.5 ka ago (D). Lines e and f mark the southeastern limits of sites with $\geqslant 2\%$ elm pollen 7.0 and 5.0 ka ago, respectively. Line g marks the southern limit of regions where *Ulmus glabra* may have been the only elm species 5 ka ago [87]. Simplified from the maps in Huntley and Birks [99].

grow best on moist fertile soils [41] this deterioration could have had an adverse effect on their performance. Moreover, the soils favoured by elm were also those favoured by Early Neolithic settlers [42,43] and there is a close association in the pollen record between the decline of elm pollen and the appearance of hemerophilous pollen types such as Cerealia, *Linum* and *Vitis* (cultural types), and *Artemisia*, Compositae, *Plantago lanceolata*, *P. major*, *Polygonum aviculare*, *Pteridium*, *Ranunculus*, *Rumex*, *Succisa* and *Urtica* [88,90,92–95]. Some palaeoecologists have attributed the elm decline directly to the activities of these first farmers, either by lopping leafy branches of elm for use as animal fodder [95,96], or by forest clearance from the agriculturally more desirable soils [88], but to achieve a reduction in pollen frequency of the magnitude encountered in the pollen diagrams over so extensive an area in so short a time-period is seen as a prodigious feat [45]. Moreover, there is increasing evidence that agriculture preceded the elm decline in parts of Europe [93,97,98], so that the association of hemerophilous pollen with the elm decline at many sites could merely have been a fortuitous consequence of the removal of the 'filtration effect' of a continuous tree canopy [27,93]. At Pawlaw Moss in the uplands of northern England there are few signs of cultivation in the pollen diagram at the time of the elm decline, but clear evidence of burning of the forest [100]; it is possible that at such sites, at the limits of Early Neolithic penetration, the old hunter–gatherer way of life was still being practised. Yet the elm decline took place.

The forested areas favoured by Early Neolithic settlers were probably dominated by lime (*Tilia* spp.) over much of Europe [88]. The gradual demise of lime through the Neoglacial was almost certainly the result of human activity. That demise was not a sudden synchronous event like that of elm, though; there was, instead, considerable variability in the timing between sites [88,101]. The differing behaviours of elm and lime when exposed to similar pressures is further evidence in support of additional stress on the elm populations that could have been caused by epidemic disease – and enhanced in its effects by the weakening of the host trees through climatic change, soil deterioration, lopping and ring-barking, and facilitated in its spread by major movements of human populations.

Over much of Europe the elm populations never recovered after the elm decline. Their place in the forests of central and southeastern Europe was taken by beech and hornbeam [99]. In northwestern Europe oak and ash increased in abundance – though in some areas the pollen evidence indicates that there was partial or even complete recovery of the elm populations [89,90]. This recovery, like that of hemlock in North America, took place only slowly.

12.3.5 Catastrophic disturbance

Various other events of low frequency but high destructive capability may influence the long-term floristic composition of certain forest communities. These events include earthquakes, volcanic eruptions, hurricanes and cyclones, and landslides and avalanches. Their probability of occurrence is markedly non-random over the earth's surface. Thus major earthquake activity is primarily confined to comparatively narrow belts corresponding with plate-margin contacts (Figure 4.1). Most volcanic activity occurs in these belts also, with 62% concentrated in the 'ring of fire' surrounding the Pacific Ocean and 14% in the Indonesian islands. Hurricanes and cyclones are most frequent between latitudes 10° and 30°; on average, about one-quarter of all events each year occur in southeast Asia and one-

Figure 12.12 Forest regions of the world most likely to be regularly perturbed today by earthquakes (larger black areas), volcanic eruptions ● and cyclonic storms (arrows). The incidence of these perturbations is shown only within potentially forested regions (shaded). Base map re-drawn from Whittaker [38].

seventh in the Caribbean and adjacent waters (source: *Encyclopedia Britannica*). Consequently, some forest regions are more prone than others to these perturbations (Figure 12.12). However, severe damage may not necessarily follow perturbation: thus only 5% of the forest situated along the path of the 1938 New England hurricane was actually destroyed [102].

The frequency of occurrence of perturbation at a particular site is known in only a few instances: either from documentary records, tree-ring records, radiocarbon-dating of engulfed forest, or painstaking examination of tree remains on the forest floor. Some examples are given in Table 12.6. It is clear that the average return frequency of perturbation, as with wildfires, is very variable, ranging from decades to many centuries. For some light-demanding tree species such infrequent perturbation may be essential for their continued survival. At least two cases are known where the canopy-dominants maintain their position in this way: the lower-altitude forests in the Valdivian Andes of Chile dominated by *Nothofagus dombeyi* and *N. alpina* (which are essentially long-lived and massive pioneer species on ground disturbed by tectonic activity [24]); and the tropical forests of the Solomon Islands dominated by *Campnosperma brevipetiolatum* (a pioneer of cyclone-devastated areas [103]). Many other cases must be suspected on the basis of the 'sensitive' regions identified in Figure 12.12.

Table 12.6 Approximate return frequencies* for various perturbations in forest regions of the world†

Location	Return frequency (yr)	Time-period (yr AD)
Samoa (C) [103]	20	1960–1970
New Hampshire (C) [104]	80	1890–1970
Northern Queensland (C) [105]	140	1867–1939
Valdivian Andes, Chile (E) [24]	170	1520–1960
Puerto Rico (E) [106]	240	1928–1976
Papua New Guinea (E) [107]	400–800	1900–1976
Massachusetts (C) [108]	2800	1400–1960
North-east China (E) [109]	2900	1400–1980
Northern New Hampshire (C) [102]	4800	1492–1976

* Return frequency = 'mean interval between perturbation' divided by 'mean proportion of area perturbed at each event'.

† C = cyclone or hurricane damage, E = earthquake and volcanic damage.

12.4 EPILOGUE: MODERN FORESTS

In this chapter we have emphasized the processes that have acted on forest vegetation during the Neoglacial. Some of the compositional changes that have taken place in the world's forests over the last 5 ka as a result of these processes were summarized in the Introduction. So pervasive are the products of perturbation today that it is tempting to ascribe the composition of modern forest communities wholly to the action of Neoglacial disturbance. To do so, however, is to ignore the earlier perturbations that have affected forests, and determined the pool of species available to respond to the vicissitudes of the Neoglacial. The very great contrasts between the modern temperate forests of Europe and eastern North America, for example, are less a consequence of the different settlement patterns of the two regions than of their Pleistocene histories. The fossil record shows clearly that certain major tree taxa became extinct in Europe, but not in North America, during the Pleistocene; as a result trees of smaller stature have been able to fill niches in Europe that are not available to them in North America. Examples of such 'replacements' are *Liriodendron* by *Carpinus*, *Liquidambar* by *Alnus*, *Carya* by *Corylus*, and *Tsuga* and *Thuja* by *Taxus* [110]. Nevertheless, although we might lament the floristic poverty of European forests, we must remember that they are made up of 'survivors' – trees that have weathered the perturbations of the past. To some extent their future is assured. Yet even these survival forests may succumb if perturbation is long-continued, and, in particular, if it acts in combination with climatic stress. Rather little has been said in the present chapter about the burgeoning impact of Man on the world's forests through the Neoglacial; though as a result of this destructive impact over the last 5 ka, the extent of non-forest communities has been substantially enlarged. The prime example of human impact acting conjointly with climatic stress is the deforestation of the Mediterranean Basin; and this event is considered further in the next chapter, after related non-forest communities and their history have been examined.

REFERENCES

1. Birks, H. J. B. (1986) Late-Quaternary biotic changes in terrestrial and lacustrine environments, with particular reference to northwest Europe, in *Handbook of Holocene Palaeoecology and Palaeohydrology* (ed. B. E. Berglund), Wiley, Chichester, pp. 3–65.
2. Von Post, L. (1946) The prospect for pollen analysis in the study of the earth's climatic history. *New Phytol.*, **45**, 193–217.

3. Kukla, G. (1980) End of the last interglacial: a predictive model of the future? *Palaeoecol. Africa*, **12**, 395–408.
4. Huntley, B. (1988) Europe, in *Vegetation History* (eds B. Huntley and T. Webb III), Kluwer Academic Publishers, Dordrecht, pp. 341–83.
5. Ritchie, J. C. and Yarranton, G. A. (1978) The Late-Quaternary history of the boreal forest of central Canada, based on standard pollen stratigraphy and principal components analysis. *J. Ecol.*, **66**, 199–212.
6. Webb, T. III (1986) Is vegetation in equilibrium with climate? How to interpret Late-Quaternary pollen data. *Vegetatio*, **67**, 75–91.
7. Webb, T. III, Cushing, E. J. and Wright, H. E. Jr (1984) Holocene changes in the vegetation of the Midwest, in *Late-Quaternary Environments of the United States*, Vol. 2, *The Holocene* (ed. H. E. Wright Jr), Longman, London, pp. 142–65.
8. Tsukada, M. (1982) Late-Quaternary development of the *Fagus* forest in the Japanese Archipelago. *Jap. J. Ecol.*, **32**, 113–8.
9. Davis, M. B., Spear, R. W. and Shane, L. C. K. (1980) Holocene climate of New England. *Quat. Res.*, **14**, 240–50.
10. Heusser, C. J. (1974) Vegetation and climate of the southern Chilean Lake District during and since the last interglaciation. *Quat. Res.*, **4**, 290–315.
11. Lees, C. M. (1986) Late Quaternary palynology of the southern Ruahine Range, North Island, New Zealand. *N. Z. J. Bot.*, **24**, 315–29.
12. McGlone, M. S. and Topping, W. W. (1977) Aranuian (post-glacial) pollen diagrams from the Tongariro region, North Island, New Zealand. *N. Z. J. Bot.*, **15**, 749–60.
13. Watts, W. A. (1979) Late-Quaternary vegetation of central Appalachia and the New Jersey coastal plain. *Ecol. Monogr.*, **49**, 427–69.
14. Moore, D. M. (1978) Post-glacial vegetation in the South Patagonian territory of the giant ground sloth *Mylodon*. *Bot. J. Linn. Soc.*, **77**, 177–202.
15. Beug, H.-J. (1982) Vegetation history and climatic change in central and southern Europe, in *Climatic Change in Later Prehistory* (ed. A. F. Harding), Edinburgh University Press, Edinburgh, pp. 85–102.
16. Dubois, A. D. and Ferguson, D. K. (1985) The climatic history of pine in the Cairngorms based on radiocarbon dates and stable isotope analysis, with an account of the events leading up to its colonization. *Rev. Palaeobot. Palynol.*, **46**, 55–80.
17. Knox, J. C. (1984) Responses of river systems to Holocene climates, in *Late-Quaternary Environments of the United States*, Vol. 2, *The Holocene* (ed. H. E. Wright Jr), Longman, London, pp. 26–41.
18. Starkel, L. (1984) The reflection of abrupt climatic changes in the relief and sequence of continental deposits, in *Climatic Changes on a Yearly to Millennial Basis* (eds N.-A. Mörner and W. Karlén), Reidel, Dordrecht, pp. 136–46.
19. Wendland, W. M. and Bryson, R. A. (1974) Dating climatic episodes of the Holocene. *Quat. Res.*, **4**, 9–24.
20. Graham, S. A. (1941) Climax forests of the Upper Peninsula of Michigan. *Ecology*, **22**, 355–62.
21. Whitmore, T. C. (1982) On pattern and process in forests, in *The Plant Community as a Working Mechanism* (ed. E. I. Newman), Blackwell, Oxford, pp. 45–59.

22. Rowe, J. S. (1961) Critique of some vegetational concepts applied to forests of northwestern Alberta. *Can. J. Bot.*, **39**, 1007–17.
23. Hanxi, Y. and Yegang, W. (1987) Tree composition, age structure and regeneration strategy of the mixed broadleaved/*Pinus koraiensis* (Korean pine) forest in Changbai Mountain Reserve, in *The Temperate Forest Ecosystem* (eds Y. Hanxi, W. Zhan, J. N. R. Jeffers and P. A. Ward), ITE, Merlewood, Cumbria, pp. 12–20.
24. Veblen, T. T. and Ashton, D. H. (1978) Catastrophic influences on the vegetation of the Valdivian Andes, Chile. *Vegetatio*, **36**, 149–67.
25. Bell, M. and Limbrey, S. (eds) (1982) *Archaeological Aspects of Woodland Ecology*, BAR International Series 146, Oxford.
26. Jones, M. (ed.) (1988) *Archaeology and the Flora of the British Isles*, Oxbow Books, Oxford.
27. Perry, I. and Moore, P. D. (1987) Dutch elm disease as an analogue of Neolithic elm decline. *Nature*, 326, 72–3.
28. Brugam, R. B. (1978) Pollen indicators of land-use change in southern Connecticut. *Quat. Res.*, **9**, 349–63.
29. Godwin, H. (1944) Neolithic forest clearance. *Nature*, **153**, 511–2.
30. Behre, K.-E. (1981) The interpretation of anthropogenic indicators in pollen diagrams. *Pollen Spores*, **23**, 225–45.
31. Delcourt, P. A., Delcourt, H. R., Cridlebaugh, P. A. and Chapman, J. (1986) Holocene ethnobotanical and paleoecological record of human impact on vegetation in the Little Tennessee River Valley, Tennessee. *Quat. Res.*, **25**, 330–49.
32. McAndrews, J. H. (1988) Human disturbance of North American forests and grasslands: the fossil pollen record, in *Vegetation History* (eds B. Huntley and T. Webb III), Kluwer Academic Publishers, Dordrecht, pp. 673–98.
33. Flenley, J. R. (1979) *The Equatorial Rain Forest: a Geological History*, Butterworths, London.
34. Hamilton, A. C. (1982) *Environmental History of East Africa. A Study of the Quaternary*, Academic Press, London.
35. Leyden, B. W. (1987) Man and climate in the Maya lowlands. *Quat. Res.*, **28**, 407–14.
36. Maloney, B. K. (1985) Man's impact on the rainforests of West Malesia: the palynological record. *J. Biogeogr.*, **12**, 537–58.
37. Walker, D. and Flenley, J. R. (1979) Late Quaternary vegetational history of the Enga Province of upland Papua New Guinea. *Phil. Trans. Roy. Soc. Lond., B*, **286**, 265–344.
38. Whittaker, R. H. (1970) *Communities and Ecosystems*, Collier-Macmillan, Toronto.
39. Roberts, N. (1982) Forest re-advance and the Anatolian Neolithic. *B.A.R. Internat. Series*, **146**, 231–46.
40. Bender, B. (1975) *Farming in Prehistory. From Hunter–Gatherer to Food-Producer*, Baker, London.
41. Iversen, J. (1960) Problems of the early post-glacial forest development in Denmark. *Danm. Geol. Unders. Series IV*, **4**, 4–32.
42. Jarman, M. R., Bailey, G. N. and Jarman, H. N. (eds) (1982) *Early European*

Agriculture. Its Foundations and Development, Cambridge University Press, Cambridge.

43. Kosse, K. (1979) Settlement ecology of the Early and Middle Neolithic Körös and Linear Pottery cultures in Hungary. *B.A.R. Internat. Series*, **64**, 1–238.
44. Evans, J. G. (1975) *The Environment of Early Man in the British Isles*, Elek, London.
45. Rowley-Conwy, P. (1982) Forest grazing and clearance in temperate Europe with special reference to Denmark: an archaeological view. *B.A.R. Internat. Series*, **146**, 199–215.
46. Turner, J. (1965) A contribution to the history of forest clearance. *Proc. Roy. Soc., Series B*, **161**, 343–54.
47. Engelmark, R. (1976) The vegetational history of the Umeå area during the past 4000 years. *Early Norrland*, **9**, 75–111.
48. Walker, D. (1982) The development of resilience in burned vegetation, in *The Plant Community as a Working Mechanism* (ed. E. I. Newman), Blackwell, Oxford, pp. 27–43.
49. Stewart, O. C. (1956) Fire as the first great force employed by man, in *Man's Role in Changing the Face of the Earth* (ed. W. L. Thomas), Chicago University Press, Chicago, pp. 115–33.
50. Wright, H. A. and Bailey, A. W. (1982) *Fire Ecology, United States and Southern Canada*, Wiley, New York.
51. Innes, J. B. and Simmons, I. G. (1988) Disturbance and diversity: floristic changes associated with pre-elm decline woodland recession in northeast Yorkshire, in *Archaeology and the Flora of the British Isles* (ed. M. Jones), Oxbow Books, Oxford, pp. 7–20.
52. Jacobi, R. M., Tallis, J. H. and Mellars, P. A. (1976) The southern Pennine Mesolithic and the ecological record. *J. Archaeol. Sci.*, **3**, 307–20.
53. Daubenmire, R. F. (1968) Ecology of fires in grasslands. *Adv. Ecol. Res.*, **5**, 209–66.
54. Wein, R. W. and MacLean, D. A. (eds) (1983) *The Role of Fire in Northern Circumpolar Ecosystems*, Wiley, Chichester.
55. Tolonen, K. (1983) The post-glacial fire record, in *The Role of Fire in Circumpolar Ecosystems* (eds R. W. Wein and D. A. MacLean), Wiley, Chichester, pp. 21–44.
56. Swain, A. M. (1973) A history of fire and vegetation in northeastern Minnesota as recorded in lake sediments. *Quat. Res.*, **3**, 383–96.
57. Winkler, M. G. (1985) A 12,000-year history of vegetation and climate for Cape Cod, Massachusetts. *Quat. Res.*, **23**, 301–12.
58. Green, D. G. (1982) Fire and stability in the postglacial forests of southwest Nova Scotia. *J. Biogeogr.*, **9**, 29–40.
59. Cwynar, L. C. (1987) Fire and the forest history of the North Cascade Range. *Ecology*, **68**, 791–802.
60. Colhoun, E. A. (1985) Pre-last glaciation maximum vegetation history at Henty Bridge, western Tasmania. *New Phytol.*, **100**, 681–90.
61. Molloy, B. P. J., Burrows, C. J., Cox, J. E. *et al.* (1963) Distribution of subfossil forest remains, eastern South Island, New Zealand, *N. Z. J. Bot.*, **1**, 68–77.
62. Cwynar, L. C. (1978) Recent history of fires and vegetation from laminated

sediment of Greenleaf Lake, Algonquin Park, Ontario. *Can. J. Bot.*, **56**, 10–21.

63. Swain, A. M. (1978) Environmental changes during the last 2000 years in north-central Wisconsin: analysis of pollen, charcoal and seeds from varved lake sediments. *Quat. Res.*, **10**, 55–68.
64. Van Wagner, C. E. (1983) Fire behaviour in northern conifer forests and shrublands, in *The Role of Fire in Northern Circumpolar Ecosystems* (eds R. W. Wein and D. A. MacLean), Wiley, Chichester, pp. 65–80.
65. Hanes, T. L. (1981) California chaparral, in *Ecosystems of the World 11. Mediterranean-type Shrublands* (eds F. di Castri, D. W. Goodall and R. L. Specht), Elsevier, Amsterdam, pp. 139–74.
66. Zackrisson, O. (1977) Influence of forest fires on the north Swedish boreal forest. *Oikos*, **29**, 22–32.
67. Heinselman, M. L. (1973) Fire in the virgin forests of the Boundary Waters Canoe Area, Minnesota. *Quat. Res.*, **3**, 329–82.
68. Brubaker, L. B. (1986) Responses of tree populations to climatic change. *Vegetatio*, **67**, 119–30.
69. Green, D. G. (1981) Time series and postglacial forest ecology. *Quat. Res.*, **15**, 265–77.
70. Cottam, G. (1949) The phytosociology of an oak woods in southwestern Wisconsin. *Ecology*, **30**, 271–87.
71. Wells, P. V. (1970) Postglacial vegetational history of the Great Plains. *Science*, **167**, 1574–82.
72. Behre, K.-E. (1988) The role of man in European forest history, in *Vegetation History* (eds B. Huntley and T. Webb III), Kluwer Academic Publishers, Dordrecht, pp. 633–72.
73. Bryson, R. A., Irving, W. N. and Larsen, J. A. (1965) Radiocarbon and soils evidence of former forest in the southern Canadian tundra. *Science*, **147**, 46–8.
74. Lamb, H. F. and Edwards, M. E. (1988) The Arctic, in *Vegetation History* (eds B. Huntley and T. Webb III), Kluwer Academic Publishers, Dordrecht, pp. 519–55.
75. Grimm, E. C. (1983) Chronology and dynamics of vegetation change in the prairie–woodland region of southern Minnesota, USA. *New Phytol.*, **93**, 311–50.
76. Rowe, J. S. (1983) Concepts of fire effects on plant individuals and species, in *The Role of Fire in Northern Circumpolar Ecosystems* (eds R. W. Wein and D. A. MacLean), Wiley, Chichester, pp. 135–54.
77. Patterson, W. A. III and Backman, A. E. (1988) Fire and disease history of forests, in *Vegetation History* (eds B. Huntley and T. Webb III), Kluwer Academic Publishers, Dordrecht, pp. 603–32.
78. Anderson, T. W. (1974) The chestnut pollen decline as a time horizon in lake sediments in eastern North America. *Can. J. Earth Sci.*, **11**, 678–85.
79. Keever, C. (1953) Present composition of some stands of the former oak–chestnut forests of the southern Blue Ridge Mountains. *Ecology*, **34**, 44–54.
80. Gibbs, J. N. (1978) Intercontinental epidemiology of Dutch Elm Disease. *Ann. Rev. Phytopath.*, **16**, 287–307.

81. Karnosky, D. F. (1979) Dutch elm disease: a review of the history. Environmental implications, cultural and research needs. *Env. Conserv.*, **6**, 311–22.
82. Allison, T. D., Moeller, R. E. and Davis, M. B. (1986) Pollen in laminated sediments provides evidence for a mid-Holocene forest pathogen outbreak. *Ecology*, **67**, 1101–5.
83. Davis, M. B. (1984) Holocene vegetational history of the eastern United States, in *Late-Quaternary Environments of the United States*, Vol. 2, *The Holocene* (ed. H. E. Wright Jr), Longman, London, pp. 166–81.
84. White, P. S. (1987) Research objectives in US National Park Service Biosphere Reserves, with special reference to Great Smoky Mountains National Park, in *The Temperate Forest Ecosystem* (eds Y. Hanxi, W. Zhan, J. N. R. Jeffers and P. A. Ward), ITE, Merlewood, Cumbria, pp. 153–63.
85. Girling, M. A. (1988) The bark beetle *Scolytus scolytus* (Fabricius) and the possible role of elm disease in the early Neolithic, in *Archaeology and the Flora of the British Isles* (ed. M. Jones), Oxbow Books, Oxford, pp. 34–8.
86. Rackham, O. (1980) *Ancient Woodland: its History, Vegetation and Uses in England*, Arnold, London.
87. Richens, R. H. (1983) *Elm*, Cambridge University Press, Cambridge.
88. Iversen, J. (1973) The Development of Denmark's Nature since the Last Glacial. *Danm. Geol. Unders. Series V*, **7-C**, 1–126.
89. Ten Hove, H. A. (1968) The *Ulmus* fall at the transition Atlanticum–Subboreal in pollen diagrams. *Palaeogeogr. Palaeoclimatol. Palaeoecol.*, **5**, 359–69.
90. Pennington, W. (1965) The interpretation of some post-glacial diversities at different Lake District sites. *Proc. Roy. Soc., Series B*, **161**, 310–23.
91. Pennington, W. (1975) The effect of Neolithic man on the environment in northwest England: the use of absolute pollen diagrams, in *The Effect of Man on the Landscape: the Highland Zone* (eds J. G. Evans, S. Limbrey and H. Cleere), Council for British Archaeology, London, pp. 74–86.
92. Beckett, S. C. and Hibbert, F. A. (1979) Vegetational change and the influence of prehistoric man in the Somerset Levels. *New Phytol.*, **83**, 577–600.
93. Groenman-van Waateringe, W. (1983) The early agricultural utilization of the Irish landscape: the last word on the Elm Decline? *B.A.R. British Series*, **116**, 217–32.
94. Scaife, R. G. (1988) The elm decline in the pollen record of southeast England and its relationship to early agriculture, in *Archaeology and the Flora of the British Isles* (ed. M. Jones), Oxbow Books, Oxford, pp. 21–33.
95. Troels-Smith, J. (1960) Ivy, mistletoe and elm. Climate indicators–fodder plants. *Danm, Geol. Unders. Series IV*, **4**, 1–32.
96. Heybroek, H. M. (1963) Diseases and lopping for fodder as possible causes of a prehistoric decline of *Ulmus. Acta bot. neerl.*, **12**, 1–11.
97. Edwards, K. J. and Hirons, K. R. (1984) Cereal pollen grains in pre-elm decline deposits: implications for the earliest agriculture in Britain and Ireland. *J. Archaeol. Sci.*, **11**, 71–80.
98. Molloy, K. and O'Connell, M. (1987) The nature of the vegetational changes at about 5000 B.P. with particular reference to the elm decline: fresh evidence from Connemara, western Ireland. *New Phytol.*, **107**, 203–20.

99. Huntley, B. and Birks, H. J. B. (1983) *An Atlas of Past and Present Pollen Maps for Europe: 0–13,000 Years Ago*, Cambridge University Press, Cambridge.

100. Sturlodottir, S. A. and Turner, J. (1985) The elm decline at Pawlaw Mire: an anthropogenic interpretation. *New Phytol.*, **99**, 323–9.

101. Turner, J. (1962) The *Tilia* decline: an anthropogenic interpretation. *New Phytol.*, **61**, 328–41.

102. Bormann, F. H. and Likens, G. E. (1979) *Pattern and Process in a Forested Ecosystem*, Springer Verlag, New York.

103. Whitmore, T. C. (1975) *Tropical Rain Forests of the Far East*, Clarendon Press, Oxford.

104. Henry, J. D. and Swan, J. M. A. (1974) Reconstructing forest history from live and dead plant material – an approach to the study of plant succession in southwest New Hampshire. *Ecology*, **55**, 772–83.

105. Webb, L. J. (1958) Cyclones as an ecological factor in tropical lowland rain forest, north Queensland. *Aust. J. Bot.*, **6**, 220–8.

106. Crow, T. R. (1980) A rain forest chronicle: a 30 year record of change in structure and composition at El Verde, Puerto Rico. *Biotropica*, **12**, 42–55.

107. Garwood, N. C., Janos, D. P. and Brokaw, N. (1979) Earthquake-caused landslides: a major disturbance to tropical forests. *Science*, **205**, 997–9.

108. Oliver, C. D. and Stephens, E. P. (1977) Reconstruction of a mixed-species forest in central New England. *Ecology*, **58**, 562–72.

109. Machida, H., Moriwaki, H. and Arai, F. (1987) Historical eruptions of the Changbai volcano resulting in large-scale forest devastation (deduced from widespread tephra), in *The Temperate Forest Ecosystem* (eds Y. Hanxi, W. Zhan, J. N. R. Jeffers and P. A. Ward), ITE, Merlewood, Cumbria, pp. 23–26.

110. Campbell, J. J. N. (1982) Pears and persimmons: a comparison of temperate forests in Europe and eastern North America. *Vegetatio*, **49**, 85–101.

CHAPTER THIRTEEN

Dryland communities

13.1 INTRODUCTION

Climates suitable for the development of forest and woodland communities currently affect about 35% of the world's land surface; a further 16% is capable of supporting sparse tree growth (Table 2.2). Nearly half of the present land surface, however, is climatically unsuitable for the growth of trees. In these areas that are too cold, too dry, or too wet for trees to grow, a range of tundra and alpine, desert and semi-desert, grassland and peatland communites now occurs (Figure 13.1). Other non-forest communities can be found where forest has been cleared over the centuries – from more than 10 million km^2 of land, principally in temperate regions. Here tree growth is prevented not by climatic limitations but by tillage, grazing, or burning. The actual extent of forest and woodland at the present day is consequently only 27.5% of the world's land surface (40.9 million km^2 [1]). However, there is clear evidence that trees have formerly grown in what are today some of the world's most inhospitable regions. Obvious woody remains, often of massive trunks and roots, have been found widely

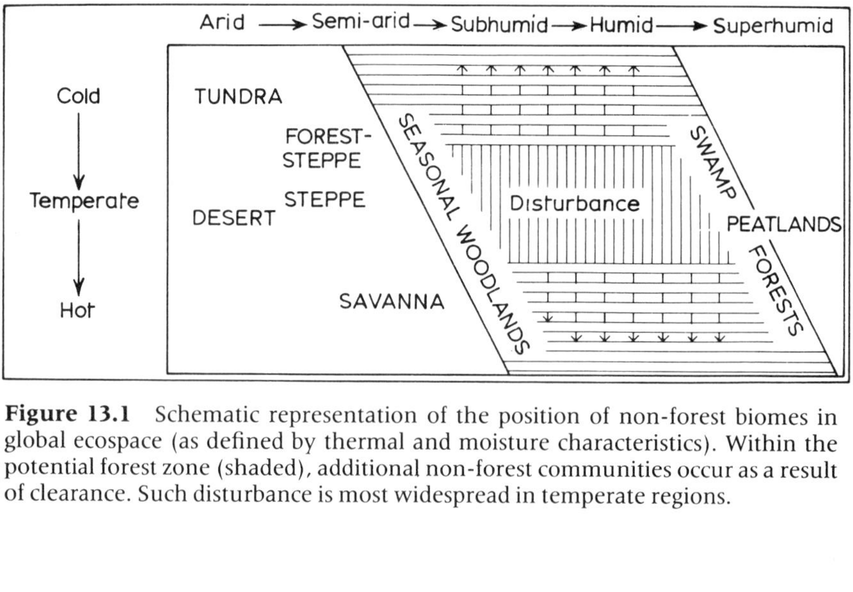

Figure 13.1 Schematic representation of the position of non-forest biomes in global ecospace (as defined by thermal and moisture characteristics). Within the potential forest zone (shaded), additional non-forest communities occur as a result of clearance. Such disturbance is most widespread in temperate regions.

in the Sahara Desert [2], the prairies of Wyoming [3], and the Canadian and Siberian tundra (section 11.4); they have also been found above the treeline in many mountain ranges of North America and Europe (section 11.4), and below the peat in many wetland areas [4,5]. Vivid and lifelike paintings and engravings of animals such as antelope, crocodile, elephant, giraffe, hippopotamus, ostrich, rhinoceros and water buffalo on rock faces in the arid central Sahara testify to a more genial climate there in the past [6–8]; sometimes the bleached bones of these animals have also been found preserved in the desert sand [9]. The ages of these incongruous plant and animal remains range from many million to a few hundred years. The plant remains are a clear demonstration of the flux between forest and non-forest communities over the long time-span considered in this book. That flux during the Tertiary Period has already been described in Chapter 9. Its continuation through the last 20 ka of the Quaternary Period, and the character of the participating floras, is examined in this chapter and the next. We consider first the floras of dryland regions, and the communities that they form (desert, steppe and savanna, and also some seasonal woodlands – Figure 13.1); the emphasis is on desert floras, as these represent the culmination of adaptation to drought. Subsequently in Chapter 14 we examine the history of the cold-adapted floras of tundra, forest–steppe and boreal peatland regions of the northern hemisphere.

13.2 THE CHARACTER OF DRYLAND FLORAS

13.2.1 Speciation in arid environments

Modern dryland floras are characterized by a higher frequency of herbaceous taxa than the forest floras that adjoin them. In regions of lowered annual precipitation, where seasonal woodland communities occur, the competitiveness of the tree cover declines progressively with decreasing precipitation. That decline is expressed in reduced interception of incident light by the tree canopy at certain times of the year, reduced stature of the individual trees and shrubs, and increased spacing between individuals, and is accompanied by an increasing frequency of herbaceous taxa [10]. The ability of herbaceous taxa to spread under these conditions is evident in regions where the competitiveness of the tree cover has been further diminished by prolonged human intervention. Thus the three climatically similar regions of Israel, central Chile and coastal California differ in the lengths of time to which their native floras have been subjected to extensive human settlement (Figure 12.5), and have very different patterns of representation of herbaceous and woody taxa (Table 13.1). Annual species are most prominent in the Israeli flora, subjected to 8–10 ka of

Table 13.1 Physiognomic characteristics of some mediterranean-type floras

		Number of species			
		Total	Woody taxa	Annual taxa	Perennial herbs
Israel	open shrubland	139	27	72	40
	closed shrubland	34	17	5	12
	grazed woodland	134	14	97	23
California	open shrubland	65	23	27	15
	closed shrubland	33	12	15	6
	grazed woodland	63	3	46	14
Chile	open shrubland	108	41	39	28

Data of Naveh and Whittaker [11].

human impact, and least common in California, with less than 250 yr of white settlement. In both these regions, species-richness is greatest in grazed communities, and is markedly reduced in closed shrubland communities [11].

Herbaceous species, and particularly annual ones, are widely believed to undergo faster rates of evolution, on average, than woody ones (Table 7.3). Moreover, arid environments generate a number of plant responses that are themselves conducive to rapid evolutionary change, so that the development of aridity is seen as a major 'evolutionary pump' mechanism [12]. Drought-stressed environments probably arose initially in the Early Tertiary as localized 'islands of unfavourability' within the enveloping forest matrix, and even when they came subsequently to occupy larger areas they must still have presented to plant colonists a small-scale mosaic of separated habitats with differing moisture availabilities – and hence with differing selection pressures that each favoured a slightly different genome. The primary sources of colonization of these developing dryland 'islands' must have been the immediately adjoining woody plants in the investing forest or woodland, and any spatial floristic heterogeneity there would have been reflected in the composition of the colonizing floras. As these floras subsequently maintained partial or complete reproductive isolation from each other, any initial floristic differences could thus have been progressively intensified. Moreover, morphological adaptation to drought stress can be achieved in a number of ways (and presumably by a number of different genetic mechanisms): by the reduction of leaf surface-areas, the development of deciduous leaves, the development of specialized epidermal coverings or sunken stomata, or the development of extensive root systems or underground perennating organs. All these, as well as the adoption of an annual habit, represent alternative and viable solutions to the problem of drought stress, so that a variety of morphological responses is possible and a variety of genetic mutants can be utilized. The likelihood of such genetic mutants becoming established is much greater in dryland communities than in forest, because of the reduced competition. In extremely arid environments competition ceases to be a limiting factor. Accordingly, aberrant individuals or populations have an inherently greater chance of survival in dryland situations [12].

Nevertheless, the most extreme dryland environments present a considerable challenge to plant growth and survival, and the chance of appropriate mutants arising and surviving has always been small. Because competition is low, the chance of establishment, not only of these mutants but also of the occasional disseminule dispersed to exceptional distances, is high. Even though long-distance dispersal is a rare event, its frequency may still exceed that of mutation. Consequently, the possibility of long-distance dispersal as a method of recruiting new individuals to dryland areas can never be discounted. Its effect will be to increase floristic similarities between geographically-separated regions.

13.2.2 Similarities between dryland floras

Similarities between modern dryland floras, whether at family, genus, or species level, may have arisen in various ways in the past. Some shared components, firstly, may be derivatives of the common ancestral stock that gave rise to the Madrean–Tethyan Geoflora in the Early Tertiary Period (section 9.2.2). Many woody genera found in both the Mediterranean Basin and California are probably derivatives of this type (Table 13.2). Other shared components may be derived from large, wide-ranging, genetically-plastic genera and families that have been a consistent source of mutant individuals throughout the Cenozoic Era. Thus the majority of desert plants belong to a rather small number of families, most of which are large and 'successful' at the present day (Table 13.3); three of the four

Table 13.2 Wide-ranging genera in dryland floras

(a)	Wide-ranging in desert floras [13] *Acacia, Capparis, Euphorbia, Heliotropium, Indigofera, Lappula, Panicum, Poa, Polygala, Senecio, Zizyphus*
(b)	Common to the Mediterranean Basin and California [14] *Acer, Aesculus, Alnus, Arbutus, Cercis, Clematis, Crataegus, Cupressus, Fraxinus, Juniperus, Lonicera, Pinus, Platanus, Populus, Prunus, Quercus, Rhamnus, Rhus, Rosa, Rubus, Smilax, Staphylea, Styrax, Viburnum, Vitis*
(c)	Wide-ranging in savanna floras [15] *Aristolochia, Aster, Cassia, Eupatorium, Euphorbia, Phaseolus, Polygala, Salvia, Senecio, Vernonia*

Table 13.3 Major families in the desert flora, and the number of warm-desert regions (ex 7*) in which they occur. Data of Schmida [13]

Chenopodiaceae	7
Compositae	7
Gramineae	7
Leguminosae	7
Cruciferae	5
Zygophyllaceae	5
Boraginaceae	4
Liliaceae	4
Scrophulariaceae	4

*Australian, South African, Irano–Turanian, Saharan, Middle Eastern, Thar and North American.

largest families in the world (Compositae, Gramineae and Leguminosae) are included. Many of the plants also belong to wide-ranging and successful genera (with species in several different habitat-types); examples are given in Table 13.2. The same is true of savanna floras (Table 13.2).

Yet other shared components may reflect former continuity of dryland habitats. Greater-than-average similarity between two dryland floras may indicate that the two regions were connected at one time. Four pairs of currently-separated dryland floras fall into this category: North and South Africa, North Africa and India, central Asia and the Great Basin of North America, and North and South America [13]. Good independent evidence for former complete or partial land linkage exists for three of these paired regions. (1) The Northern and Southern African dryland floras, firstly, have about 180 genera in common, and share at least 38 species or species pairs [13]; the similarities are mainly in the savanna and dry woodland floras, and only eight true desert species are common to the Sahara and Namib Deserts. Savanna vegetation is known to have bridged the equator during the hypothermal stages of the Pleistocene, when the rainforest was greatly contracted in extent (section 10.5), and interchange of taxa would almost certainly have occurred then. By contrast, the desert regions in the north and south of the continent have probably always remained separate from each other. (2) Similarities between the southern Sahara and the Thar Desert of northwestern India are again mainly in the savanna floras, with such shared genera as *Capparis*, *Indigofera*, *Maerua*, *Moringa*, *Salvadora* and *Zizyphus* [13]. In addition to near-continuity of vegetation-type during glacial stages, a savanna connection between the two areas may have been present as recently as the mid-Holocene, when lake levels in Arabia were high [16] and rainfall considerably above present levels in the Thar Desert

[17]. (3) The floristic links between a third pair of regions, central Asia and the Great Basin, include a number of genera that were widespread in the full-glacial steppes of Eurasia (*Artemisia*, *Atriplex* and *Eurotia*, for example), and migration via Beringia at times of lowered sea-level is clearly likely to have occurred. The land bridge, however, was forested until late in the Pliocene, so that Pleistocene interchange is more likely; and as the North American Ice Sheet would have blocked direct passage between the two areas at full glacial, interchange during the less cold periods of each glacial stage must be postulated. Direct fossil evidence is lacking. (4) Floristic links between North and South America comprise a number of important desert genera (*Larrea*, *Prosopis*, for example), some dry woodland trees (species of *Dodonaea*, *Gutierrezia*, *Hoffmanseggia* and *Menodora*) and grasses (species of *Andropogon*, *Bouteloua* and *Scleropogon*), several shared species that are confined to high latitudes in the two continents, and more than one hundred wider-ranging temperate species and genera [18]. Current opinion favours long-distance dispersal as the agent, perhaps by strong-flying migrant birds targeting on non-forest vegetation.

The best instance of long-distance dispersal between dryland habitats, however, is furnished by certain *halophytic* species, which occur in both desert and coastal habitats throughout the world – notably *Salicornia europaea*, *Salsola kali*, *Spergularia marina* and *Suaeda maritima*. In these and other ecologically similar taxa, saline desert may have been the original habitat, and coastal salt marshes were used as staging posts between one desert region and another. Many of these desert halophytes belong to the Chenopodiaceae, a family which has a concentration of species and genera in the deserts of the Near East and Central Asia. It is probable that the extensive saline deposits that formed there as the remnants of the Tethys Sea evaporated in the later Tertiary provided the stimulus for the evolution of the group [13]. From this presumed centre of origin, some taxa have subsequently spread to all the major desert regions of the world.

13.2.3 Mediterranean floras

Apart from the Chenopodiaceae, and to a lesser extent the Zygophyllaceae, Frankeniaceae, Polygonaceae and Tamaricaceae, most families (and indeed genera) are poorly represented in desert floras – a reflection of the extreme environment and the low probability of plant adaptation to it. The situation is in marked contrast to that in mediterranean-climate regions, where many large genera have their centres of distribution (Table 13.4). Active speciation still seems to be occurring in some of these genera to produce a range of closely related species, some with extremely localized ranges [19,20].

Table 13.4 Examples of large genera centred in mediterranean regions

Mediterranean Europe	*Aethionema, Cistus, Dianthus, Nigella*
California	*Arctostaphylos, Ceanothus, Clarkia, Mimulus*
Cape Province	*Cliffortia, Erica, Murattia*
Southwest Australia	*Boronia, Calythrix, Daviesia, Isopogon*

As a result, **endemism** is a notable feature of mediterranean-climate floras. In the Californian Floristic Province, for example (encompassing the west-coast region of summer-drought conditions), with a total flora of 748 genera and more than 4000 species, 6.5% of the genera and 48% of the species are found nowhere else in the world [20]. For the Mediterranean region of Europe and for southwestern Australia, the corresponding values are about 10% and 40%, and about 25% and 68% [21,22]. For the Cape region of South Africa the values may be even higher. Such data can be somewhat misleading, however, as they relate to more than one vegetation type in each region (the Californian data, for example, include taxa found in the redwood forests); nevertheless, they are in striking contrast to the values for certain other dryland communities: the North American grassland biome, for example, has only 56 endemic species [23].

A number of possible factors that might in combination have produced the richness of the modern Californian flora has been suggested [20]. These are (1) the ability of the region to provide a continuing favourable environment for a variety of taxa moving south or west in North America during the Tertiary Period as climatic conditions worsened; (2) the maintenance of that environment through the vicissitudes of the Quaternary; (3) high topographic diversity within the region; and (4) recent active speciation. The Californian flora is one assembled over a long period of time from a variety of sources [20], with a strong record of recruitment and differentiation and a low record of extinction. The poverty of endemic taxa in a region such as the Great Plains, conversely, must be seen as evidence of a new flora recently assembled in a new region (see below).

Whether the same considerations can be applied to other mediterranean-climate floras is unclear. The floras of South Africa and southwest Australia differ from those of California and Mediterranean Europe in their low percentages of annual taxa: 5% in South Africa, according to Cody and Mooney [24]. As annual species in California include a greater-than-average proportion of endemics (63%), endemism might be expected to be lower in these southern-hemisphere floras. That it is not indicates that the presence of annual taxa is not necessary to achieve high endemism. Indeed, annual taxa may be less frequent generally in southern-hemisphere dryland areas (they contribute only 14% of the Australian arid flora, for example [25]), and it is possible that the evolution of an annual habit has occurred less readily in the Gondwanic flora.

13.2.4 The antiquity of the desert flora

Geophysical evidence for the time of origin of the major desert regions of the world is ambiguous (section 9.3.3), and various features of the world's desert floras have accordingly been studied to try and derive supplementary information regarding their antiquity. Two features in particular suggest considerable antiquity for certain of the floras: palaeo-endemism and growth-form diversity. Some endemic desert taxa at least appear to be **palaeo-endemics**, with no close surviving relatives within or outside the deserts. Schmida [13] lists 13 palaeo-endemic genera (out of a total of 22 endemic ones) in the Sonoran Desert, and 6 (out of 33) in the Saharo–Arabian Desert, but does not provide data for other desert regions. The presence in a desert flora of a variety of weird growth-forms with special ecophysiological attributes (plants with stem-succulence, a rosette-shrub habit, a spiny aphyllous habit, basicarpy, or salt-secretion) is likewise believed to be the product of evolution over a long time-period [13]. Such growth-forms are widespread in the Namibian, Chile–Peruvian and Sonoran Deserts, in particular. This floristic evidence for the antiquity of the desert floras of the Sahara, the Namib and South America is in accord with the geological evidence – which would also identify the Central Asian deserts as probably being of considerable age. Palaeoecological evidence might question the antiquity of the Sonoran Desert; but it must be realized that the floristic evidence relates to desert taxa and not to desert vegetation, and these taxa, as we shall see, may have grown in rather different plant communities in the past.

13.3 THE DRYLAND COMPLEX

Extensive dryland communities occur in three geographical situations: in the subtropical arid zone, between about latitudes 15° and 45° in both hemispheres; in 'rainshadow' areas in the lee of major mountain ranges; and in the interior of large land masses, far distant from oceanic sources of moisture. Typically, dryland communities occur intercalated between forest biomes; and typically they comprise a graded physiognomic series related to decreasing moisture. A generalized sequence of communities across the subtropical arid zone, from middle to low latitudes, is: temperate forest → seasonal woodland or shrubland → semi-desert → desert → savanna → dry woodland → tropical forest. The progression towards desert in either direction involves a reduction in stature and density of the woody plants, which become increasingly scleromorphic and xeromorphic. In desert and semi-desert the only woody plants are of low stature and form a sparse ground cover – <10% in deserts and 20–50% in semi-deserts [13]. Herbaceous plants, by contrast, become increasingly prominent along the gradient to desert, though maximum species-diversity characteristically occurs part-way along, in the seasonal woodland types [26]. In continental interiors at mid-latitudes, where conditions are also too dry for tree growth, the generalized progression from west to east is: temperate forest → seasonal shrubland → semi-desert and desert → steppe → parkland → temperate forest.

The progression from forest to desert in both these sequences is associated not only with a reduction in annual precipitation (from >500 mm p.a. to <150 mm p.a.), but also with a change from year-round precipitation to strongly seasonal precipitation and then to non-seasonal and sporadic precipitation. The change results in part from seasonal movements of the global climatic belts, brought about by seasonal differences between the temperature gradients from equator to pole in the two hemispheres. Gradients (ΔT) in either hemisphere are steeper in winter, so that the climatic belts in the summer hemisphere become meridionally compressed (section 3.2.1). As a result of these seasonal movements of the climatic belts there is a transitional zone, extending over 5–10° latitude on either side of the subtropical high-pressure belt in each hemisphere, where dry conditions develop either in the winter (on the equatorial side) or in the summer (on the polar side); these transitional zones support dry tropical woodland or savanna, and mediterranean woodland or shrubland, respectively. The year-round arid subtropical belt supports desert.

In additional to these seasonal displacements, longer-term displacements of the climatic and vegetation belts have occurred as a result of the climatic oscillations of the Quaternary Period (the 'glacial/interglacial cycles'). The pattern of these displacements can be predicted to some extent from climatic models. Thus Harrison *et al.* [27] showed that maximum southward displacement of the climatic belts in the northern hemisphere could be expected on theoretical grounds to have occurred 16–18 ka ago; by 4–6 ka ago, however, the belts would have been displaced northwards of their present position (Figure 6.4).

An overall displacement of 5–7° latitude over the last 20 ka is suggested both from the history of forest communities (Chapters 10 to 12), and from the recent work of Rea and Leinen [28] on the input of dust particles to ocean sediments in the northwest Pacific Ocean over the last 24 ka. Rea and Leinen showed that the position of the northern margin of the zone of intense westerly winds in this region (currently at latitudes 39–40°N) had varied from a southernmost position 24 ka ago at 38°N to a northernmost position 6 ka ago at 44–45°N. Displacements of such magnitude (if they occurred also at lower latitudes) must have had a profound influence on the vegetation of the dryland complex.

It is important when reconstructions of former dryland vegetation patterns are attempted that the whole complex of dryland communities listed above is considered, and not its separate components. In the following pages, consideration is limited to two geographical regions where sufficient evidence exists for such an integrative approach to be adopted: western North America, and Northern Africa and the Mediterranean region. For western North America, abundant packrat-midden evidence allows good reconstructions of the vegetation of the dryland complex to be made; for Northern Africa and the Mediterranean region, however, reconstruction is hampered by scanty biological evidence, and in particular by the scarcity of well-dated pollen sequences. Even where polleniferous deposits are available for study, moreover, the sources of pollen input to the deposits are often not clear [29,30]. Consequently, use has to be made of additional non-biological evidence – changing lake levels, input of wind-blown dust, presence of palaeosols (buried soils), for example – from both terrestrial and offshore sites. As a result, the picture of vegetation change revealed is far more sketchy than for western North America. Nevertheless, taken together, the evidence from the two regions provides a picture which can now be seen to contain many of the elements discernible also in the better-documented record for forest communities that we have been considering in previous chapters.

13.4 THE LATE-QUATERNARY HISTORY OF DRYLAND COMMUNITIES

13.4.1 North America

(a) The evidence from preserved plant remains

The origin and spread of dryland communities in North America in the Tertiary Period has already been discussed in Chapter 9, and their subsequent history in the glacial stages of the Pleistocene in Chapter 10. Although most of the taxa prominent in modern deserts and grasslands in North America were already present 18 ka ago, the plant communities in which they occurred were very different. A tree cover extended over almost the entire area now occupied by dryland communities, and modern desert and prairie taxa ranged widely as understorey components, sometimes to higher altitudes than at present. Climatic changes first began to affect these parklands and woodlands 13–15 ka ago [31,32], but the major manifestations of rising temperature and declining moisture status took place from 11 ka ago onwards*. Lake levels over the entire American Southwest fell dramatically within a period of 1–2 ka (Figure 6.3), and drought conditions spread at lower altitudes. The plant response was equally dramatic. Pine and spruce populations vanished from the parklands within 2 ka, the former to higher altitudes in the Rocky Mountains and the latter to higher latitudes in Canada. From these disintegrating lower-altitude communities there emerged, in some cases gradually and in other cases more abruptly, the modern range of dryland biomes.

The broad patterns of change in western North America in the Early Holocene, as revealed by midden and pollen evidence, are as follows (Figure 13.2). Dry grasslands of modern aspect, and with a modern grazing fauna, were present in the western Great Plains by 10 ka ago [33], and perhaps earlier in northwestern Montana [34], and followed the retreating spruce populations closely northwards into southern Canada (A in Figure 13.2); prairie floras and faunas were well established in southern Alberta by 9.5 ka ago [35]. Further to the east, however, in the Lake States and westwards to Iowa and Missouri, the spruce parkland was replaced instead by deciduous forest: a mixed forest rich in *Ulmus* in the north, and oak–hickory forest to the south [23,37,38] (B in Figure 13.2). Then as drier conditions spread eastwards over the next 2 ka, these forests retreated and were replaced by oak savanna and then by prairie [23,33,37]. In the southern Great Plains, too, in central and northern Texas, mesic conditions may have prevailed in the Early Holocene [33,39].

In the western United States, the pinyon–juniper woodland that had occupied the lower ground in the Late Pleistocene retreated to higher altitudes as temperatures rose in the Early Holocene. However, some juniper populations remained, to form dry woodlands at altitudes up to at least 1500 m in the regions now occupied by the Sonoran and Chihuahuan Desert floras [40–42]. Common associates of the junipers were *Quercus pungens*, *Q. turbinella*, *Cercocarpus montanus* and *Fraxinus anomala*. These lower-altitude juniper woodlands persisted generally until about 8 ka ago; above them, pinyon–juniper woodlands of modern aspect had formed widely by 9.5 ka ago [42] (Figure 13.2(b)). Pinyon–juniper woodland also moved gradually northwards into the higher-lying Great Basin (D in Figure 13.2), to occupy the ground vacated by the subalpine conifers (section 11.3.2(b)). Massive individuals of *Juniperus scopulorum* that survive to the present day on the floor of the Great Basin below the Snake Range may be relics of this Early Holocene woodland [43].

Accompanying these movements of the tree taxa were movements of the desert flora itself. Taxa that had persisted on site through the last glacial stage were joined, as temperatures rose after 11 ka ago, by migrants from as

*Some American palaeoecologists accordingly place the start of the Holocene at 11 ka ago.

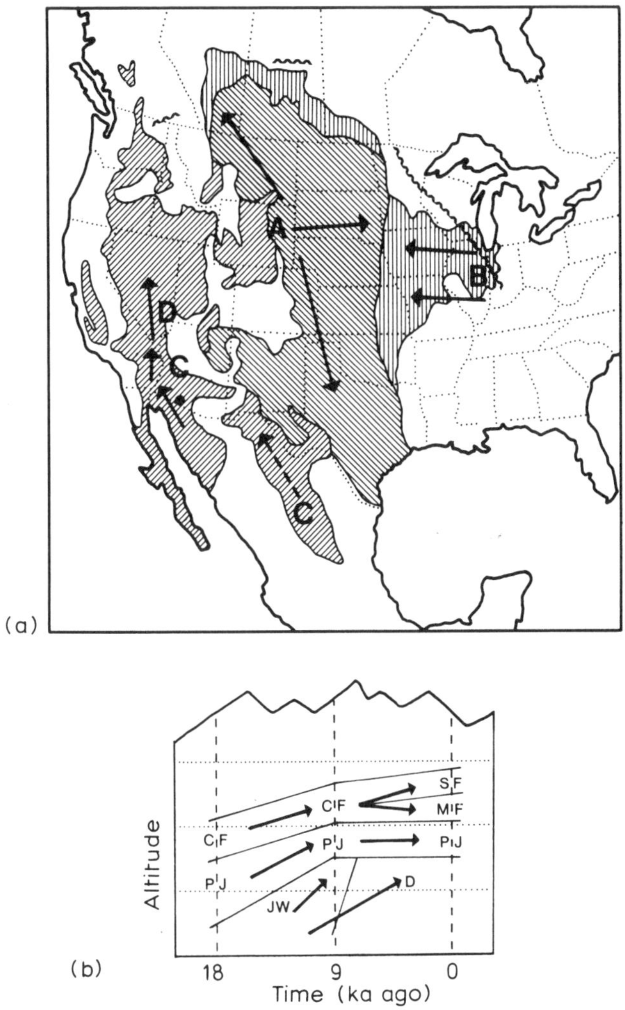

Figure 13.2 North American dryland vegetation in the Holocene. (a) Latitudinal migrations of floristic components; A–D are migration pathways in the Early Holocene. A = prairie; B = deciduous woodland; C = warm desert; D = pinyon–juniper woodland. Shaded areas show the approximate modern extent (from Daubenmire [44]) of desert and semi-desert ////; prairie \\\\; and savanna /parkland ||||||||. •, Organ Pipe Cactus National Monument Park; ~~, extent of prairie 7 ka ago. (b) Altitudinal shifts of vegetation zones in the American southwest. CF = conifer forest/parkland, SF = subalpine forest, MF = montane conifer forest, PJ = pinyon–juniper woodland, JW = juniper woodland, D = desert. The horizontal dotted lines are drawn at approximately 1000-m altitudinal intervals, but the precise altitudinal limits of the various vegetation zones varied from region to region (Figure 10.8).

yet largely undetermined refugia to the south [42]. Thus modern Mojave Desert taxa such as *Chrysothamnus*, *Coleogyne*, *Eriogonum* and *Yucca* moved north into the region, probably from the Lower Colorado River Valley [45], and were replaced in that region by *Ambrosia*, *Encelia* and *Larrea* (C in Figure 13.2). In the Sonoran Desert, *Opuntia* and *Yucca* were displaced by *Acacia greggii*, *Encelia*, *Eriogonum* and *Prosopis* [46], perhaps from refugia in Mexico. For certain desert taxa, however, the full modern range was achieved only slowly. Thus in the Great Basin region, the chenopods that are prominent in the current desert flora (*Atriplex*, *Kochia*, *Salicornia* and *Suaeda*) spread only in the mid-Holocene [47]; *Larrea divaricata* had completed its spread into the Mojave Desert before 5.4 ka ago, but was still spreading into the Sonoran and Chihuahuan Deserts 4 ka ago [46,48]; and *Ambrosia*, co-dominant today with *Larrea* and *Encelia* over large areas of the Mojave and Sonoran Deserts, was first recorded from the Lower Colorado River Valley about 700 yr ago [45].

In all dryland communities change continued beyond the Early Holocene. Prairie vegetation reached its maximum extent about 7 ka ago, when open grasslands occurred well to the north [49] and east [38] of their present limits (Figure 13.2). At the same time, *Artemisia* steppe replaced *Pinus ponderosa/contorta* forests in northeastern Washington and northwestern Idaho [47], and perhaps at mid-altitudes in the Rocky Mountains also, as forest boundaries moved upslope during the Altithermal [47,50]. If summer drought intensified then, as seems likely, the increased pollen values of *Quercus*, *Artemisia*, Chenopodiaceae and RAR (Rosaceae–Anacardiaceae–Rhamnaceae) in deposits from the Santa Barbara Basin, California, after 7.8 ka ago (and rather later at Clear Lake in the Coast Ranges) could well document the development of dry woodland and chaparral communities [47].

However, parts of the American southwest – notably the Sonoran and Chihuahuan Deserts – remained moister than at present into the mid-holocene. Thus Sonoran plants which at the present day are restricted either to riparian habitats (*Acacia greggii*, *Cercidium floridanum* and *Prosopis velutina*), or to higher altitudes (*Ferocactus acanthodes*, *F. wislizenii* and *Opuntia chlorotica*), where the moisture regime is more favourable, were widespread 9–4 ka ago [46] (Figure 13.3). These plants declined in abundance about 4 ka ago, when the present dry subtropical climatic regime became established.

About that time also there was a retraction of range limits of grassland, as trees invaded the prairie margins [37,38,47,49,51]. This reversion, variable in time between 5 and 3 ka ago, was contemporaneous with a renewed rise in lake levels in the Great Basin and California [31], and with the onset generally of Neoglacial conditions. Many North American dryland communities became recognizably modern only during the Neoglacial. Some –

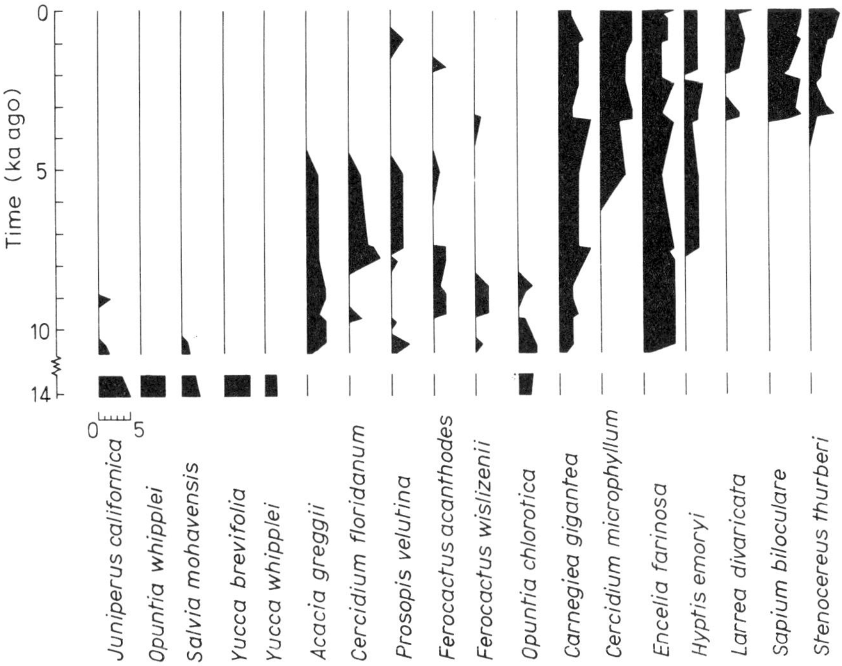

Figure 13.3 Packrat-midden record of 18 plant taxa for the last 14 ka in the Puerto Blanco Mts, southwestern Arizona. Plant remains are scored on an abundance scale of 1 to 5. Re-drawn from Van Devender [46].

those in the Sonoran and Chihuahuan Deserts [42,46] and the Laramie Basin of Wyoming [3], for example – appear to be more arid now than they ever were in the past.

The overall patterns of change that have occurred in dryland communities through the Holocene are illustrated by studies of packrat middens in the Organ Pipe Cactus National Monument Park in southwestern Arizona [46]. The frequencies of 18 of the 103 plant taxa identified there from middens extending back in time to 14.1 ka ago are shown in Figure 13.3. Of the seven taxa shown in Figure 13.3 that are prominent in the study area today, none was found in the oldest middens, and only two (*Carnegiea gigantea* and *Encelia farinosa*) were continuously present from the start of the Holocene. Three of the modern dominants (*Larrea divaricata, Sapium biloculare* and *Stenocereus thurberi*) appeared about 4 ka ago. Taxa prominent in the oldest (Pleistocene) middens (e.g., *Juniperus californica, Yucca* spp.) no longer occur in the Sonoran Desert, but are found instead at the present day 200–300 km to the northwest in the Mojave Desert. It is clear that re-sorting and migration of taxa must have occurred in dryland communities just as in forest communities.

(b) The climatic scenario

A picture of the climatic changes that directed these vegetation changes in the western United States is now emerging. A partial reconstruction of the evidence (based on lake levels, plant distributions, sediment characteristics and climatic modelling) is shown in Figure 13.4. Critical lines of evidence that define space/time boundaries in that reconstruction include the following.

1. The recognition of successive zones to the south of the Cordilleran Ice Sheet at full glacial that were affected by cold dry conditions (extending south to about latitude 45°N), winter rainfall (with the winter storm track of the Westerlies shifted to about latitude 36°N), and summer rainfall [32,52–54].
2. A dramatic change to drier conditions around 11 ka ago, manifested most strikingly in lake level and vegetation changes in the Great Basin and the Mojave Desert [42,55], and culminating in the final disappearance of juniper in those regions between 8.9 and 7.8 ka ago [54]. These changes were caused by the demise of the winter-rain regime of the full glacial as the westerly-wind storm track shifted northwards again from its full-glacial position.
3. The maintenance of dry conditions uninterruptedly from 12 ka to 6 ka ago in the northwestern Great Plains of Montana at latitude 48°N [34].
4. The time-limits of the mid-Holocene xerothermic period in the

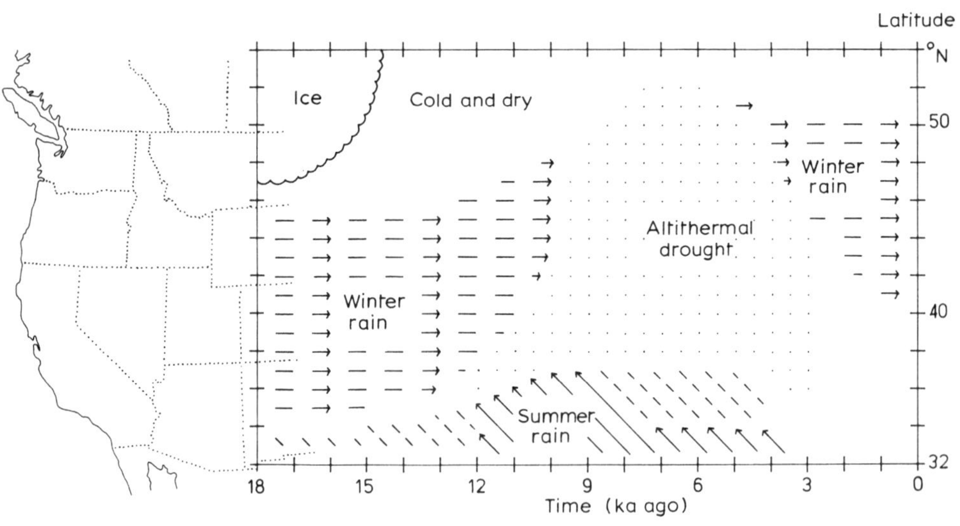

Figure 13.4 Climatic conditions in the western United States over the last 18 ka.

northern and southern Great Plains [39,56], and particularly of its most northerly expression (7.5 to 6.0 ka ago [49]).

5. The evidence for increased summer (monsoonal) rainfall in the Chihuahuan and Mojave Deserts between 12 and 4 ka ago [32,54,57].
6. Modernization of desert and grassland floras beginning about 6 ka ago [34], and most marked after about 4.5 ka ago [54,56].

These changes can be seen as a manifestation of the general trends of climatic change in the northern hemisphere over the last 18 ka discussed in sections 6.4 and 13.2. Four primary components of climatic change are involved: an overall rise in temperature to maximum levels 9–6 ka ago, and a subsequent decline; increasing seasonal extremes of temperature, with maximum summer warmth about 6 ka ago; displacement polewards of the climatic belts as the continental ice sheets melted after 15 ka ago; and variations in the amplitude of the seasonal movements of these belts consequent upon changes in the thermal gradients of the northern and southern hemispheres. In the specific context of the western United States, these changes had far-reaching effects on rainfall regimes, which at the present day are generated by the interactions between three air masses: the cold dry Arctic Air Mass to the north; the cool moist Pacific Air Mass to the west (but which becomes increasingly dry inland after passage over the western mountain ranges); and the warm wet Gulf of Mexico Air Mass to the southeast [54,58]. A strong **zonal** circulation over North America, promoted by a steep thermal gradient, draws in Pacific air from the west; **meridional** circulation, when the thermal gradient is weakened, draws in Arctic and Gulf air; and increased summer warmth, which heightens the pressure differences between land and water surfaces, draws in **monsoonal** Gulf air. Zonal circulation prevailed in the latest Pleistocene [53,58], and again probably in the Neoglacial; meridional circulation prevailed about 6 ka ago (Figure 6.6), and monsoonal circulation intensified between about 12 and 6 ka ago [53,57]. The reconstruction in Figure 13.4 reflects the changing interplay of these air masses in the western United States over the last 18 ka; however, because the reconstruction ignores east/west regional differences across the country, and seasonal variation in climate within a region, it must be regarded as a first approximation only.

13.4.2 Northern Africa and the Mediterranean region

(a) *The long-term pollen record*

The Saharo–Mediterranean region extends from latitudes 10° to 44°N and from longitudes 18°W to 40°E (Figure 13.5), and includes part of the largest continuous desert region in the world (the Saharo–Arabian Desert). The

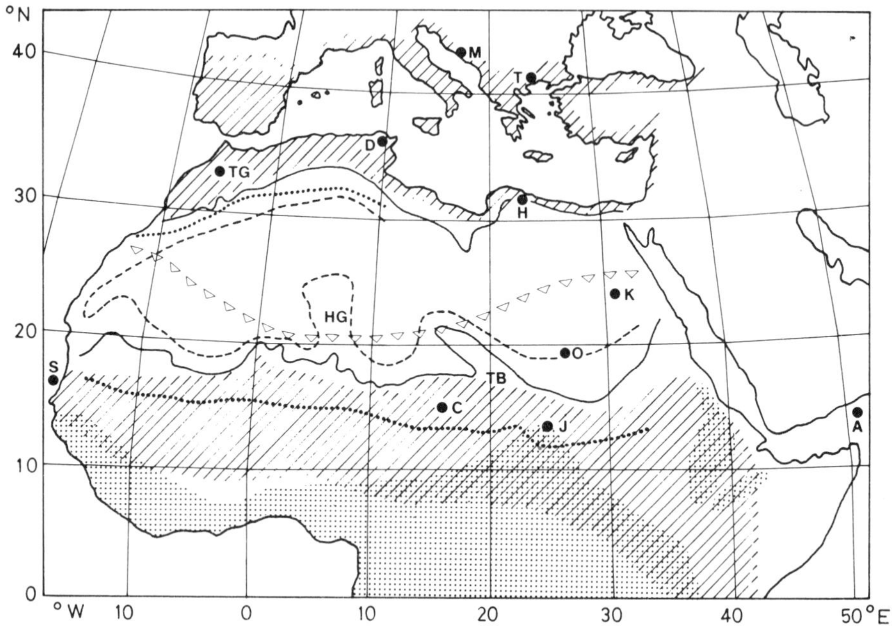

Figure 13.5 The Saharo–Mediterranean region, showing: (1) Sites mentioned in the text. A, Arabian Sea; C, Chad basin; D, Dar Fatma; H, Haua Fteah; HG, Hoggar; J, Jebel Marra; K, Kharga oasis; M, Malo Jezero; O, Oyo Depression; S, Senegal core; T, Tenaghi Philippon; TG, Tigalmamine. (2) Modern vegetation and environment: ⁓ present limit of desert region; //// regions with strongly seasonal rain. (3) Some proposed limits of the desert region in the past: --- 6 ka ago; .... 18 ka ago. (4) Southern limit of region with high lake levels 18 ka ago (▽▽▽). (5) Probable extent of tropical rainforest 8 ka ago ::::::. Sources of information: [22,59–64].

broad pattern of pollen changes in this region through the last glacial stage and the Holocene is illustrated by the summary diagrams in Figure 13.6, for an offshore sediment in the Arabian Sea and terrestrial sediments in Tunisia and northern Greece. The Arabian Sea and Greek diagrams both cover a time-period of 100–150 ka (as judged by C-14 datings back to 50 ka ago for the latter diagram, and by correlation with the appropriate oxygen-isotope marine stages for the former). The Tunisian core is undated below the 19.6-ka level, but probably extends over a rather similar time-period. The summary diagrams for the two terrestrial cores are both based on total pollen. The diagram for the marine core considers only the less common pollen types, and apportions them to three probable geographical source areas (equivalent to modern floristic provinces): the Sudano–Guinean (characterized today by tropical forest and woodland), the Sahelian (predominant savanna and thorn-scrub) and the Saharan (desert and semi-desert). Typical pollen taxa derived from these, and other adjoining, floristic provinces are listed in Table 13.5.

The varying representation of these pollen components in the Arabian Sea core can be explained as reflecting differing influences of the dry northeast Trade Winds (transporting pollen from the Arabian and Iranian Deserts) and the moist Southwesterlies (bringing savanna and forest pollen from equatorial Africa [30]. The summary diagram for the Arabian Sea core shows a peak of tropical pollen (i.e., an expanded influence of the

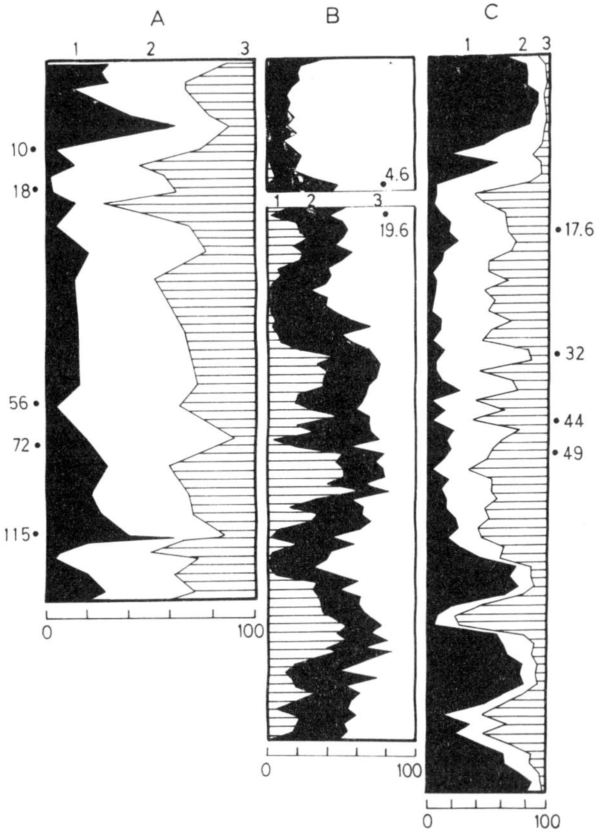

Figure 13.6 Summary pollen diagrams for the Holocene and the last 'glacial' stage. A, ocean core MD76135, off the south coast of Saudi Arabia (1 = Sudano–Guinean, 2 = Sahelian, 3 = Saharan; Chenopodiaceae, Amaranthaceae, Gramineae and Cyperaceae are not included in the pollen sum). B, Dar Fatma, Tunisia (1 = *Alnus*, 1 + 2 = arboreal pollen, 3 = non-arboreal pollen). C, Tenaghi Philippon, Greece (1 = arboreal pollen, 2 = *Artemisia*, 2 + 3 = non-arboreal pollen), •56, dated level, ka ago. Re-drawn from Wijmstra [65], Ben Tiba and Reille [66] and Van Campo, Duplessy and Rossignol-Strick [30].

Table 13.5 Geographical elements in Saharan terrestrial and offshore pollen spectra [30,67,68]

A. Temperate (forest)	*Betula, Corylus*
B. Mediterranean (woodland)	*Alnus, Cupressus, Pinus, Quercus, Olea, Vitis*
C. Saharan (desert, steppe)	*Artemisia**, *Calligonum*, Chenopodiaceae, *Ephedra, Moltkia, Zygophyllum*
D. Sahelian (savanna)	*Acacia, Blepharis, Commiphora, Maerua, Phyllanthus, Tribulus, Zizyphus*
E. Sudano–Guinean (tropical forest and woodland)	*Alchornea*, Celastraceae, Combretaceae, *Elaeis, Eugenia*, Meliaceae, *Sizygium, Uapaca*
F. Tropical montane	*Podocarpus, Taxus*

Data of Van Campo *et al.* [67], Rossignol-Strick and Duzer [68] and Van Campo, Duplessy and Rossignol-Strick [30].

* Found in high steppes of the Atlas Mts at the present day.

southwesterly monsoon) early in the Holocene (about 9 ka ago) and in the last interglacial (115 ka ago), and a peak of Saharan pollen at the last glacial maximum (15–18 ka ago). The time-period covered by these episodes, however, is relatively short, and for much of the last cold stage (the hypothermal) the Sahelian component was predominant. As the proportions of the three pollen components 3–6 ka ago were rather different from what they are today, a rather different vegetation situation must accordingly be envisaged – the higher proportions of the Saharan pollen component and the reduced values of Cyperaceae pollen indicate more arid conditions.

Both the Arabian Sea core and a similar core off the Senegal coast of West Africa [68] show a period of maximum aridity between 20 and 12.5 ka ago, which probably corresponded with the time of maximum extension of active desert dunes mapped by Sarnthein [69] (Figure 5.1). Possible limits of the desert zone 18 ka ago are shown in Figure 13.5.

The Tunisian pollen diagram in Figure 13.6(B) provides clear evidence that north of these cold-stage deserts was a zone where forest was widespread. The major tree pollen taxa at the Dar Fatma site throughout the full glacial were *Quercus* (deciduous) and *Alnus*, with smaller amounts of *Cedrus, Pinus* and *Salix*. These forests must have been very different in character from those surviving at the present day in Tunisia, where the major oak is the evergreen *Quercus suber*, and *Alnus* is very local. The full-glacial forest zone extended across Tunisia, Algeria and Morocco [70], and probably also through northern Libya and southern Lebanon [71]. To the south may have been a zone of mediterranean woodland, scrub and steppe extending discontinuously to the isolated Saharan highlands of Hoggar, Tassili and Tibesti [22], while to the north lay the arid steppe

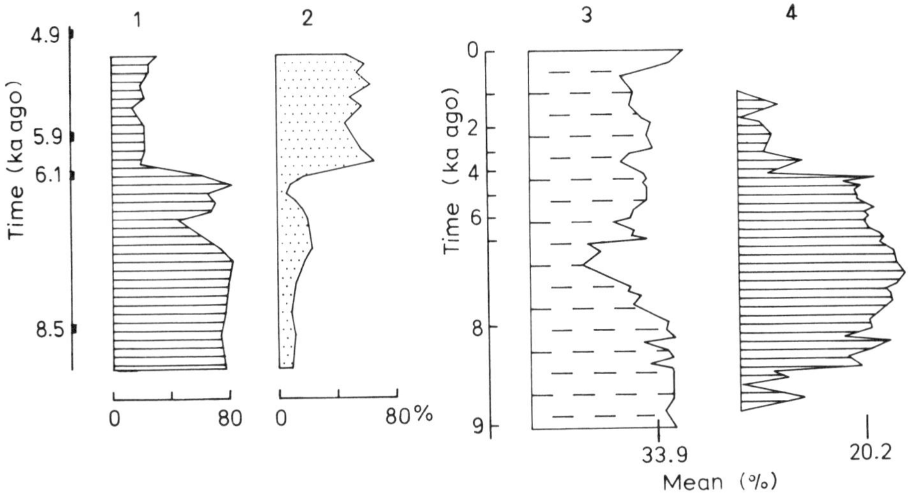

Figure 13.7 Relative proportions of pollen of the Sudanian (1) and Saharan (2) components in sediments from the Oyo Depression, Sudan, and of the Sahelian (3) and Sudano–Guinean (4) components in sediments from the Chad Basin, West Africa. Values in (1) and (2) are percentage of total tropical pollen [76]; values in (3) and (4) are expressed relative to the mean of all samples, and are plotted on a log scale [84].

vegetation of southern and central Europe discussed in section 10.3 (Figure 10.5). The Greek pollen diagram indicates that this European steppe vegetation persisted relatively unchanged throughout the 40–50 ka of the full glacial (Figure 13.6), but the North African forests (and perhaps the scrub to the south) were clearly less stable. The periodic reductions in *Alnus* pollen shown in Figure 13.6(B) probably resulted from recurring drier intervals of climate within the last glacial cycle [66], the latest of which began about 20 ka ago and reached peak aridity about 10 ka ago [70].

(b) Holocene changes in the Saharan belt

The arid period within the Saharan region was terminated by a major increase in precipitation in the tropics, which was manifested by the deposition of black organic muds (sapropels) in the Eastern Mediterranean Basin, dated to between 11.8 and 8.0 ka ago [72], and resulting from increased discharge by the River Nile. The heavy precipitation led to a rise in lake levels, first apparent in equatorial latitudes [73], and peaking between 10 and 8 ka ago around the southern margins of the Sahara [74,75]. The rise was delayed progressively northwards [75], with maximum lake levels in the central and eastern Sahara being reached only between 8.5 and 6.1 ka ago [9,76]. Temperate forest trees such as *Acer*,

Alnus, Juglans, Quercus ilex and *Tilia* were probably growing in the Saharan highlands at this time [22]. Conditions remained generally dry in North Africa during this period [70], as evidenced by the animal bones recovered from the Haua Fteah cave in northern Libya [77], and by the cessation of usage of the Kharga Oasis in Egypt [78] – though at higher altitudes in the Atlas Mountains of Morocco mixed evergreen and deciduous oak forests replaced cold steppe vegetation about 8.5 ka ago, at a time of rapidly rising lake levels [79]. There may have been a short-lived drier period in the southern Sahara after about 7.5 ka ago [73,80], but conditions became wetter again around both the northern and southern margins of the Sahara about 6.5 ka ago [61,74], and persisted until nearly 4 ka ago [70,80–83]. The animal paintings and engravings from the central Sahara date to this time-period [7,8].

These changes in climatic wetness were accompanied by progressively increasing temperatures. The combined effects of more favourable moisture and warmth in the desert core can be seen in the pollen analyses of lake sediments of the Chad Basin [84] and the Oyo Depression, Sudan [76]. The dominant taxa in both pollen spectra were Gramineae, Cyperaceae and *Typha*, presumably growing locally, but many of the less common pollen taxa could be apportioned to the phytogeographical source areas listed in Table 13.5. The relative proportions of the Sudano–Guinean and Sahelian components in the Chad spectra, and of the Sudanian and Saharan components in the Oyo spectra, are shown in Figure 13.7. At Oyo the Sudanian component was predominant from about 8.5 to 6.1 ka ago, when the lake was probably surrounded by deciduous savanna (with the trees *Grewia* and *Piliostigma*), of a type now found 500 km to the south. Thorn scrub with *Acacia* succeeded the savanna, and only after 4.5 ka ago (when the lake dried up completely) did the present desert vegetation become established [76]. Leaf-impressions of the oil palm *Elaeis guineensis* were recorded by Wickens [85] from Jebel Marra in the western Sudan (Figure 13.5), some 600 km north of its present-day distribution limits, and these fossils could belong to the same Holocene warm moist phase – though a Middle Pleistocene provenance has also been suggested [22]. At Chad, in the sediments accumulated between about 8.5 and 4.5 ka ago, the Sudano–Guinean component was far commoner than in modern sediments, whereas the Sahelian component was less abundant from about 7.5 to 1 ka ago [84]. A similar pattern was found by Rossignol-Strick and Duzer [68] in marine sediments off the West African coast.

Trees were still growing on the Hoggar and Tibesti highlands of the central Sahara between about 6.5 and 4.3 ka ago. Pollen of a variety of woody taxa has been found in cave and rock-shelter deposits in these regions, both of tropical (*Acacia, Balanites*) and of Mediterranean provenance (*Cedrus, Celtis, Cupressus, Juglans, Juniperus, Olea, Pinus, Pistacia,*

Quercus ilex and *Tamarix*) [7,8,22]. Woody plants also grew during the Holocene in parts of Tunisia now too arid for tree growth [70,86].

The plant and animal remains found in the rock shelters of the central Sahara led Hays [8] to describe the region as consisting of 'grasslands associated with lakes fed by rivers draining forested highlands'. Whilst it would undoubtedly be erroneous to envisage the whole Saharan region as being of this character, it is clear that 4–6 ka ago the arid desert nucleus was considerably contracted, and the fringing semi-desert and savanna communities supported more luxuriant plant growth, especially along river valleys draining from the Saharan highlands where a more persistent water flow was maintained than today (Figure 13.5). The current vegetation situation is the result of a shift to more arid conditions only 3.5–4.0 ka ago. By 2.8 ka ago the Mediterranean elements of the Saharan highlands had almost entirely disappeared, to be replaced by savanna species of *Acacia* and then by desert taxa [22].

(c) The Mediterranean Basin

Reconstructions of Saharan-zone vegetation are based on very limited biological evidence. Rather more sites have been investigated pollen-analytically in the Mediterranean Basin to the north, but very few provide a continuous pollen record through the Holocene, and few have sufficient C-14 dates to sustain a firm timescale. Nevertheless, broad patterns of change throughout the region are discernible. At a number of sites pollen changes indicative of improving temperature are recognizable between about 14.5 and 11 ka ago [108], but aridity probably remained a limiting factor to tree growth [87], and it was not until the start of the Holocene that significant expansion of tree pollen occurred: at 10.2–10.7 ka ago in Greece and Iran, and perhaps as early as 11.4 ka ago in western Syria [87]. The pollen spectra indicate widespread forest vegetation throughout the Mediterranean Basin in the Early Holocene, with deciduous species of oak (chiefly *Quercus pubescens*, but also probably *Q. canariensis* and *Q. faginea*) forming the main component. The dominant oaks were accompanied by *Corylus*, *Fagus* and *Pinus* in Yugoslavia [88]; evergreen oak and *Pinus* in Greece, Crete and Spain [89–91]; *Juniperus* and *Pinus* in southern France [91]; and *Abies*, *Cedrus* and *Pinus* in Tunisia [66]. The impression is of forest vegetation developed under more favourable moisture conditions than at the present day: summer drought might well have been less pronounced, though the presence of small amounts of pollen of typical 'Mediterranean' taxa such as *Fraxinus ornus*, cf. *Ostrya*, *Phillyrea* and *Pistacia* [92] suggests some seasonality of rainfall.

Pollen values of deciduous oaks decline at all sites in the mid-Holocene (Figure 13.8), and those of evergreen oaks (notably *Quercus ilex*) rise. The time of this changeover, however, is firmly dated at only a few sites. Available dates range from 6.3 to 6.7 ka ago [91,93] in the northern Mediterranean, but perhaps later in Tunisia [66]. The increasing *Q. ilex* pollen is almost always accompanied by charcoal in the sediments, and many ecologists believe that this vegetation change was the direct result of increased demographic pressure following the establishment of settled agriculture from about 8 ka ago onwards [94–97]. Major deforestation is certainly apparent in some pollen diagrams by about 5 ka ago [89,90]. At some sites, however, earlier pollen changes occur before the main *Q. ilex* rise (about 7.5 ka ago [89,93]); the taxa involved (*Juniperus, Phillyrea, Pistacia*), like *Quercus ilex*, are all capable of surviving under lower, and more strongly seasonal, rainfall conditions at the present day than the deciduous oaks (Figure 13.9). The possibility of climatic change being involved in the rather later expansion of the evergreen oaks cannot therefore be ignored – especially as at Tigalmamine, at 1626 m altitude in the Atlas Mountains of Morocco, a rise in *Cedrus* pollen about 4.0 ka ago probably documents a change to a more strongly winter-rain climate [79].

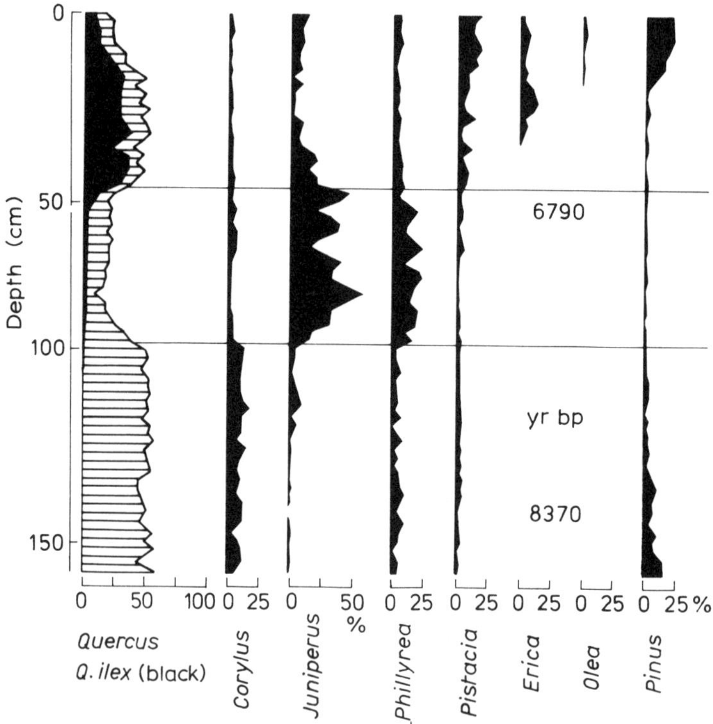

Figure 13.8 Pollen diagram from Malo Jezero on the Dalmatian coast of Yugoslavia. Re-drawn from Beug [93].

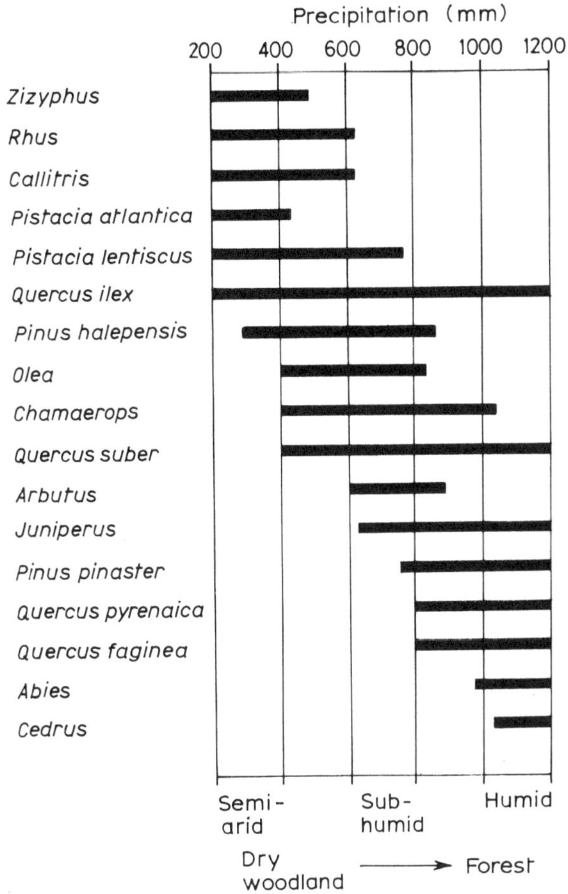

Figure 13.9 Present-day distribution of woody taxa in northern Morocco in relation to annual precipitation. Modified from Mikesell [99].

Recent climatic modelling certainly suggests decreased summer rainfall in southern Europe in the last 6 ka [98].

(d) The climatic scenario

Some parallelism in the timing of events can be seen in the Saharan and Mediterranean regions; the onset of the warm moist phase in the Sahara, around 6.5 ka ago, in particular, matches closely with the shift towards *Quercus ilex* dominance in the Mediterranean region. Such parallelism should not be surprising, as both regions are included within the same climatic complex (Figure 13.5), and both should have been affected similarly by any displacements of climatic belts over the last 20 ka. The best documented expression of these climatic shifts is the changing location of the West African savanna zone, reconstructed by Hooghiemstra [29] from the pollen content of offshore sediments at latitude 21°N. Hooghiemstra showed that the northern margin of the savanna moved northward through about 9.5° of latitude between 12.5 and 7.8 ka ago, and then retreated 6.5° to its current position. This reconstruction forms the basis of Figure 13.10, which collates available information for the vegetation of the western Saharo–Mediterranean region over the last 20 ka, and attempts to locate other vegetation boundaries during the same period. In the southern

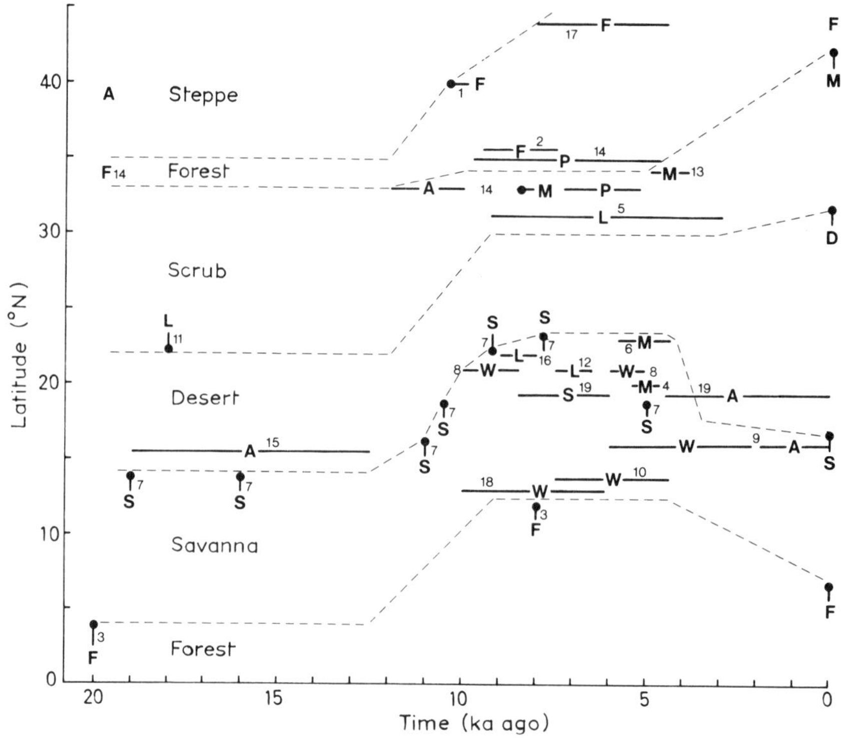

Figure 13.10 Summary of evidence relating to environmental and vegetational changes in northern Africa and the western Mediterranean Basin over the last 20 ka. ---, reconstructed vegetation boundary; —, time-span of environmental or vegetational event; •—, range limit. A = arid, D =desert, F = forest, L = high lake level, M = Mediterranean scrub or woodland, P = palaeosol, S = savanna, W = woodland. Sources of information: 1 = Bottema [87]; 2 = Bottema [89]; 3 = Bradley [60]; 4 = Butzer and Twidale [7]; 5 = Gasse *et al.* [81]; 6 = Hays [8]; 7 = Hooghiemstra [29]; 8 = Jäkel [73]; 9 = Lézine [100]; 10 = Maley [84]; 11 = Peterson *et al.* [62]; 12 = Petit-Maire and Riser [9]; 13 = Ritchie [86]; 14 = Rognon [70]; 15 = Rossignol-Strick and Duzer [68]; 16 = Street and Grove [75]; 17 = Vernet and Thibault [91]; 18 = Wickens [22]; 19 = Ritchie *et al.* [76].

part of the region the pattern of change conforms to theoretical predictions (of e.g., Harrison *et al.* [27] – Figure 6.4), but in the north the picture is less clear – with a muted response of the southern limit of forest to the climatic optimum of the mid-Holocene, and little evidence for the expected southward displacement of vegetation zones in the last 4 ka. Nevertheless, Figure 13.10 shows clearly that between about 8 and 4 ka ago (the North American Altithermal period) all vegetation zones were displaced far to the north. The desert region then appears to have been affected by rains from both the north and the south. Possible mechanisms are discussed by Jäkel [73], Maley [82,84] and Van Zinderen Bakker [64]; basically they involve a weakened westerly wind circulation in the mid-Holocene [27], which allowed tropical depressions to move into the Sahara region and bring sporadic rains to it. For much of the Early Holocene the Mediterranean region was apparently influenced by rain-bearing westerly winds, but the progressive northward migration of the subtropical arid zone (defined approximately by the movement of the savanna/desert boundary in Figure 13.10) brought the region increasingly into a climatic regime of summer drought. The lack of any vegetation changes comparable to those in the southern Sahara that might be attributed to the climatic 'recession' of the last 4 ka (which should have enhanced the moisture status of the Mediterranean region) could be the result of inadequate data, the complicating presence of the Mediterranean Sea, or the burgeoning impact of human settlement in the Mediterranean region after 8 ka ago. This aspect is considered further in section 13.5.

The Holocene records for the drylands of North America and the Saharo–Mediterranean region that we have been considering are largely complementary. The broad patterns of climatic change are perhaps most evident in the latter region, whereas a picture of vegetation flux rather similar to that for forest communities emerges from recent work in North America. Three features of note are common to both regions: modern climatic regimes have probably become established only within the last 4–6 ka; summer (monsoonal) precipitation affected substantially greater areas between 12 and 6 ka ago; and the magnitude of recorded range displacements of individual taxa (200–600 km) over the last 18 ka is broadly similar to that for forest trees. We must hope that these features apply also to other parts of the world where documentation of the history of dryland complexes is negligible.

13.5 EPILOGUE: THE GLORY THAT WAS ROME

The evidence that we have been considering in this chapter testifies to the reality of climatic change in dryland regions of the world in recent geological time. The incongruous Saharan plant and animal records mentioned in the Introduction are clear evidence of this change. Further obvious incongruities can be found in the lands bordering the Mediterranean Sea. At El Djem in Tunisia, for example, the remains of a Roman amphitheatre built to seat 60 000 people tower above scattered modern Arab settlements in a now-barren desert region [101]; at Ephesus in Turkey the Roman harbour is isolated from the sea at the present day by 5 km of flat marshland; and in the dry steppes of northwestern Syria the ruins of 'a hundred dead cities' lie exposed [102]. In the same region the site of Antioch, formerly a city of 400 000 inhabitants, is buried beneath 6 m of soil washed down from the uplands to the east [103]. On the barren western slopes of the mountains of northern Lebanon (from which some of this soil at Antioch has come) there are Roman inscriptions carved on rock faces that record the former limits of imperial forests in which cedar, fir, pine and cypress were protected species [103]; only vestiges now remain [104] of what was once described as 'a vast forest whose branches hide the sky' [103]. All these incongruities date back to Greek and Roman times, when the Mediterranean basin was the centre of the civilized world and 50–60% of the land was under cultivation [105]. It is tempting to see them as further evidence of climatic change, but current opinion favours human mismanagement as the cause. The Graeco–Roman civilizations practised a sophisticated land-use system, with the lowland plains bordering the Mediterranean Sea largely given over to cereal-growing, and the terraced hillslopes to vineyards and orchards extending up to the upper limit of cultivation of the olive (to at least 1000 m altitude). In the drier climates of North Africa and the Near East, irrigation was widely used. At the highest altitudes were mountain forests and summer pastures. This land-use system had a pedigree stretching back to the early farming communities that developed in the 'Fertile Crescent' of the Middle East 9–10 ka ago, and spread from there throughout the Mediterranean basin by 7.5 ka ago [96,106]. The system maintained a high level of productivity in the countries of the Mediterranean basin for many centuries, but it is doubtful whether the level could have been maintained indefinitely. In the event it was destroyed gradually by a series of internal and external perturbations: by successive 'barbarian' invasions (Seljuks, Turkomans, Ottomans, Afjars, Mongols, Uzbeks, Qazars, Baluchies, Iranians, Tartars, Goths, Vandals and Arabs), by internal strife, by plague, by famine and by earthquake. After each perturbation a new land-use system was developed that almost invariably represented a lower level of productivity than the one it replaced. The present degraded Mediterranean landscape is thus attributable to a combination of effects, in which man is believed to have played a major role and climatic change only a minor one [99,101,103,105,107,108].

The impact of human settlement in the Mediterranean basin is clearly evident by 4–5 ka ago in pollen diagrams [89,90], when the first documentary evidence also becomes available. Documentary evidence for the three millennia from 2600 BC to 400 AD indicates, on the one hand, the much greater productivity of the Mediterranean lowlands and the mountain forests then than now, and, on the other hand, the increasing pressures on land, and on timber resources in particular, by an expanding urban population. Even by the fifth century BC Plato was describing parts of southern Greece as 'like the skeleton of a sick man, all the fat and soft earth wasted away and only the bare framework of the land being left' [103]. Nevertheless, the major deterioration of the Mediterranean economy probably post-dates the collapse of the Roman Empire. Among the effects that have been singled out are those resulting from deforestation, grazing and soil erosion. Documentary and pollen evidence indicates that the higher ground fringing the Mediterranean basin was formerly covered with extensive forests of deciduous and evergreen oaks, pines, cedar, fir and cypress; the cedar forests of southern Turkey (the Taurus Mountains), Lebanon (Mount Lebanon) and North Africa (the Atlas Mountains) were particularly valued for their timber. From prehistoric times the Mediterranean civilizations were linked, in peace and in war, by the sea, and shipbuilding remained a vital industry, and a major consumer of timber, right through to the sixteenth century AD, when indigenous supplies finally ran out [107]. Wood also remained the principal, and often the only, source of fuel for Mediterranean peoples through to the present day – when it is estimated that more than half the total annual production of forest and scrub is consumed as firewood [105]. As the forests were felled, so the higher ground formerly occupied by them was increasingly taken over as year-round grazing for sheep and goats, and firing of the vegetation to encourage new growth has been regularly practised for at least 4 ka [105]. Setting fire to the forests has also been a common form of political protest and of warfare for many centuries [36,103]. The woody plants that remain today are thus those adapted to persistent browsing and periodic firing; of them, *Quercus coccifera* is probably the most notable and widespread.

The major animals in the Graeco–Roman economy were cattle, pigs and horses, which are normally part of an integrated farming system. After the collapse of the Roman Empire, self-contained, and often nomadic, grazing developed, based on sheep and goats, and these animals have continued to dominate Mediterranean land-use up to the present day. Associated paticularly with Moslem peoples, these economies rely on common grazing on uncultivated land, and, where wealth is measured in terms of numbers of animals rather than their productivity, can result in massive overgrazing. For many ecologists, it is the huge numbers of free-range goats, introduced initially to North Africa by Arab pastoralists in the seventh century AD, that have been the scourge of Mediterranean vegetation over the centuries [103].

The consequences of deforestation, political instability, burning and overgrazing have been erosion over large areas of the Mediterranean basin; and although plant diversity has increased as a result of human impact (Table 13.2), so too has the extent of man-made desert, steppe and scrubland. The degraded state of the Mediterranean lands today might thus be seen as a stark indictment of Man's failure in his stewardship of the earth's resources.

REFERENCES

1. FAO (1986) *FAO Production Yearbook*, Vol. 40. Food and Agriculture Organization of the United Nations, Rome.
2. Aubréville, A. (1970) La flore tropicale tertiaire du Sahara. *Adansonia*, Series 2, **10**, 9–14.
3. Wells, P. V. (1970) Postglacial vegetational history of the Great Plains. *Science*, **167**, 1574–82.
4. Godwin, H. (1975) *History of the British Flora*, 2nd edn, Cambridge University Press, Cambridge.
5. Tallis, J. H. and Switsur, V. R. (1983) Forest and moorland in the South Pennine uplands in the mid-Flandrian period. I. Macrofossil evidence of the former forest cover. *J. Ecol.*, 585–606.
6. Brentjies, B. (1969) *African Rock Art* (transl. A. Dent), Dent, London.
7. Butzer, K. W. and Twidale, C. R. (1966) Deserts in the past, in *Arid Lands. A Geographical Appraisal* (ed. E. S. Hills), Methuen, London, pp. 127–44.
8. Hays, T. R. (1975) Neolithic settlements of the Sahara as it relates to the Nile Valley, in *Problems in Prehistory: North Africa and the Levant* (eds F. Wendorf and A. Marks), Southern Methodist University Press, Dallas, TX, pp. 193–204.
9. Petit-Maire, N. and Riser, J. (1981) Holocene lake deposits and palaeoenvironments in central Sahara, northeastern Mali. *Palaeogeogr., Palaeoclimatol., Palaeoecol.*, **35**, 45–61.
10. Whittaker, R. H. and Niering, W. A. (1965) Vegetation of the Santa Catalina Mountains, Arizona: a gradient analysis of the south slope. *Ecology*, **46**, 429–52.
11. Naveh, Z. and Whittaker, R. H. (1979) Structural and floristic diversity of shrublands and woodlands in northern Israel and other mediterranean areas. *Vegetatio*, **41**, 171–90.
12. Stebbins, G. L. (1952) Aridity as a stimulus to plant evolution. *Am. Nat.*, **86**, 33–44.
13. Schmida, A. (1985) Biogeography of the desert flora, in *Ecosystems of the World 12A. Hot Deserts and Arid Shrublands* (eds M. Evenari, I. Noy-Meir and D. W. Goodall), Elsevier, Amsterdam, pp. 23–77.

14. Raven, P. H. (1971) The relationships between 'Mediterranean' floras, in *Plant Life of South-West Asia* (eds P. H. Davis, P. C. Harper and I. C. Hedge), Botanical Society of Edinburgh, Edinburgh, pp. 119–34.
15. Sarmiento, G. (1983) The savannas of tropical America, in *Ecosystems of the World 13. Tropical Savannas* (ed. F. Bourlière), Elsevier, Amsterdam, pp. 245–88.
16. McClure, H. A. (1976) Radiocarbon chronology of Late Quaternary lakes in the Arabian Desert. *Nature*, **263**, 755–6.
17. Singh, G., Joshi, R. D., Chopra, S. K. and Singh, A. B. (1974) Late Quaternary history of vegetation and climate of the Rajasthan Desert, India. *Phil. Trans. Roy. Soc. Lond. B*, **267**, 467–501.
18. Raven, P. H. (1963) Amphitropical relationships in the floras of North and South America. *Q. Rev. Biol.*, **38**, 151–77.
19. Axelrod, D. I. and Raven, P. H. (1978) Late Cretaceous and Tertiary vegetation history of Africa. *Monogr. Biol.*, **31**, 77–130.
20. Raven, P. H. and Axelrod, D. I. (1977) Origin and relationships of the California flora. *Univ. Calif. Publs Bot.*, **72**, 1–134.
21. Nelson, E. C. (1981) Phytogeography of southern Australia, in *Ecological Biogeography of Australia*, Vol. 1 (ed. A. Keast), Junk, The Hague, pp. 733–59.
22. Wickens, G. E. (1976) *The Flora of Jebel Marra (Sudan Republic) and its Geographical Affinities*, HMSO, London.
23. Axelrod, D. I. (1985) Rise of the grassland biome, central North America. *Bot. Rev.*, **51**, 163–201.
24. Cody, M. L. and Mooney, H. A. (1978) Convergence versus nonconvergence in mediterranean-climate ecosystems. *Ann. Rev. Ecol. Syst.*, **9**, 265–321.
25. Beadle, N. C. W. (1981) The vegetation of the arid zone, in *Ecological Biogeography of Australia*, Vol. 1 (ed. A. Keast), Junk, The Hague, pp. 695–716.
26. Whittaker, R. H. (1977) Evolution of species diversity in land communities. *Evol. Biol.*, **10**, 1–67.
27. Harrison, S. P., Metcalfe, S. E., Street-Perrott, F. A. *et al.* (1984) A climatic model of the last Glacial/Interglacial transition based on palaeotemperature and palaeohydrological evidence, in *Late Cainozoic Palaeoclimates of the Southern Hemisphere* (ed. J. C. Vogel), Balkema, Rotterdam, pp. 21–34.
28. Rea, D. K. and Leinen, M. (1988) Asian aridity and the zonal westerlies: Late Pleistocene and Holocene record of eolian deposition in the Northwest Pacific Ocean. *Palaeogeogr., Palaeoclimatol., Palaeoecol.*, **66**, 1–8.
29. Hooghiemstra, H. (1988) Changes of major wind belts and vegetation zones in NW Africa 20,000–5000 yr BP, as deduced from a marine pollen record near Cap Blanc. *Rev. Palaeobot. Palynol.*, **55**, 101–40.
30. Van Campo, E., Duplessy, J. C. and Rossignol-Strick, M. (1982) Climatic conditions deduced from a 150-kyr oxygen isotope–pollen record from the Arabian Sea. *Nature*, **296**, 56–9.
31. Smith, G. I. and Street-Perrott, A. (1983) Pluvial lakes of the western United States, in *Late-Quaternary Environments of the United States*, Vol. 1, *The Late Pleistocene* (ed. S. C. Porter), Longman, London, pp. 190–212.
32. Spaulding, W. G. and Graumlich, L. J. (1986) The last pluvial climatic episodes in the deserts of southwestern North America. *Nature*, **320**, 441–4.

33. Semken, H. A. Jr (1984) Holocene mammalian biogeography and climatic change in the eastern and central United States, in *Late-Quaternary Environments of the United States*, Vol. 2, *The Holocene* (ed. H. E. Wright Jr), Longman, London, pp. 182–207.
34. Barnosky, C. W. (1989) Postglacial vegetation and climate in the northwestern Great Plains of Montana. *Quat. Res.*, **31**, 57–73.
35. Harris, S. A. (1985) Evidence for the nature of the early Holocene climate and palaeogeography, High Plains, Alberta, Canada. *Arctic Alpine Res.*, **17**, 49–67.
36. Trabaud, L. (1981) Man and fire: impacts on Mediterranean vegetation, in *Ecosystems of the World 11. Mediterranean-Type Shrublands* (eds F. di Castri, D. W. Goodall and R. L. Specht), Elsevier, Amsterdam, pp. 523–37.
37. King, J. E. (1981) Late Quaternary vegetational history of Illinois. *Ecol. Monogr.*, **51**, 43–62.
38. Webb, T., Cushing, E. J. and Wright, H. E. Jr (1984) Holocene changes in the vegetation of the Midwest, in *Late-Quaternary Environments of the United States*, Vol. 2, *The Holocene* (ed. H. E. Wright Jr), Longman, London, pp. 142–65.
39. Holliday, V. T. (1989) Middle Holocene drought on the Southern High Plains. *Quat. Res.*, **31**, 74–82.
40. Spaulding, W. G. (1983) Late Wisconsin macrofossil records of desert vegetation in the American South-west. *Quat. Res.*, **19**, 256–64.
41. Van Devender, T. R. (1977) Holocene woodlands in the Southwestern deserts. *Science*, **198**, 189–92.
42. Van Devender, T. R., and Spaulding, W. G. (1979) Development of vegetation and climate in the southwestern United States. *Science*, **204**, 701–10.
43. Wells, P. V. (1983) Paleobiogeography of montane islands in the Great Basin since the last glaciopluvial. *Ecol. Monogr.*, **53**, 341–82.
44. Daubenmire, R. (1978) *Plant Geography, with Special Reference to North America*, Academic Press, New York.
45. Cole, K. L. (1986) The Lower Colorado River Valley: a Pleistocene desert. *Quat. Res.*, **25**, 392–400.
46. Van Devender, T. R. (1987) Holocene vegetation and climate in the Puerto Blanco Mountains, southwestern Arizona. *Quat. Res.*, **27**, 51–72.
47. Baker, R. G. (1984) Holocene vegetational history of the western United States, in *Late-Quaternary Environments of the United States*, Vol. 2, *The Holocene* (ed. H. E. Wright Jr), Longman, London, pp. 109–27.
48. Thompson, R. S. (1988) Western North America, in *Vegetation History* (eds B. Huntley and T. Webb III), Kluwer Academic Publishers, Dordrecht, pp. 415–58.
49. Ritchie, J. C. and Yarranton, G. A. (1978) The Late-Quaternary history of the boreal forest of central Canada, based on standard pollen stratigraphy and principal components analysis. *J. Ecol.*, **66**, 199–212.
50. Cole, K. (1985) Past rates of change, species richness, and a model of vegetational inertia in the Grand Canyon, Arizona. *Am. Nat.*, **125**, 289–303.
51. Reider, G., Huckleberry, G. A. and Frison, G. C. (1988) Soil evidence for postglacial forest–grassland fluctuation in the Absaroka Mountains of northwestern Wyoming, USA. *Arctic Alpine Res.*, **20**, 188–98.
52. Barnosky, C. W., Anderson, P. M. and Bartlein, P. J. (1987) The north-

western US during deglaciation: vegetational history and paleoclimatic implications, in *North America and Adjacent Oceans during the Last Deglaciation* (eds W. F. Ruddiman and H. E. Wright Jr), Geological Society of America, Boulder, CO, pp. 289–321.

53. Kutzbach, J. E. and Guetter, P. J. (1986) The influence of changing orbital parameters and surface boundary conditions on climatic simulations for the past 18,000 years. *J. Atmos. Sci.*, **43**, 1726–59.
54. Van Devender, T. R., Thompson, R. S. and Betancourt, J. L. (1987) Vegetational history of the deserts of southwestern North America; the nature and timing of the Late Wisconsin–Holocene transition, in *North America and Adjacent Oceans during the Last Deglaciation* (eds W. F. Ruddiman and H. E. Wright Jr), Geological Society of America, Boulder, CO, pp. 323–52.
55. Forester, R. M. (1987) Late Quaternary paleoclimate records from lacustrine ostracods, in *North America and Adjacent Oceans during the Last Deglaciation* (eds W. F. Ruddiman and H. E. Wright Jr), Geological Society of America, Boulder, CO, pp. 261–76.
56. Winkler, M. G., Swain, A. M. and Kutzbach, J. E. (1986) Middle Holocene dry period in the northern Midwestern United States: lake levels and pollen stratigraphy. *Quat. Res.*, **25**, 235–50.
57. Kutzbach, J. E. (1987) Model simulations of the climate patterns during the deglaciation of North America, in *North America and Adjacent Oceans during the Last Deglaciation* (eds W. F. Ruddiman and H. E. Wright Jr), Geological Society of America, Boulder, CO, pp. 425–46.
58. Knox, J. C. (1984) Responses of river systems to Holocene climates, in *Late-Quaternary Environments of the United States*, Vol. 2, *The Holocene* (ed. H. E. Wright Jr), Longman, London, pp. 26–41.
59. Ayyad, M. A. and Ghabbour, S. I. (1986) The deserts of Egypt and the Sudan, in *Ecosystems of the World 12B. Hot Deserts and Arid Shrublands* (eds M. Evenari, I. Noy-Meir and D. W. Goodall), Elsevier, Amsterdam, pp. 149–202.
60. Bradley, R. S. (1985) *Quaternary Paleoclimatology. Methods of Paleoclimatic Reconstruction*, Allen and Unwin, Winchester, MA.
61. Lauer, W. and Frankenberg, P. (1980) Modelling of climate and plant cover in the Sahara for 5500 BP and 18,000 BP. *Palaeoecol. Africa*, **12**, 307–14.
62. Peterson, G. M., Webb, T., Kutzbach, J. E. *et al.* (1979) The continental record of environmental conditions at 18000 BP: an initial evaluation. *Quat. Res.*, **12**, 47–82.
63. Rognon, P. and Williams, M. A. J. (1977) Late Quaternary climatic changes in Australia and North Africa: a preliminary interpretation. *Palaeogeogr., Palaeoclimatol., Palaeoecol.*, **21**, 285–327.
64. Van Zinderen Bakker, E. M. (1980) Comparison of late-Quaternary climatic conditions in the Sahara and the Namib–Kalahari region. *Palaeoecol. Africa*, **12**, 381–94.
65. Wijmstra. T. A. (1969) Palynology of the first 30 metres of a 120 m deep section in northern Greece. *Acta bot. neerl.*, **18**, 511–27.
66. Ben Tiba, B. and Reille, M. (1982) Recherches pollenanalytiques dans les montagnes de Kroumirie (Tunisie septentrionale): Premiers résultats. *Ecol. Mediterr.*, **8**, 75–86.

67. Van Campo, M., Guinet, Ph., Cohen, J. and Dutil, P. (1967) Contribution à l'étude du peuplement végétale Quaternaire des montagnes sahariennes. III. Flore de l'Oued Outoul (Hoggar). *Pollen Spores*, **9**, 107–20.
68. Rossignol-Strick, M. and Duzer, P. (1979) West African vegetation and climate since 22,500 BP from deep-sea cores palynology. *Pollen Spores*, **21**, 105–34.
69. Sarnthein, M. (1978) Sand deserts during glacial maximum and climatic optimum. *Nature*, **272**, 43–6.
70. Rognon, P. (1987) Late Quaternary climatic reconstruction for the Maghreb (North Africa). *Palaeogeogr., Palaeoclimatol., Palaeoecol.*, **58**, 11–34.
71. Butzer, K. W. (1978) The late prehistoric environmental history of the Near East, in *The Environmental History of the Near and Middle East since the last Ice Age* (ed. W. C. Brice), Academic Press, London, pp. 5–12.
72. Rossignol-Strick, M., Nesteroff, W., Olive, P. and Vergnaud-Grazzini, C. (1982) After the deluge: Mediterranean stagnation and sapropel formation. *Nature*, **295**, 105–10.
73. Jäkel, D. (1977) Run-off and fluvial formation processes in the Tibesti Mountains as indicators of climatic history in the central Sahara during the Late Pleistocene and Holocene. *Palaeoecol. Africa*, **11**, 13–44.
74. Flohn, H. and Nicholson, S. (1980) Climatic fluctuations in the arid belt of the 'Old World' since the Last Glacial Maximum; possible causes and future implications. *Palaeoecol. Africa*, **12**, 3–21.
75. Street, F. A. and Gove, A. T. (1979) Global maps of lake-level fluctuations since 30,000 yr BP. *Quat. Res.*, **12**, 83–118.
76. Ritchie, J. C., Eyles, C. H. and Haynes, C. V. (1985) Sediment and pollen evidence for an early to mid-Holocene humid period in the eastern Sahara. *Nature*, **314**, 352–5.
77. Farrand, W. R. (1971) Late Quaternary paleoclimates of the Eastern Mediterranean area, in *The Late Cenozoic Glacial Ages* (ed. K. K. Turekian), Yale University Press, New Haven, CT, pp. 529–64.
78. Brooks, C. E. P. (1926) *Climate through the Ages*, Ernest Benn, London.
79. Lamb, H. F., Eicher, U. and Switsur, V. R. (1989) An 18,000-year record of vegetation, lake-level and climatic change from Tigalmamine, Middle Atlas, Morocco. *J. Biogeogr.*, **16**, 65–74.
80. Hastenrath, S. (1985) *Climate and Circulation in the Tropics*, Reidel, Dordrecht.
81. Gasse, F., Fontes, J. C., Plaziat, J. C. *et al.* (1987) Biological remains, geochemistry and stable isotopes for the reconstruction of environmental and hydrological changes in the Holocene lakes from North Sahara. *Palaeogeogr. Palaeoclimatol. Palaeoecol.*, **60**, 1–46.
82. Maley, J. (1982) Dusts, clouds, rain types and climatic variations in tropical North Africa. *Quat. Res.*, **18**, 1–16.
83. Monod, Th. (1986) The Sahel zone north of the Equator, in *Ecosystems of the World 12A. Hot Deserts and Arid Shrublands* (eds M. Evenari, I. Noy-Meir and D. W. Goodall), Elsevier, Amsterdam, pp. 203–43.
84. Maley, J. (1977) Palaeoclimates of central Sahara during the early Holocene. *Nature*, **269**, 573–7.
85. Wickens, G. E. (1975) Quaternary plant fossils from the Jebel Marra volcanic

complex and their palaeoclimatic interpretation. *Palaeogeogr., Palaeoclimatol., Palaeoecol.*, **17**, 109–22.

86. Ritchie, J. C. (1984) Analyse pollinique de sédiments holocènes supérieurs des hauts plateaux du Maghreb oriental. *Pollen Spores*, **26**, 489–96.
87. Bottema, S. (1978) The Late Glacial in the Eastern Mediterranean and the Near East, in *The Environmental History of the Near and Middle East since the last Ice Age* (ed. W. C. Brice), Academic Press, London, pp. 15–28.
88. Brande, A. (1973) Untersuchungen zur postglazialen Vegetationsgeschichte im Gebiet der Neretva-Niederungen (Dalmatien, Herzegowina). *Flora, Jena*, **162**, 1–44.
89. Bottema, S. (1980) Palynological investigations on Crete. *Rev. Palaeobot. Palynol.*, **31**, 193–217.
90. Greig, J. R. A. and Turner, J. (1974) Some pollen diagrams from Greece and their archaeological significance. *J. Archaeol. Sci.*, **1**, 177–94.
91. Vernet, J.-L. and Thiebault, S. (1987) An approach to northwestern Mediterranean recent prehistoric vegetation and ecologic implications. *J. Biogeogr.*, **14**, 117–29.
92. Huntley, B. and Birks, H. J. B. (1983) *An Atlas of Past and Present Pollen Maps for Europe: 0–13,000 Years Ago*, Cambridge University Press, Cambridge.
93. Beug, H.-J. (1967) On the forest history of the Dalmatian coast. *Rev. Palaeobot. Palynol.*, **2**, 271–9.
94. Behre, K. E. (1988) The role of man in European forest history, in *Vegetation History* (eds B. Huntley and T. Webb III), Kluwer Academic Publishers, Dordrecht, pp. 633–72.
95. Di Castri, F. (1981) Mediterranean-type shrublands of the world, in *Ecosystems of the World 11. Mediterranean-Type Shrublands* (eds F. Di Castri, D. W. Goodall, and R. L. Specht), Elsevier, Amsterdam, pp. 1–52.
96. Guilaine, J. (1979) The earliest Neolithic in the West Mediterranean: a new appraisal. *Antiquity*, **53**, 22–30.
97. Pons, A. (1981) The history of the Mediterranean shrublands, in *Ecosystems of the World 11. Mediterranean-Type Shrublands* (eds F. di Castri, D. W. Goodall and R. L. Specht), Elsevier, Amsterdam, pp. 131–8.
98. Huntley, B. (1988) Europe, in: *Vegetation History* (eds B. Huntley and T. Webb III), Kluwer Academic Publishers, Dordrecht, pp. 341–83.
99. Mikesell, M. W. (1960) Deforestation in northern Morocco. *Science*, **132**, 441–8.
100. Lézine, A. M. (1988) New pollen data from the Sahel, Senegal. *Rev. Palaeobot. Palynol.*, **55**, 141–54.
101. Murphey, R. (1951) The decline of North Africa since the Roman Occupation: climatic or human? *Ann. Ass. Am. Geogr.*, **41**, 116–32.
102. Lowdermilk, W. C. (1940) Erosion control lessons from Old World experience. 3. Erosion at its worst, and a hundred dead cities. *Soil Conserv.*, **5**, 157–62.
103. Thirgood, J. V. (1981) *Man and the Mediterranean Forest. A History of Resource Depletion*, Academic Press, London.
104. Beals, E. W. (1965) The remnant cedar forests of Lebanon. *J. Ecol.*, **53**, 679–94.

105. Le Houérou, H. N. (1981) Impact of man and his animals on Mediterranean vegetation, in *Ecosystems of the World 11. Mediterranean-Type Shrublands* (eds F. di Castri, D. W. Goodall and R. L. Specht), Elsevier, Amsterdam, pp. 479–521.
106. Reed, C. A. (1977) *Origins of Agriculture*, Chicago University Press, Chicago.
107. Attenborough, D. (1987) *The First Eden. The Mediterranean World and Man*, Collins, London.
108. Van Zeist, W. and Bottema, S. (1982) Vegetational history of the eastern Mediterranean and the Near East during the last 20,000 years. *B.A.R. Internat. Series*, **133**, 277–321.

CHAPTER FOURTEEN

Coldland communities

14.1 INTRODUCTION

Land surfaces at higher latitudes in the northern hemisphere support a range of forest, scrub, tundra and peatland communities at the present day that may collectively be called the 'coldland complex'. Physiognomically and floristically similar communities also occur at higher altitudes on mountains further south and in the southern hemisphere (Figure 14.1). Both sets of communities share a common resistance to extreme low temperature. Several physiognomic types can be recognized [1,2]: coniferous forest; tall-shrub thicket; heath (with a predominance of dwarf shrubs); sedge–moss carpet; and steppe–desert (with a discontinuous cover of low-growing perennial herbs). At high latitudes these types form successive zones extending northwards towards the pole – the boreal forest, forest–tundra, and low-Arctic, middle-Arctic and high-Arctic zones of some ecologists – and they may also form a smaller-scale topographic mosaic within each zone. The forest–tundra represents an **ecotone**, often of considerable extent, between forest and tundra, and is bounded to the north and south by the tree-limit (treeline) and the forest-border, respectively (section 8.3.1). In mountain regions the tree-limit and the forest-border are usually more nearly coincident; here heath and steppe–desert communities comprise the low-Alpine and middle-Alpine zones, respectively*. The cold climate at high latitudes and altitudes, coupled with the scleromorphic nature of many of the larger plants and the prevalence of acidic soils, can lead to retarded rates of decomposition of organic matter; where loss of water from flatter ground is checked by a high regional water table, by permafrost in the soil, or by low evaporation throughout the year, then waterlogging ensues, and may result in the accumulation of peat. Peat can form below any of the physiognomic vegetation types mentioned earlier except steppe–desert, though in the arctic zone peat accumulation is largely confined to the sedge–moss communities. Peatlands form a major

* Considerable confusion exists over the terminology of coldland communities [3]. 'Subarctic' and 'subalpine', in particular, have been used in conflicting senses by some ecologists. Their usage as part of any vegetation-zonation system is avoided here (except in specific contexts; cf. Chapter 10).

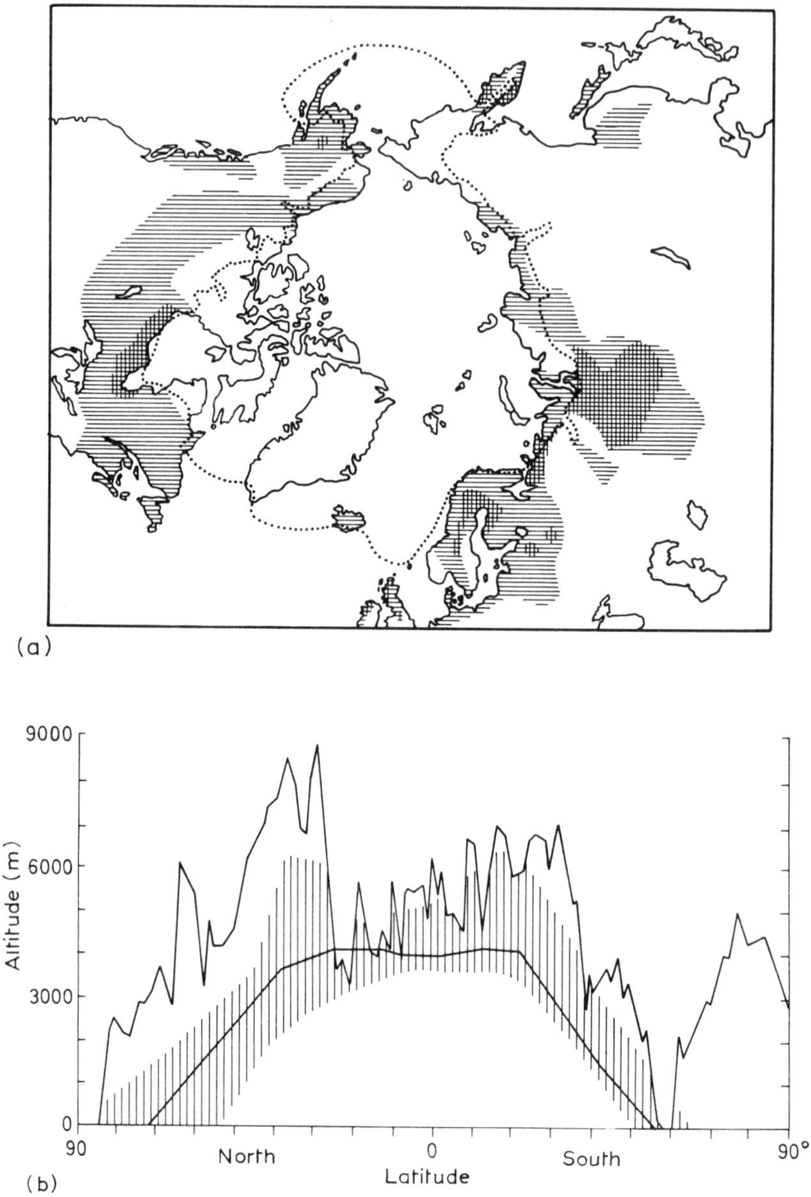

Figure 14.1 The occurrence of coldland and wetland communities. (a) Approximate position of the modern treeline in the northern hemisphere ...., and the occurrence of areas with >10% cover ≡ and dominance # of wetland communities. Re-drawn from Sjörs [4] and Gore [5]. (b) Schematic profile along a north/south transect of the highest mountain summits, showing the position of the maximum recorded modern treeline ⁄‾, and the greatest theoretical extent of alpine communities |||||||| between the extreme recorded positions of the snowline and the treeline at the present day. Re-drawn from Barry and Ives [3].

element in the coldland landscape between about latitudes 50° and 70°N (Figure 14.1). The vegetation of these 'boreal mires' comprises a complex mosaic of forest, heath and sedge–moss communities at predominantly low altitudes, apart from some 'blanket mires'. At the highest latitudes and altitudes within the coldland zone plant life is effectively absent, and landscapes are dominated by permanent snow and ice.

The coldland complex is largely a Pleistocene creation, and occupies those regions that were most profoundly affected by the climatic vicissitudes of the Pleistocene. The present plant communities clearly bear the impress of these vicissitudes, and are the product of a highly changeable environment. Their history can be regarded as a history of the interactions between the four basic components of the coldland complex – forest, tundra, peatland and ice. The timescale over which the interactions can be followed is shorter than that for temperate and tropical forest communities, being confined essentially to the last glacial/interglacial cycle, but the magnitude of change is just as great; because of the widespread occurrence of ice and peat deposits within the coldland zone, with their contained archives of past climates and vegetation, the nature of the change is well documented. Nevertheless, after more than a century of interest and research, many questions still remain unanswered. The current state of knowledge of coldland dynamics, and its relevance to plant community history in general, is considered in the following pages.

14.2 THE ORIGINS OF COLDLAND COMMUNITIES

The low temperatures characteristic of modern coldland habitats were almost certainly absent from the Eocene world. Appropriate conditions developed only late in the Tertiary, and reached their culmination during the Pleistocene. The Tertiary origins of certain parts of the coldland complex (and particularly the so-called boreal mires) are still obscure, but from a knowledge of other parts of the complex we can begin to reconstruct a partial picture. Two parts are well documented (though the evidence is strongly biased towards North America), and provide the dating framework: permanent glacier ice and boreal conifer forest.

North polar ice – over Greenland and Baffin Land – formed during the Pliocene, some time between 3.5 and 2.4 Ma ago [6–9]. Mountain glaciers almost certainly formed earlier, spreading progressively southwards from high-altitude centres in Alaska about 9 Ma ago to the newly-elevated Sierra Nevada about 3 Ma ago [6,10]. A similar dating range probably applies to other northern-hemisphere mountain ranges also. Boreal coniferous forest developed by segregation of cold-resistant taxa (species of *Abies*, *Larix*, *Picea* and *Pinus*) from the diverse mixed coniferous forests that were present at higher altitudes in the Middle Tertiary (sections 9.3 and 9.4). The earliest records of species-poor coniferous forest are from high northern mountains in Alaska and Japan about 15 Ma ago [11,12]; these forests spread subsequently to mountain ranges further south as they were elevated in the Late Miocene and Pliocene [12]. 'Ancestral' mixed coniferous forest at low altitudes was present in the Alaskan lowlands 10–14 Ma ago [11], and true boreal forest derived from it was widespread over Alaska, northeast Siberia and northern Japan by the Pliocene [12–14]. The extreme dates of 15 Ma ago and 2.4 Ma ago for the inception of boreal forest and permanent ice, respectively, must thus bracket the time of formation of the coldland complex in the northern hemisphere.

It is the herbaceous tundra and peatland communities, however, that are most characteristic of the coldland complex, and for these the fossil record is much less good. Almost all the more precise identifications of fossil plant material from the Tertiary Period are of woody taxa; pollen records of herbaceous taxa are largely non-specific. Although it might be logical to assume that tundra floras first evolved in the Arctic, as deteriorating climate in the Late Tertiary forced trees to evacuate high latitudes, the fossil record indicates that coniferous forest did not generally begin to pull back from the shores of the Arctic Ocean until the Pleistocene; at only a few sites in Alaska and Siberia is there evidence of open forest or incipient tundra even in the Late Pliocene [12,13,15]. However, there is almost no fossil documentation of Late Tertiary vegetation from further north, from the islands of the Canadian Arctic; if indeed tundra communities were present in the northernmost parts of Greenland, Ellesmere Island and Axel Heiberg Island (beyond latitude 80°N) in the Late Miocene, as Murray [2] suggests, then coldland floras could have been widespread in those islands by the close of the Tertiary. The problem is as yet unresolved.

There is better fossil evidence of non-forest vegetation at high altitudes on mountain ranges further south during the Tertiary, above the developing conifer forests. It is likely that herbaceous floras radiated extensively in the new alpine habitats created by orogenesis in the Miocene and Pliocene. Fusion of these initially separate arctic and alpine floras occurred only during the cold stages of the Pleistocene, as low-altitude treeless habitats spread to temperate latitudes.

The sources of cold-adapted herbaceous plants are obscure. We might surmise that in the temperate and boreal forests of the Late Tertiary there was a comparable range of subordinate herbaceous plants to that found in similar forests today, and that some of those herbaceous taxa remained behind at high latitudes when the trees had gone, or colonized the newly-elevated high-altitude ecospace to form the nucleus of coldland floras. However, modern conifer forests are floristically poor in undergrowth species; as forests of this type probably abutted on early tundra habitats, the developing arctic floras of the Late Tertiary and Early Pleistocene might also have been of restricted diversity – and dominated perhaps by ericads such as *Empetrum, Ledum* and *Vaccinium*. Sources of colonizing plants for newly-uplifted alpine habitats were probably more varied, and might well have included temperate (nemoral) forests if orogenesis were rapid. The resulting alpine floras could then have been more diverse [16]. We might thus envisage the modern low-arctic flora as derived largely from boreal forest sources, and the middle- and high-arctic floras as including substantial contributions from precursor alpine floras. The sources of the boreal peatland flora are unknown. Some taxa (for example, the ericads) might have been drawn from the boreal forests; others (for example, sphagna) might have originated from more generalist wetland taxa, which were undoubtedly present in local waterlogged habitats throughout the Tertiary. Because of the dryness of the Late Miocene climate, however (section 9.3.3), it is unlikely that peatland floras evolved extensively before the Pliocene.

Consolidation and diversification of coldland floras was concentrated into the Pleistocene Epoch. Major range expansions of many coldland plants occurred during the cold stages, when continuity must have been established between the previously-separated high-altitude and high-latitude floras; much of the land area shown in Figure 14.1 (a), for example, was treeless during these stages, and the coldland complex of Eurasia was connected to that of Alaska by a land bridge up to 1500 km wide ('Beringia' [17]). During the intervening warm stages, contraction and fragmentation of range ensued. We must presume, by analogy with the Pleistocene scenario for forests (Figure 10.1), that some losses of coldland taxa occurred when their ranges suffered contraction during the warm stages. The available evidence indicates that such extinction was low. Thus of the 444 species recorded in the fossil record from the last glacial stage in the British Isles*, probably only twelve no longer occur there (Table 14.1).

The large-scale displacements of range that repeatedly affected coldland taxa through the Pleistocene must have meant that they were exposed to great changes in seasonal temperature patterns and in day-length, precipitation and wind regimes (changes equivalent to the total range of conditions experienced by arctic and alpine floras at the present day). Physiological adaptation to these differing regimes must have been an ongoing process during the migrations of coldland plants [19]. Indeed, a case can be made out that coldland plants have been among the most successful constituents of the world's flora. The range of conditions to which they were subjected through the Quaternary was extreme, the

* Including the lateglacial period.

Table 14.1 Species recorded from Middle and Late Weichselian (last glacial and late glacial) deposits in the British Isles that are now extinct [18]

Carex capitata
Corispermum sp.
Dianthus carthusianorum
Erysimum cheiranthoides
Papaver Section *Scapigerum*
Pedicularis hirsuta
Pedicularis lanata
Ranunculus aconitifolius
Ranunculus hyperboreus
Salix polaris
Silene furcata
cf. *Stellaria crassipes*

distances over which they migrated were immense, and the physiological adaptations they were required to make as a result of these migrations were considerable. Moreover, migrations at the onset of each glacial stage were accomplished in the face of resistance from an already-established (and often tree-dominated) vegetation [20]. The occurrence at the present day on tropical mountains of such typical arctic genera as *Arabis, Astragalus, Cerastium, Silene* and *Veronica* is testimony of the ability of coldland plants in the past to adapt to the very different diurnal and seasonal climatic patterns at high and low latitudes; the presence on southern-hemisphere mountains of northern polar species such as *Carex echinata, C. lachenalii, Koenigia islandica* and *Trisetum spicatum*, or their close relatives [16], testifies to their high dispersal capabilities [21]. Because coldland plants inhabit areas which have been highly changeable through time, their characteristics of adaptability and dispersibility must be seen as responses to this changeableness. These characteristics may not always be apparent at the present day, when the ranges of many coldland communities are greatly contracted under near-maximum interglacial conditions: the evolutionary success of coldland plants cannot be judged solely on present-day performance and range.

14.3 THE DYNAMICS OF COLDLAND COMMUNITIES

14.3.1 Cold-stage communities

Although there is a variety of information relating to different facies of the coldland complex prior to the last glacial maximum of the Pleistocene at 18–20 ka ago, it is the detailed reconstructions of the ice-age world for that time-period that form a realistic starting point for a consideration of the history of coldland communities. These reconstructions show a coldland complex very different in character and composition from that at the present day. Environmental conditions, discussed in Chapters 5 and 10, were not equally favourable to all components of the coldland complex. Permanent ice and snow in the northern hemisphere was greatly expanded in extent – 29.67 as against 1.99 million km^2 (Table 5.1) – and covered large parts of the area now occupied by coldland communities. In heavily glaciated regions it is possible that all vascular plants were exterminated, so that survival was possible only by major displacements of range southwards (the '*tabula rasa* hypothesis'). However, there is some geological evidence, particularly for the Canadian Arctic, that residual areas within the ambit of the major ice sheets at glacial maximum remained ice-free. These areas comprised certain mountain summits that projected as unglaciated 'nunataks' through the investing ice, and some coastal regions close to the edges of the continental shelf. The geological evidence for Europe is more ambiguous [22]. Nevertheless, it is possible that some hardy plants survived through the glacial maximum in such localized ice-free areas – albeit under extreme climatic stress. Disjunct plant distributions at the present day, particularly in Fennoscandia, Iceland and Greenland, have been used by some ecologists to reconstruct the locations of possible 'ice-age refugia' [16,22]. Examples of such reconstructions are shown in Figure 14.2, and the whole question of 'perglacial survival' is discussed further in section 14.4.

To the south of the ice sheets in all northern-hemisphere land masses stretched a very extensive region where herbaceous plants were widespread and often predominant (Figures 10.5, 10.9 and 10.11). These herbaceous, and presumably cold-adapted, plants are found today in a variety of different habitats, and combined in assemblages quite unlike the full-glacial ones. Specific examples of such full-glacial plants and their modern habitats were given in section 10.3. Dwarf shrubs were also widespread, particularly shrubby willows and birches (*Salix* and *Betula* spp.). The character and composition of these full-glacial coldland communities have been discussed in sections 10.3 and 10.4. Many of the component plants must have had very extensive geographical ranges. Nevertheless, it is important to realize that major discontinuities also occurred. Thus the coldland communities of the United States were separated from those of Alaska and Eurasia by the North American Ice Sheet, whilst smaller segregated populations could have survived in ice-free refugia at higher latitudes and on high mountains at lower latitudes. During the earlier parts of each glacial stage of the Pleistocene, when climatic conditions were less severe, discontinuities of range may have been more pronounced.

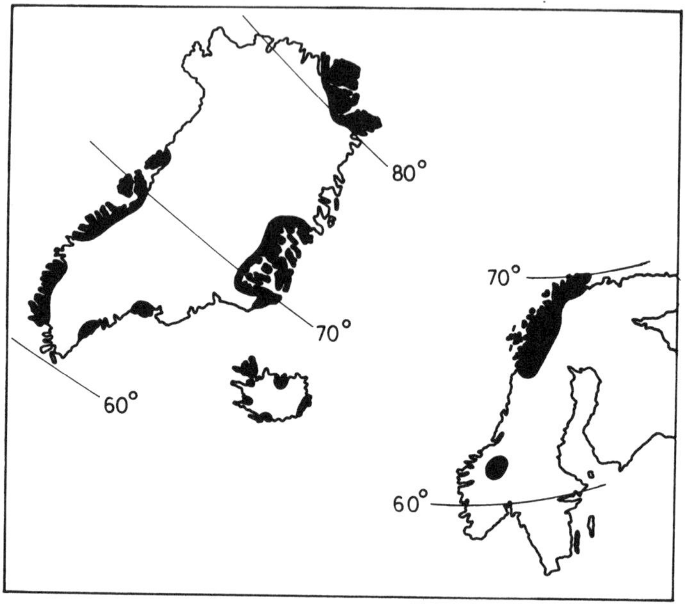

Figure 14.2 Areas with concentrations of disjunct populations of arctic-alpine species at the present day that have been suggested as located close to 'ice-age refugia'. Based on maps in Gjaerovoll [23], Steindórsson [24], Ives [22] and Funder [25].

Other components of the coldland complex, however, were markedly restricted in extent at full glacial. The arid conditions to the south and east of the ice sheets in Eurasia, and to a lesser degree in North America, were inimical to the survival of forest, heath and peatland communities over vast areas. Cold-hardy trees, it is true, occurred widely as scattered individuals or in small groves in parkland and forest–steppe communities in both land masses, but continuous coniferous forest stands were confined to limited refugia in mountainous regions of Eurasia and North America (Figures 10.7 and 10.11) and in warmer oceanic regions – western and south-western Europe [26], the western United States (section 10.4), the Caucasus, and probably Kamchatka and around the Sea of Okhotsk in eastern Asia [27]. These refugia almost certainly sheltered different populations of conifers: *Larix decidua, Picea abies* and *Pinus sylvestris* in southern and eastern Europe; *Larix sibirica, Picea obovata* and *Pinus sibirica* in western Asia; *Larix dahurica* and *Pinus pumila* in eastern Asia; *Picea glauca, P. rubens* and *Pinus banksiana* in the eastern United States; and *Picea engelmannii, Pinus flexilis* and *P. longaeva* in the western United States, for example. It is likely that these same refugia were the main areas of survival of ericads [26] and boreal mire plants, but no collated information exists. What is clear is that for these components of the coldland complex the full glacial was a time of moisture-stress and consequent range contraction; in this respect their history differs strikingly from that of the 'arctic–alpine' component.

14.3.2 The lateglacial period

The glacial maximum of 18–20 ka ago in the northern hemisphere was succeeded by a prolonged period of slow and hesitant amelioration of climate before temperatures rose rapidly at the start of the Holocene (section 6.1). The time-interval from 18 to 10 ka ago, the latter part of which is often referred to as the 'lateglacial period' – when the pace of change quickened somewhat – was characterized by extreme instability of climate: an instability expressed most clearly in complex fluctuations of the margins of the major ice sheets. Simplistically, the period was one in which the heterogeneity of the coldland ecospace was substantially enhanced. The partial and episodic melting of the ice sheets, accentuated after 15 ka ago, led to the exposure of bare and sometimes waterlogged substrates, the periodic irrigation of certain areas by snow-melt waters, the impounding of drainage streams, and at least partial mitigation of the full-glacial aridity. These effects were variable both in space and in time, but the net result was to transform some of the vast areas of relatively uniform climate and vegetation of the full glacial into a much smaller-scale landscape mosaic that offered opportunities for the spread of many of the components of the coldland complex. In Europe, the lateglacial and Early Holocene periods were characterized by a large number of geographically-restricted vegetation units [28]. The lateglacial period was the heyday of the coldland complex, as during it the extent of permanent ice and snow was substantially diminished yet temperate forest had not spread much from its full-glacial locations by the close. Within their expanded area of occupancy, plants of the coldland complex flourished, aided in their spread by the widespread occurrence of immature soils of high base-status and low humus content, and by the absence of competition from trees and shrubs [29]. Fossil records show clearly that many of the plants occupied more extensive ranges than they do at the present day [18]. A number of attempts have been made to reconstruct the various plant communities making up the vegetation during this period of time, by seeking analogies with modern arctic and alpine communities; three examples of such reconstructions are shown in Table 14.2, for central Minnesota, the British Isles and northwest Europe. Most ecologists, however, agree that many of these lateglacial plant assemblages have no precise modern analogues. Birks [30], for example, could find approximate modern matches (i.e.,

Table 14.2 Lateglacial plant communities: suggested modern analogues for the pollen and macrofossil assemblages found in (a) central Minnesota [30], (b) the British Isles [18] and (c) north-west Europe [32] between 12 and 15 ka ago. Some characteristic fossil taxa are shown*

	Central Minnesota	British Isles	Northwest Europe
Snow-bed communities	O,S	L,S	+
Screes and boulders	F		+
Fellfield (tundra barrens)	A,D	A,R	+
Moss and lichen communities			+
Steppe–tundra	A,Ch,R	Ch	
Herb meadows		L,Sg	+
Tall-herb communities		Fi,T	+
Heath		E,J	+
Scrub	P,S	B,H,S	+
Copses and woodland		B	+
Springs and flushes		+	+
Marsh and fen	C,J,Sg	C,J	
Sedge mires			+
Aquatic communities	+	+	

* A = *Artemisia*; B = *Betula*; C = Cyperaceae; Ch = Chenopodiaceae; D = *Dryas*; E = *Empetrum*; F = ferns; Fi = *Filipendula*; H = *Hippophaë*; J = *Juncus*; L = *Lycopodium*; O = *Oxyria*; P = *Populus*; R = *Rumex*; S = *Salix*; Sg = *Selaginella*; T = *Trollius*.

from surface samples) for only about 12 of the 58 pollen spectra of lateglacial age that he examined at Wolf Creek, Minnesota.

Pollen diagrams from lateglacial deposits typically show complex patterns of change through time that reflect the expansions and contractions of plant ranges in response to the constantly varying environment. The pollen record, however, is an amalgam culled from a variety of sources, so that any landscape mosaic that might have existed is difficult to detect. The 'pollen assemblage zones' customarily used to subdivide such pollen diagrams do not therefore correspond to the reconstructed plant communities derived from modern analogues. Nevertheless, in regions that have been intensively studied, such as the British Isles, some correspondence of the more important pollen assemblage zones with broad community types can be recognized [31]. The overall pattern of change in the pollen diagrams can be seen as a general trend away from assemblages dominated by pollen of herbaceous plants and towards ones dominated by pollen of woody plants; however, particularly in northwestern Europe, a major reversion towards herbaceous assemblages is apparent between 11 and 10 ka ago [18,29,31,33,34]. This reversion was caused by a temporary shift to colder conditions, and was accompanied by glacier expansion at higher latitudes and altitudes (section 6.1).

The general trend through the lateglacial period towards scrub or woodland communities was paralleled by an overall increase in the range and abundance of the boreal conifers. The increase was manifested both as the progressive closing of the tree canopy locally in existing parkland areas and as an establishment of outlying populations further to the north. In central Europe and Minnesota these changes were first apparent about 13.5 ka ago; in the western United States the spread started rather earlier (section 11.3.1). Almost certainly there was a concomitant increase in ericads, but the process is poorly documented. In Europe appreciable amounts of Ericaceae pollen are recorded by 10 ka ago only in deposits along the Atlantic coasts from southern Spain to southwestern Norway [26], and conditions in the interior of the continent might still have been too dry generally at that time for the spread of ericads. Dwarf shrub tundra was certainly becoming established in East Beringia by 14 ka ago [35]. The status of the boreal mire flora during the lateglacial period is even more obscure, but the success of both ericads and mire plants in responding rapidly to the climatic amelioration of the Early Holocene suggests that by the end of the lateglacial they had established a series of outlying populations in favourable microhabitats over a large part of the coldland zone of the northern hemisphere.

14.3.3 Boreal forest in the Holocene

Russian palaeoecologists in particular have stressed the lack of any marked north/south zonation in the lateglacial vegetation of the USSR. A climate with short hot summers, long cold winters and little precipitation allowed plants characteristic today of tundra, steppe, forest and desert communities to intermingle throughout a periglacial zone that extended up for to 1500 km south of the ice sheets [36]. A similar, though less extreme, picture has been reconstructed for the British Isles [18,31]. The subsequent rapid rise in temperature in the Early Holocene caused an equally rapid restructuring of this lateglacial vegetation, and a segregation of the various floristic elements into separate vegetation zones. A broad belt of birch–conifer forest formed across middle latitudes in North America and Eurasia, between tundra vegetation to the north and steppe and desert to the south; it was this boreal forest zone that responded most quickly to the Early Holocene temperature rise, advancing northwards into recently deglaciated ground and into periglacial tundra–steppe, compressing the tundra zone before it. By 9 ka ago trees had reached the arctic coasts of northwestern Canada and northern Siberia, where their fossilized remains persist among the modern tundra communities – birches, and some larch, in Eurasia, and spruce and larch in Canada [36,37]. In central and eastern Canada [38] and in northern Europe [39,40] the advance was slower, presumably delayed by the melting residues of the Laurentide and Fenno-scandian Ice Sheets, but by the mid-Holocene forest had everywhere become established at higher latitudes and altitudes than at the present day (200–400 km and 70–150 m, respectively, beyond current limits). In North America the present-day dominants of the boreal forest (*Picea glauca, P. mariana, Pinus banksiana, Larix laricina*) were prominent almost from the start of the forest advance, but in Eurasia the pattern was more complicated. In western Siberia *Picea obovata* dominated the forests, but in Europe pine forest (of *Pinus sylvestris*) succeeded birch, and spruce (*Picea abies*) was a late immigrant (Figure 14.3). During the mid-Holocene (i.e., the Altithermal period) *Picea obovata* extended its range eastwards in central Siberia far beyond its present limits, and broadleaved forest (of *Quercus, Ulmus* and *Tilia*) expanded in the European USSR into areas now occupied by spruce forest [36]. Subsequently in the Neoglacial, beginning about 4.5 ka ago (Table 14.3), there was a marked recession of all arctic and alpine treelines (section 8.3.1), and a shift in the composition of the boreal forests towards their modern condition. Spruce (*Picea abies*) spread gradually in Europe, at the expense of birch, pine and (southwards) the broadleaved trees, and in central and western Siberia the more drought-tolerant *Larix dahurica* and *Pinus sibirica* replaced *Picea obovata* over large areas [36] (Figure 14.3). Although climatic deterioration was undoubtedly the major cause of these changes, the role of fire cannot be ignored in preventing the regeneration of forest at its latitudinal and altitudinal limits [41,42]. The climatic deterioration also accelerated the growth of peatlands in the boreal zone, and led to further reductions in the extent of forest. The history of these boreal mires is considered in the next section.

Table 14.3 Approximate dates (ka ago) when treelines began to retreat southwards or to lower altitudes from their maximum Holocene positions

A. *Latitudinal treeline*	
Northwestern Canada [37]	4.0
Central Canada [43,44]	3.4
Labrador [38,45]	3.1
Siberia [36]	2.5
B. *Altitudinal treeline*	
Cairngorms, Scotland [46,47]	4.6
Sierra Nevada, California [48]	3.5
White Mts, California [49]	3.1
Scandes, Sweden [50]	3.0

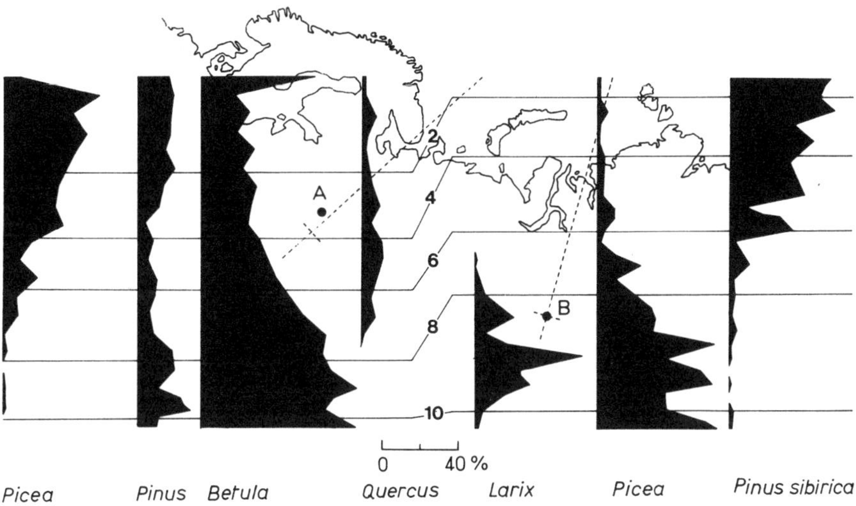

Figure 14.3 The course of Holocene forest history at two sites in the Russian taiga (A = Polovetsko–Kupanskoye mire, B = Nizhnevartovskoye Bog). Only certain pollen components are shown (all as percentage of total tree pollen). The continuous horizontal lines are synchronous time-horizons, the ages of which (ka ago) are shown at the centre. The broken lines are longitude 40°E, latitude 55°N (on left) and longitude 80°E, latitude 60°N (on right). Re-drawn from Khotinskiy [36].

14.3.4 The boreal mires

The extensive accumulations of peat that accompany the spruce and pine forests of the boreal zone are a consequence of the cold humid climate, poor drainage, and large areas of subdued relief in that zone. The continental ice sheets of the Pleistocene played a major role in developing such conditions, by abrading irregularities in the land surfaces that they overlay, by modifying drainage by the deposition of till, and, in their decay, by releasing immense quantities of meltwater that flooded over the lower-lying ground and stagnated in the flattest areas. Drainage of water was further hampered by widespread permafrost in the Early Holocene. At some sites the first organic material to accumulate, either in the lateglacial period or subsequently, was in confined lake basins, and from some of these initial foci peat then spread out over the surrounding terrain [51–53]. However, most boreal peatlands developed by direct **paludification**, with organic material building up immediately above waterlogged mineral soil. The date of this first-formed peat varies enormously (Figure 14.4), with part of the variability resulting from differences in regional climates and part from progressive paludification within a region. The gradual spread of peat over the landscape has been clearly

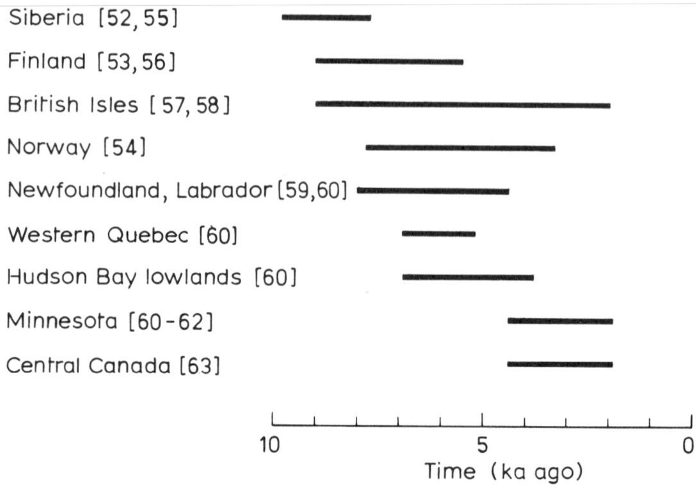

Figure 14.4 Time-periods of peat initiation in different regions of the boreal zone.

demonstrated in several different countries by radiocarbon dates and pollen analyses at nearby sites in differing topographic situations. Thus in central Norway peat spread both upslope and downslope from initial foci on drumlin plateaux and in the intervening valleys, over a 4.5-ka time-period from 7.8 to 3.3 ka ago [54]; at a site in south-eastern Labrador peat spread upslope a distance of 120 m between 4.5 and 0.8 ka ago [51]; and in northern England blanket peat spread progressively to steeper slopes and exposed hilltops over a 4-ka period [64]. The overall differences in date of the earliest peat accumulations in northern Europe (7.5–9.0 ka ago) and North America (less than 7.0 ka ago) are a consequence of the drier conditions over much of North America in the first half of the Holocene; in the continental interior there paludification did not begin before 4.5 ka ago [60,61,65]. The oldest peats in the boreal zone are in western Siberia, where large rivers swollen with meltwaters from the glaciers of the Altai range meandered northwards in the Early Holocene over vast perma-frosted plains, and flooded out each spring over the lower-lying areas [66]; peat has been accumulating in these landscapes throughout the duration of the Holocene [52].

Much of this early peat was **minerotrophic** – i.e., formed by sedge–moss communities under the influence of groundwater. By contrast, extensive areas of the boreal peatlands at the present day are accumulating **ombrotrophic** peat that relies on atmospheric precipitation to remain water-saturated. In the blanket mires of the British Isles and western Norway this ombrotrophic peat extends almost to the base of the deposits, but in most other regions the ombrotrophic peat is underlain by minerotrophic peat. The time of changeover is variable, and was clearly affected by the progressive nature of the paludification, but two main episodes of ombrotrophic peat formation can be recognized: beginning 7–8 ka ago, and beginning 1–3 ka ago. The first episode, which coincided with the onset of the wetter climate of the Atlantic period (Table 6.1) over much of northern Europe, is well represented in Britain [67], western Norway [54], southern Finland [68], western Siberia [52], and perhaps also in maritime northeastern North America [60]. The second episode, which is associated with Neoglacial deterioration of climate, has been recorded from boreal mires in Minnesota [60], central Canada [63], southeastern Labrador [51] and western Estonia [69], and from some blanket mires in the British Isles [57]. In the British Isles, however, there is increasing evidence that the inception of blanket peat accumulation was also aided by deliberate forest clearance by prehistoric man [57,70–72].

The peatlands of the boreal zone, together with the blanket mires of the British Isles, account for 85% of all the world's peatlands [73]. Nearly all of them show pronounced surface patterning, expressed in the distribution over the surface of pools of open water, hummocks and ridges of drier peat, lines of preferred water flow ('water tracks'), and (where present) trees. The patterning, most clearly visible from the air, is closely related to the contours of the mire surface, and varies in detail from one geographical region to another. Various morphological types of boreal and blanket mire have been recognized, and their distribution related to climatic factors [59,60,74–77]. Where peatlands cover extensive areas in a landscape, more than one morphological type may be present, and the landscape is then a complex mosaic of ombrotrophic and minerotrophic mires, each exhibiting characteristic surface patterning. Such **mire complexes** are landforms as well as vegetation types [1], and their history is intimately linked to the landscape evolution of the boreal zone. Consequently there has been considerable interest in how mire complexes develop (shown, for example, by a series of studies on the landscape evolution of northern Minnesota [61,62,78–82]), and in how and why surface patterning originates [51,53,59,60,83,84]. Discussion of such work is outside the scope of this chapter. However, nearly all the studies show that surface patterning is a neoglacial phenomenon, and has developed at some sites only within the last 1000 years [51,84].

The development of surface patterning led to a diversification and compartmentalization of the boreal mire flora – which thus might be said to be largely a neoglacial creation, though its participant species are of considerable antiquity [76]. Plants were probably drawn in from a variety of sources, so that the present flora consists of species capable of growing on both organic and inorganic soils, and under wet and dry conditions, and includes species that are widespread in the boreal zone and others that are widespread in the arctic zone. Many of the species are circumboreal (Table

14.4), though, as with the boreal forest, the peatland trees are more strongly segregated between the different land masses: *Pinus sylvestris*, *Picea abies* and *Betula pubescens* in northern Europe, *Larix laricina*, *Picea mariana* and *Betula glandulosa* in North America, for example.

14.3.5 Tundra communities in the Holocene

A feature of the modern coldland complex in most parts of the northern hemisphere is a broad zone of dwarf-shrub prominence that extends on both sides of of the treeline. This zone is present both at high latitudes (where it comprises forest–tundra and low-arctic communities) and at high altitudes (so-called subalpine forest and low-alpine communities). Typical wide-ranging species or species-pairs include *Betula nana/glandulosa*, *Empetrum nigrum/hermaphroditum*, *Juniperus communis*, *Ledum palustre*, *Salix glauca*, *Vaccinium myrtillus*, *V. uliginosum* and *V. vitis-idaea*. Many of these species appear to have been rare in full-glacial communities, but spread rapidly in the lateglacial and earliest Holocene wherever ice-free areas developed. In the far northwest of Canada, for example, major increases in pollen influx of *Betula* (presumably *B. glandulosa*) and ericads (*Cassiope*, *Empetrum*, *Ledum* and *Vaccinium*) occurred between about 11 and 9 ka ago (Figure 14.5), but in the northeast, where Laurentide ice persisted on the central Labrador plateau until 6.5–7.0 ka ago [85], the expansion of dwarf shrubs was considerably delayed. In pollen diagrams from northwestern Europe, peaks of *Betula nana*, *Empetrum* and *Juniperus* record a similar dwarf-shrub expansion in the earliest Holocene [18]. The spread of dwarf shrubs created a zone of low-arctic or low-alpine tundra that was subsequently invaded by trees – usually birch initially, and then conifers (section 14.3.3). What is not clear, however, is whether this dwarf-shrub

Table 14.4 Some circumboreal species in the boreal mire flora

A. Ombrotrophic mires	
Dry hummocks	*Andromeda polifolia, Betula nana, Empetrum nigrum, Ledum palustre, Rubus chamaemorus, Sphagnum fuscum.*
Wet hollows	*Carex limosa, Rhynchospora alba, Scheuchzeria palustris, Sphagnum papillosum.*
Bare wet peat	*Eriophorum vaginatum, Trichophorum cespitosum.*
B. Minerotrophic mires	
Poor-fen 'lawns'	*Carex rostrata, Eriophorum angustifolium, Menyanthes trifoliata, Potentilla palustris.*
Pools ('flarks')	*Campylium stellatum, Calliergon sarmentosum, Scorpidium scorpioides, Sphagnum lindbergii.*

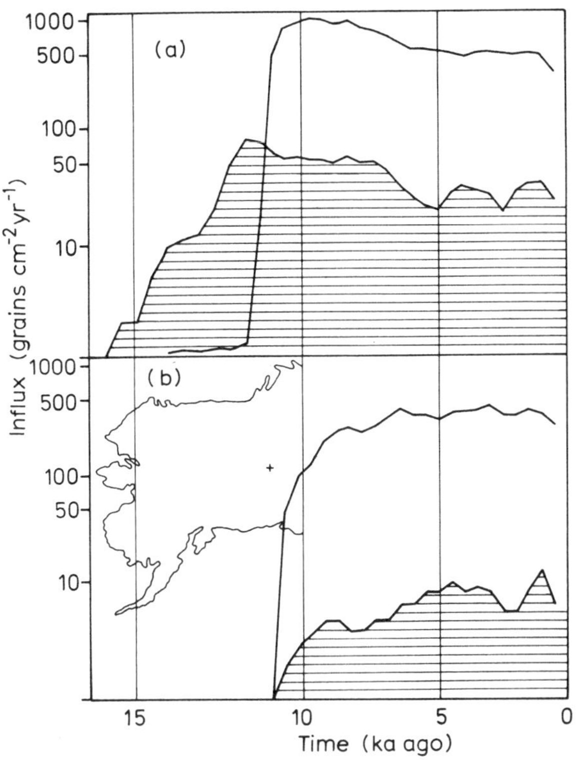

Figure 14.5 Pollen influx to sediments accumulated over the last 16 ka at Lateral Pond, Yukon, showing the initial development of dwarf-shrub tundra (a) and the subsequent invasion by conifer forest (b). Influx values in (a) are for dwarf willow, *Salix* (shaded) and dwarf birch, *Betula*; influx values in (b) are for Ericales (shaded) and spruce, *Picea*. Influx values for *Picea* comparable to those in modern forest–tundra were achieved about 9 ka ago. Re-drawn from Ritchie [37].

zone was a necessary precursor to forest at higher latitudes and altitudes. Once the zone was established, certainly, it became the principal battleground of the treeline through the Holocene.

The low-arctic zone throughout the northern hemisphere at the present day is also characterized by extensive areas of so-called 'tussock tundra' dominated by *Eriophorum vaginatum*, and with associated Cyperaceae (e.g., *Carex lugens*) and Ericaceae (e.g., *Ledum palustre, Vaccinium uliginosum*). Tussock tundra occupies poorly-drained ground in the lowlands, and is often underlain by peat deposits several metres thick. Little peat is forming today, however, and these arctic peatlands (at least in Canada) appear to be largely fossil structures developed under warmer and wetter conditions [65]. The earliest expression of those conditions, leading to the initiation of tussock tundra communities, was probably in East Beringia between 14 and 12 ka ago [41]. In the far northwest of Canada, appropriate conditions pertained between about 10.5 and 5.6 ka ago [86]; in central and northeastern Canada, slow-melting ice and a dry altithermal period delayed peat build-up until after 5 ka ago [44,65], and accumulation ceased again 600–900 yr ago [43,44].

Dwarf shrubs are far less prominent in the modern tundra communities of central and northeastern Siberia, where the extreme cold and dryness of the full glacial could have been inimical to the survival of some ericads. The areas occupied by these modern tundra communities were almost completely ice-free during the last cold stage of the Pleistocene, and accordingly we might suppose that an indigenous component (of cold-stage taxa) would be present in their floras (as opposed to an entirely adventive modern flora in the glaciated regions of North America and northern Europe). Soviet phytogeographers have equated this indigenous component with plants of localized stands of **tundra–steppe** vegetation that occur today in north-central and northeastern Siberia, on well-drained south-facing slopes on the higher terraces and mountain slopes where winter snow does not accumulate. Characteristic plants include *Calamagrostis purpurascens, Carex duriuscula, Festuca lenensis* and *Helictotrichon krylovii* [87,88]. Rather similar communities occur in northwestern North America, on south-facing knolls and escarpments with coarse-textured soils [37]. In north-central Siberia ('Yakutia') these tundra–steppe stands are almost entirely contained within the modern taiga zone (of *Larix dahurica, Pinus pumila* and *Betula middendorffii*), but in northeastern Siberia ('Chukotka') they occur beyond the treeline (Figure 14.6). Pollen analyses show clearly that the Yakutian stands were early engulfed by the northward-advancing forest, but could have persisted locally throughout the Holocene; the Chukotkan landscape has never been forested in the Holocene, but thickets of *Populus suaveolens, Pinus pumila, Betula middendorffii* and *Alnaster fruticosa* developed widely in the lowlands [36]. The modern Yakutian tundra–steppe stands may thus be true 'relicts', and the Chukotkan stands may contain true indigenous elements. However, the main components of the modern Chukotkan flora are plants that were probably uncommon in the cold-stage vegetation of 18–20 ka ago, as they require either wet and poorly-drained soils (*Carex lugens, Eriophorum vaginatum, Ledum palustre*) or lengthened snow-lie in winter (*Betula nana, Empetrum nigrum, Juniperus sibiricus, Salix pulchra, Vaccinium vitis-idaea*). The tundra–steppe stands, in fact, are but one element in a modern landscape mosaic of woods and thickets, semi-prostrate scrub (*Betula–Salix*), tussock tundra (*Eriophorum–Carex*), moss–lichen tundra and montane heath (*Dryas octopetala, Loiseleuria procumbens, Rhododendron aureum*) [87,88], that might have started to develop around 14 ka ago [41].

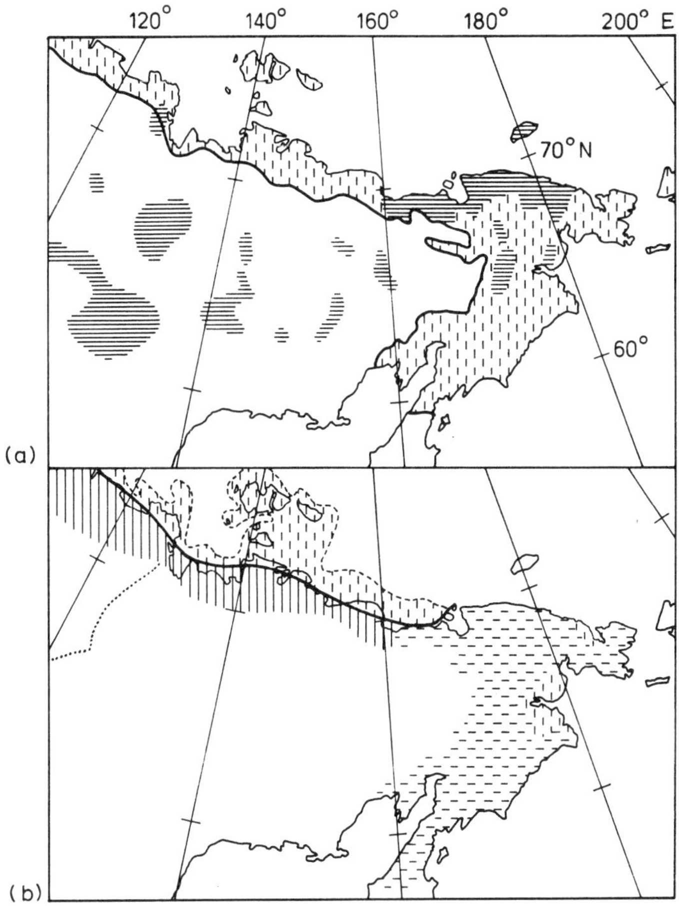

Figure 14.6 Tundra–steppe in Siberia. (a) Modern tundra–steppe stands ≡ in relation to the position of the larch treeline ⁀ and the total extent of tundra communities ¦¦¦. (b) Reconstructed vegetation map for about 8 ka ago, when the land surface extended north beyond the present coastline (to the position shown by the interrupted line ---) and much of eastern Siberia was covered by scrublands -‐-. The northern treeline ⁀ and forest–tundra |||||| are shown. The dotted line ...... within the coniferous forest zone in (b) separates the western spruce forests from the eastern larch forests. Re-drawn from Khotinskiy [36].

A similar landscape mosaic in the Central Brooks Range of Alaska has recently been shown to contain higher proportions of 'Beringian' taxa in certain of the vegetation types: dry steppe on south-facing bluffs, tundra–steppe on windswept ridges, and also snowbed areas [89]. The Holocene history of this Alaskan landscape mosaic is as yet poorly understood, but it is likely that the sources of the various elements in the mosaic should again be sought in lateglacial diversification of the vegetation. Indeed, Cooper [89] suggests that the full spectrum of lateglacial vegetation types may still survive in the Brooks Range.

In Europe the lateglacial flora became segregated early in the Holocene between high-latitude and high-altitude habitats that had been covered by ice at full glacial, and many plants widespread in the European lowlands in the lateglacial are still to be found in modern arctic and alpine floras. The contribution from plants surviving in refugia close to the ice sheets is still disputed. In North America, on the other hand, three quite distinct species-reservoirs can be seen to have contributed to the re-vegetation of areas covered by the Laurentide Ice Sheet: the full-glacial parklands to the south, the varied herbaceous communities of Beringia to the northwest, and refugial stands in the arctic islands to the north [20]. Plant populations in these reservoirs were originally isolated from one another by glacial ice, and only gradually merged as they spread following the melting of the ice. Although many details still remain to be worked out, it is clear that modern arctic floras in all three continents (and indeed the whole coldland complex) have a more involved derivation than mere displacement northwards of the full-glacial vegetation types.

14.4 MODERN DISTRIBUTION PATTERNS

Many coldland plant species (trees excepted) range widely at the present day through the boreal and arctic regions of the northern hemisphere. The circumboreal character of the modern boreal mire flora has already been noted (section 14.3.4). In the arctic flora there are nearly 300 **circumpolar** species, and another 107 taxa that are represented by closely-related species-pairs [90]. During the Pleistocene, North American and Eurasian populations of these species were linked across the Bering Straits at times of low sea level, and distributions such as those of *Rhododendron lapponicum* and *Luzula piperi* (Figure 14.7) might be interpreted as preserving a memory of this ancient land connection. More difficult to explain are the close floristic links between Arctic land masses that are now separated by major sea barriers that developed before non-forested habitats became widespread (before 15 ma ago [91]). Over 200 species show **amphi-atlantic** ranges similar to that of *Salix herbacea* (Figure 14.7), occurring in North America, Greenland, Iceland and northern Europe [90]. About 120 of the 160 species currently found in Svalbard (Spitsbergen and Bear Island) occur also on the Scandinavian mainland [92] (Figure 2.2). The conclusion would seem to be inescapable that long-distance dispersal has played, and still continues to play, a major role in the evolution of coldland floras. Such a conclusion is borne out by the varied geographical affinities

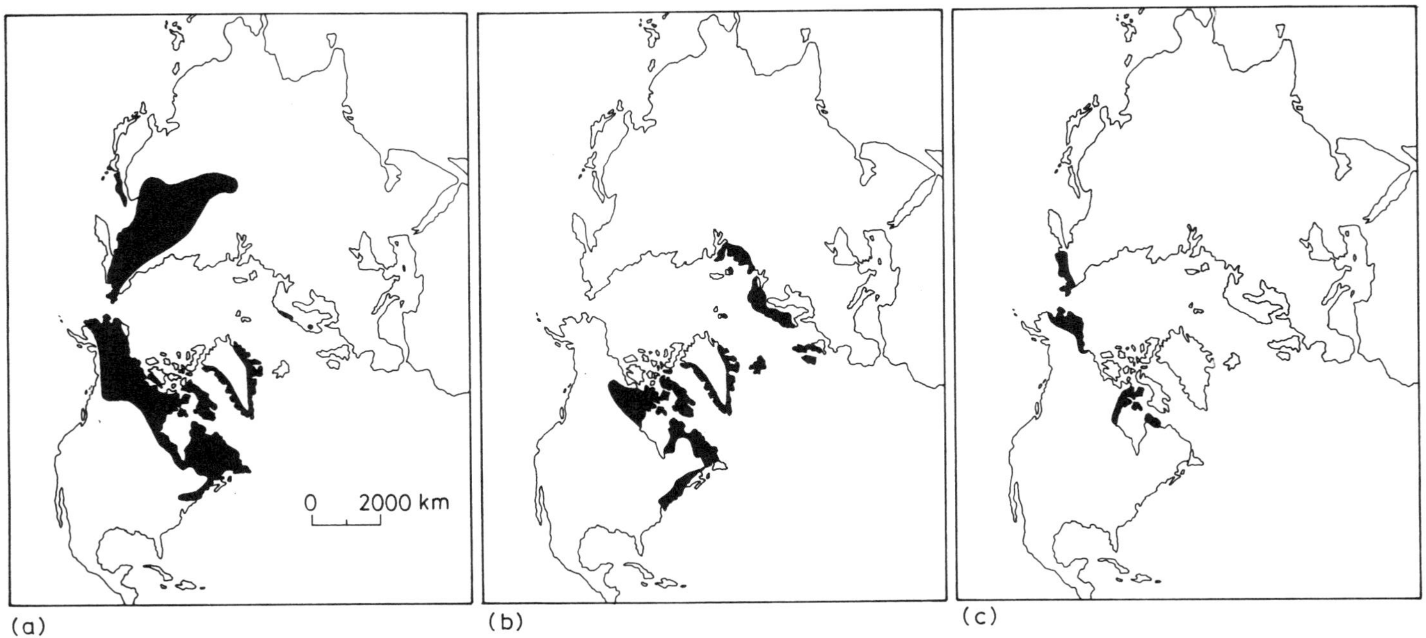

Figure 14.7 Modern ranges of (a) *Rhododendron lapponicum*, (b) *Salix herbacea*, (c) *Luzula piperi*. Re-drawn from maps in Gjaerevoll [23], Seddon [93] and Ives [22].

(and hence the presumed varied sources) of the afroalpine floras of the East African mountains: 10% of the species have their closest modern relatives in South Africa, 7% in the Mediterranean region, 2% in the Himalayas, 3% in the southern-hemisphere temperate zone and 15% in the northern-hemisphere temperate zone [94]. A wider-ranging analysis of tropicalpine floras [95] suggests that up to 83% of the modern genera may have originated as immigrants from outside the tropical zone.

The existence of large numbers of circumpolar species with ranges now subdivided by major water barriers indicates that speciation has proceeded rather slowly in some high-latitude floras. Why this should be so is not clear, especially as there is evidence (see below) of much faster speciation in some alpine floras. One possibility is that the high incidence of polyploidy in high-latitude taxa [16] helps to retard or to mask morphological differentiation. Another is that there is much genetical differentiation that has not yet been recognized. Thus of the species listed in Table 14.4, at least three show variation between North American and Eurasian populations that some taxonomists believe to be worthy of subspecific or even specific recognition: *Empetrum nigrum* (sspp. *nigrum* and *hermaphroditum*), *Eriophorum vaginatum* (sspp. *vaginatum*, *spissum* and *fauriei*) and *Ledum palustre* (sspp. *palustre* and *groenlandicum*).

Many coldland taxa show present-day ranges that are clearly disjunct *within* the major land masses (Figure 14.7). Such distribution patterns might be seen as the products of fragmentation of formerly continuous ranges, with elimination of populations over large areas during the Holocene by rising temperatures [96], accumulation of soil organic matter, and competition from more vigorously-growing plants [18,97]. In a few instances, pollen analyses at sites close to modern relict populations of certain of these taxa have shown conclusively that the populations have been continuously present at the sites since the lateglacial period – when the taxa were more widely distributed [98–100].

Where modern relict populations occur in areas in the northern hemisphere that were actually within the ambit of major glacial-stage ice sheets, then an alternative explanation is possible: that these populations survived through the last glacial stage at or near the sites they now occupy, in locally unglaciated areas. In Norway, disjunct populations of 62 species occur in one or both of the two 'refugial' areas shown in Figure 14.2 [22,23]; in Iceland, about 100 species occur in one or more of six refugial areas [24]. Similar localized areas with concentrations of disjunct populations are found in the British Isles, but most ecologists now reject the idea that these areas are 'ice-age refugia', and point to the clear fossil evidence that at least nine arctic–alpine species with localized modern distributions in Britain were formerly more widespread: *Betula nana, Draba incana, Dryas octopetala, Koenigia islandica, Oxyria digyna, Salix herbacea, Saussurea alpina, Saxifraga oppositifolia* and *Thalictrum alpinum* [18]. Nevertheless, ice-age survival has not been disproved elsewhere, and could have formed an alternative strategy to migration by which certain coldland populations persisted through at least the last Pleistocene cold stage [101]. Until conclusive fossil evidence is found which shows for taxa such as *Rhododendron lapponicum* (Figure 14.7) that their cold-stage ranges were, or were not, formerly more extensive, the question of perglacial survival in ice-age refugia must remain an open one. In Greenland, the delayed appearance in the Holocene pollen record of *Betula nana*, *Cassiope tetragona* and *Salix arctica* has been cited as evidence against perglacial survival [25], but persistence of these species in localized refugia some distance away from the sampling sites cannot be ruled out.

In some coldland taxa, genetic or chromosomal differentiation between separated small populations is clearly occurring. The best-documented case is that of the alpine poppies in Norway. Four closely-related species, which are high polyploid derivatives of the wider-ranging diploid *Papaver alpinum* ($2n = 14$), occur in Norway (Figure 14.8). All are polymorphic, and in *P. radicatum* at least ten spatially-segregated subspecies have been recognized which differ chromosomally from one another; six occur in the southern

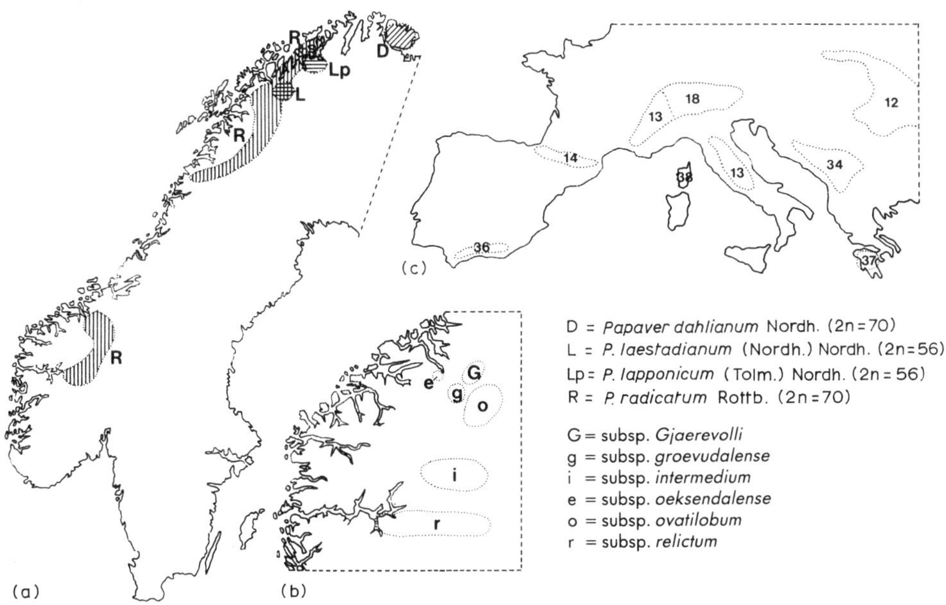

Figure 14.8 (a) Modern ranges of four alpine poppies in Norway (re-drawn from Nannfeldt [102]); (b) southern Norwegian refugial area enlarged, to show modern ranges of six subspecies of *Papaver radicatum* (re-drawn from Nannfeldt [102]); (c) percentages of endemic species in the modern alpine floras of central and southern Europe (re-drawn from Favarger [103]).

refugial area of the Dovre–Jotunheimen Mountains (Figure 14.8). The poppies are very rare plants, and even at subspecific level individuals occur as small disjunct populations in the mountains; these too are chromosomally distinct [102]. 'Speciation' thus appears to be taking place at three levels within the *Papaver alpinum* complex in Norway – though the time-scale over which such differentiation has occurred is unclear.

If we consider European alpine floras as a whole, then speciation to a more extreme degree is apparent. All the major mountain masses in central and southern Europe have substantial numbers of endemic alpine species [103], and the proportion increases southwards (Figure 14.8). Even higher endemism is found in the mountains of East Africa and Japan, where >80% and >50%, respectively, of the alpine species are found nowhere else [94,104]. Few endemic genera, however, are found in alpine floras: even in East Africa only four (or possibly five) are known [94]. This pattern can be explained most easily by assuming isolation of alpine populations for varying lengths of time. Those at the highest altitudes on the East African mountains have probably remained genetically segregated from one another ever since the mountains were first formed in the Miocene and Pliocene Epochs [94,105]. Although the European alpine floras developed at much the same time, and high-altitude populations were initially isolated from one another by the wide-ranging Tertiary forests, gene flow between them, at least in part, must have been regularly established during the cold stages of the Pleistocene. The high species-endemism in the alpine floras of southern Europe (southern Spain, Corsica and the Balkans), as compared with that of central Europe (Figure 14.8), suggests that speciation could have been active in the earlier Pleistocene, when coldland communities probably did not spread as widely in Europe as in the later Pleistocene (Figure 10.5). Given this time-scale, the patterns of differentiation in the Scandinavian alpine flora must then be seen as of Late Pleistocene and Holocene age. A possible scenario for the alpine poppies, for example, might be of chromosomally-distinct populations of the four polyploid species surviving the last glaciation in coastal refugia (but see Paus [106]), and then colonizing high-altitude habitats in the adjacent mountains as they became ice-free in the Holocene, as a series of small isolated subpopulations which are now diverging further from one another. Better quantification of this scenario might help towards a more realistic assessment of the patterns of change proposed for tropical floras and faunas in the Pleistocene (sections 8.5 and 10.5). Similar proportions of endemic species in the alpine floras of southern Europe and the Andean paramo [107] certainly support the idea of Pleistocene speciation within fragmented populations in tropical regions.

REFERENCES

1. Damman, A. W. H. and French, T. W. (1987) *The Ecology of Peat Bogs of the Glaciated Northeastern United States: a Community Profile*, US Dept of the Interior, Washington.
2. Murray, D. F. (1978) Vegetation, floristics and phytogeography of northern Alaska, in *Vegetation and Production Ecology of an Alaskan Arctic Tundra* (ed L. L. Tieszen), Springer-Verlag, New York, pp. 19–36.
3. Barry, R. G. and Ives, J. D. (1974) Introduction, in *Arctic and Alpine Environments* (eds J. D. Ives and R. G. Barry), Methuen, London, pp. 1–13.
4. Sjörs, H. (1963) Amphi-Atlantic zonation, nemoral to arctic, in *North Atlantic Biota and their History* (eds Á. Löve and D. Löve), Pergamon, Oxford, pp. 109–25.
5. Gore, A. J. P. (ed.) (1983) *Ecosystems of the World 4A. Mires: Swamp, Bog, Fen and Moor*, Elsevier, Amsterdam.
6. Flohn, H. (1984) Climate evolution in the Southern Hemisphere and the equatorial region during the Late Cenozoic, in *Late Cainozoic Palaeoclimates of the Southern Hemisphere* (ed. J. C. Vogel), Balkema, Rotterdam, pp. 5–20.
7. Hughes, T. J. (1985) The great Cenozoic ice sheet. *Palaeogeogr. Palaeoclimatol. Palaeoecol.*, **50**, 9–43.
8. Rea, D. K. and Schrader, H. (1985) Late Pliocene onset of glaciation: ice-rafting and diatom stratigraphy of North Pacific DSDP cores. *Palaeogeogr. Palaeoclimatol. Palaeoecol.*, **49**, 313–25.
9. Ruddiman, W. F. (1987) Synthesis; the ocean/ice sheet record, in *North America and Adjacent Oceans during the Last Deglaciation* (eds W. F. Ruddiman and H. E. Wright Jr), Geological Society of America, Boulder, CO, pp. 463–78.
10. Curry, R. R. (1966) Glaciation about 3,000,000 years ago in the Sierra Nevada. *Science*, **154**, 770–1.
11. Wolfe, J. A. (1969) Neogene floristic and vegetational history of the Pacific Northwest. *Madroño*, **20**, 83–110.
12. Wolfe, J. A. and Leopold, E. B. (1967) Neogene and Early Quaternary vegetation of northwestern North America and northeastern Asia, in *The Bering Land Bridge* (ed. D. M. Hopkins), Stanford University Press, Stanford, CA, pp. 193–206.
13. Petrov, O. M. (1967) Paleogeography of Chukotka during Late Neogene and Quaternary time, in *The Bering Land Bridge* (ed. D. M. Hopkins), Stanford University Press, Stanford, CA, pp. 144–71.
14. Tanai, T. (1972) Tertiary history of vegetation in Japan, in *Floristics and Palaeofloristics of Asia and Eastern North America* (ed. A. Graham), Elsevier, Amsterdam, pp. 235–55.
15. Nelson, R. E. and Carter, L. D. (1985) Pollen analysis of a Late Pliocene and early Pleistocene section from the Gubik formation of arctic Alaska. *Quat. Res.*, **24**, 295–306.
16. Löve, Á. and Löve, D. (1974) Origin and evolution of the arctic and alpine floras, in *Arctic and Alpine Environments* (eds J. D. Ives and R. G. Barry), Methuen, London, pp. 571–603.

17. Hopkins, D. M. (ed.) (1967) *The Bering Land Bridge*, Stanford University Press, Stanford, CA.
18. Godwin, H. (1975) *History of the British Flora*, 2nd edn, Cambridge University Press, Cambridge.
19. Billings, W. D. (1974) Arctic and alpine vegetation: plant adaptations to cold summer climates, in *Arctic and Alpine Environments* (eds J. D. Ives and R. G. Barry), Methuen, London, pp. 403–43.
20. Löve, D. (1959) The postglacial development of the flora of Manitoba: a discussion. *Can. J. Bot.*, **37**, 547–85.
21. Löve, D. (1963) Dispersal and survival of plants, in *North Atlantic Biota and their History* (eds Á. Löve and D. Löve), Pergamon, Oxford, pp. 189–205.
22. Ives, J. D. (1974) Biological refugia and the nunatak hypothesis, in *Arctic and Alpine Environments* (eds J. D. Ives and R. G. Barry), Methuen, London, pp. 605–36.
23. Gjaerovoll, O. (1963) Survival of plants on nunataks in Norway during the Pleistocene glaciation, in *North Atlantic Biota and their History* (eds Á. Löve and D. Löve), Pergamon, Oxford, pp. 261–83.
24. Steindórsson, S. (1963) Ice Age refugia in Iceland as indicated by the present distribution of plant species, in *North Atlantic Biota and their History* (eds Á. Löve and D. Löve), Pergamon, Oxford, pp. 303–20.
25. Funder, S. (1979) Ice-age plant refugia in East Greenland. *Palaeogeogr. Palaeoclimatol. Palaeoecol.*, **28**, 279–95.
26. Huntley, B. and Birks, H. J. B. (1983) *An Atlas of Past and Present Pollen Maps for Europe: 0–13,000 Years Ago*, Cambridge University Press, Cambridge.
27. Grichuk, V. P. (1984) Late Pleistocene vegetation history, in *Late Quaternary Environments of the Soviet Union* (ed. A. A. Velichko), Longman, London, pp. 155–78.
28. Huntley, B. (1988) Europe, in *Vegetation History* (eds B. Huntley and T. Webb III), Kluwer Academic Publishers, Dordrecht, pp. 341–83.
29. Iversen, J. (1973) The Development of Denmark's Nature since the Last Glacial. *Danm. Geol. Unders., Series V*, **7-C**, 1–126.
30. Birks, H. J. B. (1976) Late-Wisconsinan vegetational history at Wolf Creek, central Minnesota. *Ecol. Monogr.*, **46**, 395–429.
31. Pennington, W. (1977) The Late Devensian flora and vegetation of Britain. *Phil. Trans. Roy. Soc. Lond. B*, **280**, 247–71.
32. Birks, H. J. B. (1986) Late-Quaternary biotic changes in terrestrial and lacustrine environments, with particular reference to northwest Europe, in *Handbook of Holocene Palaeoecology and Palaeohydrology* (ed. B. E. Berglund), Wiley, Chichester, pp. 3–65.
33. Godwin, H. and Willis, E. H. (1959) Radiocarbon dating of the Late-glacial Period in Britain. *Proc. Roy. Soc., Series B*, **150**, 199–215.
34. Watts, W. A. (1980) Regional variation in the response of vegetation to Lateglacial climatic events in Europe, in *Studies in the Lateglacial of North-West Europe* (eds J. J. Lowe, J. M. Gray and J. E. Robinson), Pergamon, Oxford, pp. 1–21.
35. Ager, T. A. (1982) Vegetational history of western Alaska during the Wisconsin glacial interval and the Holocene, in *Paleoecology of Beringia* (eds D. M.

Hopkins, J. V. Matthews Jr, C. E. Schweger and S. B. Young), Academic Press, New York, pp. 75–93.

36. Khotinskiy, N. A. (1984) Holocene vegetation history, in *Late Quaternary Environments of the Soviet Union* (ed. A. A. Velichko), Longman, London, pp. 179–200.
37. Ritchie, J. C. (1984) *Past and Present Vegetation of the Far Northwest of Canada*, University of Toronto Press, Toronto.
38. Short, S. K. and Nichols, H. (1977) Holocene pollen diagrams from subarctic Labrador–Ungava: vegetational history and climatic change. *Arctic Alpine Res.*, **9**, 265–90.
39. Reynaud, C. (1974) The pioneer vegetational phase in Central Lapland (Finland) based on new palynological data. *Ann. Bot. Fenn*, **11**, 197–201.
40. Tipping, R. M. (1987) The prospects for establishing synchroneity in the early postglacial pollen peak of *Juniperus* in the British Isles. *Boreas*, **16**, 155–63.
41. Lamb, H. F. and Edwards, M. E. (1988) The Arctic, in *Vegetation History* (eds B. Huntley and T. Webb III), Kluwer Academic Publishers, Dordrecht, pp. 519–55.
42. Payette, S. and Gagnon, R. (1985) Late Holocene deforestation and tree regeneration in the forest–tundra of Quebec. *Nature*, **313**, 570–2.
43. Nichols, H. (1967) The postglacial history of vegetation and climate at Ennadai Lake, Keewatin, and Lynn Lake, Manitoba. *Eiszeitalter Gegenw.*, **18**, 176–97.
44. Nichols, H. (1970) Late Quaternary pollen diagrams from the Canadian Arctic barren grounds at Pelly Lake, Northern Keewatin, N.W.T. *Arctic Alpine Res.*, **2**, 43–61.
45. Lamb, H. F. (1985) Palynological evidence for postglacial change in the position of tree limit in Labrador. *Ecol. Monogr.*, **55**, 241–58.
46. Dubois, A. D. and Ferguson, D. K. (1985) The climatic history of pine in the Cairngorms based on radiocarbon dates and stable isotope analysis, with an account of the events leading up to its colonization. *Rev. Palaeobot. Palynol.*, **46**, 55–80.
47. Pears, N. V. (1967) Present tree-lines of the Cairngorm Mountains, Scotland. *J. Ecol.*, **55**, 815–30.
48. Scuderi, L. A. (1987) Late-Holocene upper timberline variation in the southern Sierra Nevada. *Nature*, **325**, 242–4.
49. Lamarche, V. C. and Mooney, H. A. (1967) Altithermal timberline advance in western United States. *Nature*, **213**, 980–2.
50. Kullman, L. (1988) Holocene history of the forest–alpine tundra ecotone in the Scandes Mountains (central Sweden). *New Phytol.*, **108**, 101–10.
51. Foster, D. R. and King, G. A. (1984) Landscape features, vegetation and developmental history of a patterned fen in south-eastern Labrador, Canada. *J. Ecol.*, **72**, 115–43.
52. Neustadt, M. I. (1984) Holocene peatland development, in *Late Quaternary Environments of the Soviet Union* (ed. A. A. Velichko), Longman, London, pp. 201–5.
53. Seppälä, M. and Koutaniemi, L. (1985) Formation of a string and pool

topography as expressed by morphology, stratigraphy and current processes on a mire in Kuusamo, Finland. *Boreas*, **14**, 287–309.

54. Solem, T. (1986) Age, origin and development of blanket mires in Sør-Trøndelag, central Norway. *Boreas*, **15**, 101–15.
55. Frenzel, B. (1983) Mires – repositories of climatic information or self-perpetuating ecosystems? in *Ecosystems of the World 4A. Mires: Swamp, Bog, Fen and Moor* (ed. A. J. P. Gore), Elsevier, Amsterdam, pp. 35–65.
56. Kivinen, E. and Tolonen, K. (1972) *4th International Peat Congress, Otaniemi, Finland 1972. Excursion Guide. F. Joensuu. Virgin Peatlands, Forestry, Cultivation.* Helsinki.
57. Moore, P. D., Merryfield, D. L. and Price, M. D. R. (1984) The vegetation and development of blanket mires, in *European Mires* (ed. P. D. Moore), Academic Press, London, pp. 201–35.
58. Taylor, J. A. (1983) The peatlands of Great Britain and Ireland, in *Ecosystems of the World 4B. Mires: Swamp, Bog, Fen and Moor* (ed. A. J. P. Gore), Elsevier, Amsterdam, pp. 1–46.
59. Foster, D. R. and Glaser, P. H. (1986) The raised bogs of southeastern Labrador, Canada: classification, distribution, vegetation and recent dynamics. *J. Ecol.*, **74**, 47–71.
60. Glaser, P. H. and Janssens, J. A. (1986) Raised bogs in eastern North America: transitions in landforms and gross stratigraphy. *Can. J. Bot.*, **64**, 395–415.
61. Glaser, P. H. (1987) *The Ecology of Patterned Boreal Peatlands of Northern Minnesota: a Community Profile*. US Dept of the Interior, Washington.
62. Glaser, P. H., Wheeler, G. A., Gorham, E. and Wright, H. E. Jr (1981) The patterned mires of the Red Lake Peatland, northern Minnesota: vegetation, water chemistry and landforms. *J. Ecol.*, **69**, 575–99.
63. Nichols, H. (1968) Chronology of peat growth in Canada. *Palaeogeogr. Palaeoclimatol. Palaeoecol.*, **6**, 61–5.
64. Tallis, J. H. (1985) Erosion of blanket peat in the southern Pennines: new light on an old problem, in *Geomorphology of North-West England* (ed. R. H. Johnson), Manchester University Press, Manchester, pp. 313–36.
65. Zoltai, S. C. and Pollett, F. C. (1983) Wetlands in Canada: their classification, distribution and use, in *Ecosystems of the World 4B. Mires: Swamp, Bog, Fen and Moor* (ed. A. J. P. Gore), Elsevier, Amsterdam, pp. 245–68.
66. Walter, H. (1977) The oligotrophic peatlands of Western Siberia – the largest peino-helobiome in the world. *Vegetatio*, **34**, 167–78.
67. Tallis, J. H. (1964) The pre-peat vegetation of the southern Pennines. *New Phytol.*, **63**, 363–73.
68. Tolonen, K. (1966) Stratigraphic and rhizopod analyses of an old raised bog, Varrassuo, in Hollola, South Finland, *Ann. Bot. Fenn.*, **3**, 147–66.
69. Botch, M. S. and Masing, V. V. (1983) Mire ecosystems in the USSR, in *Ecosystems of the World 4B. Mires: Swamp, Bog, Fen and Moor* (ed. A. J. P. Gore), Elsevier, Amsterdam, pp. 95–152.
70. Innes, J. B. and Simmons, I. G. (1988) Disturbance and diversity: floristic changes associated with pre-elm decline woodland recession in northeast Yorkshire, in *Archaeology and the Flora of the British Isles* (ed. M. Jones), Oxbow Books, Oxford, pp. 7–20.

71. Jacobi, R. M., Tallis, J . H. and Mellars, P. A. (1976) The southern Pennine Mesolithic and the ecological record. *J. Archaeol. Sci.*, **3**, 307–20.
72. Moore, P. D. (1975) Origin of blanket mires. *Nature*, **256**, 267–9.
73. Kivinen, E. and Pakarinen, P. (1981) Geographical distribution of peat resources and major peatland complex types in the world. *Ann. Acad. Scient. Fenn., Series A*, **132**, 1–28.
74. Moore, P. D. and Bellamy, D. J. (1974) *Peatlands*, Elek, London.
75. Ruuhijärvi, R. (1983) The Finnish mire types and their regional distribution, in *Ecosystems of the World 4B. Mires: Swamp, Bog, Fen and Moor* (ed. A. J. P. Gore), Elsevier, Amsterdam, pp. 47–67.
76. Sjörs, H. (1961) Surface patterns in boreal peatland. *Endeavour*, **20**, 217–24.
77. Sjörs, H. (1983) Mires of Sweden, in *Ecosystems of the World 4B. Mires: Swamp, Bog, Fen and Moor* (ed. A. J. P. Gore), Elsevier, Amsterdam, pp. 69–94.
78. Foster, D. R., King, G. A., Glaser, P. H. and Wright, H. E. Jr (1983) Origin of string patterns in boreal peatlands. *Nature*, **306**, 256–8.
79. Heinselman, M. L. (1963) Forest sites, bog processes, and peatland types in the glacial Lake Agassiz region, Minnesota. *Ecol. Monogr.*, **33**, 327–74.
80. Heinselman, M. L. (1970) Landscape evolution, peatland types, and the environment in the Lake Agassiz Peatlands Natural Area, Minnesota. *Ecol. Monogr.*, **40**, 235–61.
81. Siegel, D. I. (1983) Ground water and the evolution of patterned mires, Glacial Lake Agassiz peatlands, northern Minnesota. *J. Ecol.*, **71**, 913–21.
82. Siegel, D. I. and Glaser, P. H. (1987) Groundwater flow in a bog–fen complex, Lost River peatland, northern Minnesota. *J. Ecol.*, **75**, 743–54.
83. Boatman, D. J. (1983) The Silver Flowe National Nature Reserve, Galloway, Scotland. *J. Biogeogr.*, **10**, 163–274.
84. Hulme, P. D. (1986) The origin and development of wet hollows and pools on Craigeazle Mire, southwest Scotland. *Int. Peat J.*, **1**, 15–28.
85. Gray, J. T. and Lauriol, B. (1985) Dynamics of the Late Wisconsin ice sheet in the Ungava Peninsula interpreted from geomorphological evidence. *Arctic Alpine Res.*, **17**, 289–310.
86. Zoltai, S. C. and Tarnocai, C. (1975) Perennially frozen peatlands in the western arctic and subarctic of Canada. *Can. J. Earth Sci.*, **12**, 28–43.
87. Yurtsev, B. A. (1972) Phytogeography of northeastern Asia and the problem of trans-Beringian floristic interrelations, in *Floristics and Palaeofloristics of Asia and Eastern North America* (ed. A. Graham), Elsevier, Amsterdam, pp. 19–54.
88. Yurtsev, B. A. (1982) Relics of the xerophytic vegetation of Beringia in northeastern Asia, in *Paleoecology of Beringia* (eds D. M. Hopkins, J. V. Matthews Jr, C. E. Schweger and S. B. Young), Academic Press, New York, pp. 157–77.
89. Cooper, D. J. (1989) Geographical and ecological relationships of the arctic–alpine vascular flora and vegetation, Arrigetch Peaks region, central Brooks Range, Alaska. *J. Biogeogr.*, **16**, 279–95.
90. Hultén, E. (1963) Phytogeographical connections of the North Atlantic, in *North Atlantic Biota and their History* (ed Á. Löve and D. Löve), Pergamon, Oxford, pp. 45–72.

91. Eldholm, O. and Thiede, J. (1980) Cenozoic continental separation between Europe and Greenland. *Palaeogeogr. Palaeoclimatol. Palaeoecol.*, **30**, 243–59.
92. Rönning, O. I. (1963) Phytogeographical problems in Svalbard, in *North Atlantic Biota and their History* (eds Á. Löve and D. Löve), Pergamon, Oxford, pp. 99–107.
93. Seddon, B. (1971) *Introduction to Biogeography*, Duckworth, London.
94. Hedberg, O. (1961) The phytogeographical position of the afroalpine flora. *Recent Adv. Bot.*, **1**, 914–9.
95. Smith, J. M. B. and Cleef, A. M. (1988) Composition and origins of the world's tropicalpine floras. *J. Biogeogr.*, **15**, 631–45.
96. Dahl, E. (1951) On the relation between summer temperatures and the distribution of alpine vascular plants in the lowlands of Fennoscandia. *Oikos*, **3**, 22–52.
97. Pigott, C. D. and Walters, S. M. (1954) On the interpretation of the discontinuous distributions shown by certain British species of open habitats. *J. Ecol.*, **42**, 95–116.
98. Godwin, H. (1955) Vegetational history at Cwm Idwal: a Welsh plant refuge. *Svensk bot. Tidskr.*, **49**, 35–45.
99. Hutchinson, T. C. (1966) The occurrence of living and subfossil remains of *Betula nana* L. in Upper Teesdale. *New Phytol.*, **65**, 351–7.
100. Turner, J., Hewetson, V. P., Hibbert, F. A. *et al.* (1973) The history of the vegetation and flora of Widdybank Fell and the Cow Green reservoir basin, Upper Teesdale. *Phil. Trans. Roy. Soc. Lond. B*, **265**, 327–408.
101. Crawford, R. M. M. (1988) *Studies in Plant Survival. Ecological Case Histories of Plant Adaptation to Adversity*, Blackwell, Oxford.
102. Nannfeldt, J. A. (1963) Taxonomic differentiation as an indicator of the migratory history of the North Atlantic flora with especial regard to the Scandes, in *North Atlantic Biota and their History* (eds Á. Löve and D. Löve), Pergamon, Oxford, pp. 87–97.
103. Favarger, C. (1972) Endemism in the montane floras of Europe, in *Taxonomy, Phytogeography and Evolution* (ed. D. H. Valentine), Academic Press, London, pp. 191–204.
104. Shimizu, T. (1988) An outline of the flora of Japan. *Veröff. Geobot. Inst., Zürich*, **98**, 129–40.
105. Hedberg, O. (1969) Evolution and speciation in a tropical high mountain flora. *Biol. J. Linn. Soc.*, **1**, 135–48.
106. Paus, A. (1988) Late Weichselian vegetation, climate and floral migration at Sandvikvatn, North Rogaland, southwestern Norway. *Boreas*, **17**, 113–39.
107. Van der Hammen, T. (1988) South America, in *Vegetation History* (eds B. Huntley and T. Webb III), Kluwer Academic Publishers, Dordrecht, pp. 307–37.

PART FIVE

Looking to the Future

CHAPTER FIFTEEN

Past, present and future

15.1 INTRODUCTION

> 'Time present and time past are both perhaps present in time future, and time future contained in time past.'
>
> (T. S. Eliot)

Preceding chapters in this book have been concerned almost entirely with the world in the past. The palaeoecological scenarios that we have been discussing, however, have relevance also to the world we live in today. Indeed, it can be argued that a palaeoecological perspective is essential to a proper understanding of man's use and management of the world's resources. Accordingly, this final chapter considers the contributions that a knowledge of the world in the past can make to the way we look at contemporary plant communities, how we manage them, and how they are likely to change in the future. It examines the relevance of the palaeoecological perspective to basic ecological theory (and in particular to ideas about stability and flux in plant communities), conservation issues (the evaluation and selection of plant communities for conservation), and predictions of what the world in the 21st century will be like.

15.2 STABILITY AND FLUX IN PLANT COMMUNITIES

15.2.1 Community flux in the fossil record

Much modern ecological theory, and in particular much of what we may call the conservation ethic, is founded on the principles of plant succession and the climax state laid down by Frederic Clements and Sir Arthur Tansley in the first part of this century. At the centre is the ecosystem concept, and underpinning all is a belief in *stability*: a belief that in a stable environment the plant cover in any given locality will ultimately develop an equilibrium state of constant composition, which is then capable of being maintained without further change over many generations of the constituent species. Destabilization of the plant cover is seen as the result of perturbation, and the main agent of perturbation at the present day as Man. Although community stability may be achievable in the short term by limiting human impact, in the longer term it clearly has no validity. The history of the global plant cover over the last 65 Ma, as revealed in the various scenarios that we have been discussing in preceding chapters, is characterized by **flux**, not stability, and that flux is likely to continue into the foreseeable future. Our understanding of the history is patently incomplete, because of the large areas of *terra incognita* for which there are few documented plant fossils, and because of the small number of points in time for which there is specific information. Nevertheless, the main strands of the story seem clear. The flux has involved changes in environmental conditions, concomitant changes in plant ranges and abundances, and changes in genotype affecting a plant's ability to respond to the environmental changes. Particular assemblages of species have come together opportunistically as new ecospace has arisen through the Cenozoic Era, and have then been re-ordered as conditions continued to change. The magnitude of those changes has been greater than anything we can observe or deduce for recent centuries. The palaeoecological record for forest communities, in particular, reveals the transitoriness of those communities in time and space. Defined assemblages of tree species have not responded in concert to environmental change for any prolonged period of time. Each taxon instead has responded individualistically and idiosyncratically. As a result, modern forest assemblages have no long lineages that can be followed back through the fossil record. The same could be true for other biomes also.

On the other hand, that scenario may be to some extent an inevitable consequence of the nature of the fossil record, where the only entity that can be followed reliably through time is the individual plant taxon. Collections of taxa that are equivalent to a modern plant community do not exist in the fossil record (except in certain macrofossil assemblages – e.g., packrat middens). Those collections that do occur may derive their component plants from several adjoining communities; and even if they do not, information on the full species-content and spatial organization of the community is normally lacking. Yet that does not mean we should dismiss the community as being of little account in the long-term picture. Individual species-responses are not manifested in a vacuum, but in the presence of a wide range of interacting plant populations. We must therefore expect some interdependence of species-responses to environmental change in the past, even if that interdependence is not readily apparent in the fossil record.

The transient assemblages that we *can* detect in the fossil record each consist of three components: certain species which persist unchanged (at least in the short term) as a relatively stable core; other species recently added to the assemblage; and yet others which will soon be lost from it. Loss occurs by emigration, extinction and evolution: by species being driven from the area by changing environmental conditions, or dying out altogether, or evolving into new species. Gain occurs by immigration and evolution: by new species entering the area when environmental conditions become more favourable, or as genetic derivatives of existing species. The emphasis to date in palaeoecological studies has been on the immigrants and emigrants. However, the stable core must not be forgotten. In forest communities (for which most information is available) the assemblage of subordinate herbaceous species associated with that stable core of woody taxa is almost completely unknown. It is therefore possible that a more coordinated community than we can detect in the fossil record has existed in some circumstances, at least in the short term. Supporting evidence for that belief is considered in the next section.

15.2.1 The boreal mires as a model of community flux

In one system the integrity of the plant community through time is maintained quite clearly in the fossil record. That system is the **mire ecosystem**, where the substrate for the contemporary plant cover is the peat accumulated from the remains of earlier generations of plants that grew at the site. Because of these on-site archives, mire ecosystems afford a unique opportunity for studying the changing interactions of the plant community with its environment through time, and for identifying consistent features that might be applicable to other ecosystems also. Throughout much of the world, however, mires are only a minor element in the landscape, as the prevailing climates are unfavourable for the accumulation of peat above the regional water table. Peat deposits there are restricted to circumscribed basins [1]. The **boreal** (and blanket) **mires** (section 14.3.4), on the other hand, form in climates where continued build-up of peat *above* the regional water table is favoured, and where mire vegetation is accordingly a major landscape element. Their peat deposits show variations in floristic composition and in the degree of humification of the preserved plant remains that demonstrate the close control exercised by the climate on the rate of peat accumulation and the nature of the contemporary mire vegetation. That control is apparent both as major stratigraphic horizons within the peat – the so-called 'Grenzhorizont' and related 'recurrence surfaces' [1,2] – and as smaller-scale fluctuations in the abundance of particular peat components [3]. The peat stratigraphy shows clearly that boreal mires have functioned as very sensitive barometers of change in the past, responding in various ways to climatic shifts of differing magnitudes and duration. These responses include large-scale landscape modification (the progressive spread of peat over the landscape [4,5]), whole-community change (widespread flooding or drying-out of the mire surface as a result of pronounced climatic change [3,6,7]) and more localized change (the restricted expansion and contraction of hummocks and pools over the mire surface [3]).

Over the time-span of the Holocene, many boreal mires can be seen, moreover, to have developed an increasingly complex structural organization, expressed in progressive diversification of the mire surface and the formation of elaborate surface patterning (section 14.3.4). These changes are a product both of climatic change and of geomorphological evolution of the landscape, and typify the processes of ongoing flux within mire systems. In many higher-altitude blanket mires this flux has led to degradation and erosion. The erosion that is widespread particularly in the British uplands today has been ascribed to a variety of causes: natural instability of the peat mass, climatic change, deforestation of surrounding hillslopes, overgrazing by sheep, moor-burning and air pollution [8–10]. The proposed date of onset of erosion is also variable: as early as 3000 yr ago [11], or as recent as 200 yr ago [12]; some eroded peat areas are known to have been formed this century by wildfires or by freak weather events [13,14]. All these aspects of erosion can be regarded as manifestations of the sensitivity of mire ecosystems to perturbation by a variety of agencies, and emphasize the need to consider perturbation as an important factor in the shaping of their modern condition.

Boreal mires have been chosen for special consideration in this final chapter because the need for a palaeoecological perspective in understanding their ecology is clearly recognized. The three aspects that we have emphasized – sensitivity to climatic change, ongoing structural development, and sensitivity to perturbation – are particularly well shown by the boreal mires, but it is unlikely that these aspects are unique to that ecosystem. Their role in the history of other plant communities has been identified sufficiently in previous chapters perhaps to justify the belief that they operate universally. How particular biomes respond to climatic change and perturbation and exhibit structural change depends on their 'resilience' [15] – their inherent capacity to resist change; and mire communities are perhaps less resilient than most communities by virtue of the rather unstable substrate that they generate. What is clear is that although the individual plant species of the mire system respond differently to climatic change, perturbation, or structural evolution, nevertheless the effects of these responses are expressed within the framework of the whole community.

15.2.3 Flux in forest systems

Much modern ecological research and theory has focused on forests, as representing in some way the optimal integration of organism and environment – the ultimate ecosystem. Major concepts of plant ecology – plant succession, mineral cycling, the niche, gap regeneration – were developed in the context of forest communities. These studies generated a working model of a forest community that reconciled the conflicting properties of long-term stability and short-term flux through cycles of change [16]. Subsequent work incorporated these concepts into models of forest growth that could be used to study the likely outcome of these processes on forest composition over time-periods of several centuries. One computer model that has been used successfully in a number of situations is the JABOWA model of Botkin, Janak and Wallis [17]. This model simulates the growth and survival of a mixed population of individual trees in a 100-m^2 plot, under constant or changing environmental conditions. The program utilizes a minimum of seven soil and climate variables, and handles up to forty different tree species (whose growth and performance are defined by nine variables). The model predicts that following severe disturbance (simulated by the removal of all mature individuals from the plot) a period of 600 yr of stable environment is required for the mixed forest to attain a stable composition. In many parts of the world perturbations are known to occur at more frequent intervals than this (section 12.3), and accordingly forest communities in these regions may lack long-term stability. Detailed examination of decayed and burnt remains on the forest floor of an upland forest in New Hampshire suggested, in support of the model's prediction, that the composition of the forest 300 yr ago was very different from what it is today, as a result of perturbation by hurricane and fire in the intervening period [18]. The JABOWA model can also be used to predict whether changes in environmental conditions will result in long-term changes in forest composition. The model predicts that compositional shifts will only follow temperature changes of at least 1°C overall, maintained for a period of at least 50 yr; a change of 2°C for a 200-yr period will alter the pattern of dominance completely, but the results will be noticeable only after a lag of about 200 yr [19]. As climatic changes of this magnitude have certainly occurred within historic times, any long-term stability of forest composition is unlikely on theoretical grounds.

The model thus reinforces the evidence from the fossil record of the transitory nature of many plant communities. If that is so, then the validity of certain ecological concepts is clearly suspect but whilst it may be useful to retain the terms 'ecosystem' and 'climax' to denote short-term functional units of the global plant cover, their continued use as a basis for determining long-term management and conservation strategies is questionable. The plant communities we seek to conserve are not stable systems – and conservation strategies should be based rather on a knowledge of how plant communities have changed in the past, and on predictions of likely change in the future. These aspects are considered further in the next section.

15.3 IMPLICATIONS FOR CONSERVATION

The last glacial stage in the northern hemisphere left behind a legacy at higher latitudes of ground stripped bare by glacial action or covered with unsorted skeletal soils, and of plant communities confined in many cases to localized and separated refugia. The subsequent development of a continuous forest cover occurred only after a prolonged protocratic phase during which the soils gradually became suitable for the growth of trees, and the refugial plant populations gradually spread and coalesced. The last 8 ka have seen progressive soil maturation and the formation of increasingly complex and diversified plant communities. However, in certain parts of the world glacial-stage conditions are effectively being re-created by man's activities many millennia in advance of their expected resurgence: the soil cover is being destroyed by erosion, and the natural plant cover is becoming fragmented. Nowhere is this more apparent than in the Mediterranean Basin, where the remnant cedar forests and bare hillslopes of Lebanon typify the premature 'ice age' brought on by man's mismanagement.* Even if the climate there were to become more favourable, many plant populations would be unable to respond quickly because of fragmented ranges and unsuitable soils, whilst if the climate were to worsen towards another true ice age, trees retreating southwards might no longer find appropriate refugial situations, as the habitats had been destroyed. These considerations emphasize the need for habitat maintenance as part of any long-term conservation strategy – and in particular the identification of habitats that might be important refugia in the future. The palaeoecologist still has many answers to find concerning the location of areas that served as refugia during the last glacial stage, but certain temperate forest areas in Europe and North America (those discussed in Chapter 10) are strong candidates for urgent long-term conservation. These areas are characterized by their floristic diversity – and floristic diversity has been used by some ecologists as a criterion for locating possible refugia. Yet from the evidence presented in earlier chapters it is

* Even before the current internecine catastrophes.

clear that present-day communities that are floristically diverse can have developed in many ways. Some, such as the rainforests of New Caledonia and the laurel forests of the Canary Islands [20–23], owe their diversity to long periods of evolution in isolation; others, such as the Californian chaparral, are still actively diversifying [24]; yet others may indeed occupy sites that were glacial-stage refugia. Diversity is already recognized as a major conservation criterion, principally in the context of the maintenance of existing genetic and habitat variability. The palaeoecological perspective would endorse the value of conserving diverse plant communities, but more as maximized reservoirs of species capable of exploiting any future environmental changes – and in some cases as potential refugial communities into which species populations could contract if environmental conditions worsened. This principle is already being utilized in the selection of rainforest areas for conservation [25], and needs to be extended, particularly to mediterranean-type communities.

Maintenance of diversity (both floristic and habitat) must thus be seen as a prime conservation strategy. The palaeoecological perspective suggests that this strategy should be pursued in the context of certain premises that can be deduced from past patterns of change in plant communities and their environment. These premises are:

1. Climatic change is a fundamental feature of the global environment, and is manifested over long ($\geqslant 10^5$ yr), medium (10^3–10^5 yr) and short 10^2–10^3 yr) time-periods. This change is to some extent predictable.
2. Parts of the global environment are also subjected to sporadic, localized and less predictable perturbations such as hurricanes, earthquakes and wildfires. The frequency of these perturbations varies spatially and temporally, often over time-periods of 10^1–10^3 yr.
3. The long-term composition of plant communities is affected by climatic change and perturbation, but there may be a considerable time-lag (50–200 yr) before plant responses to a new environmental regime are apparent.
4. The 'unit of response' to climatic change and perturbation is the plant population, not the plant community; because most plant species respond individualistically to environmental change, particular assemblages of species are unlikely to persist in the long term without loss of some existing species and gain of other new ones.
5. Current plant ranges may not be in equilibrium with current climates, because of slow rates of spread occasioned by inefficient seed-dispersal.
6. Global changes in climate have undoubtedly occurred this century (Figure 15.1), and are continuing (section 15.4). Plant responses to these changes, however, may not yet have been manifested or recognized.

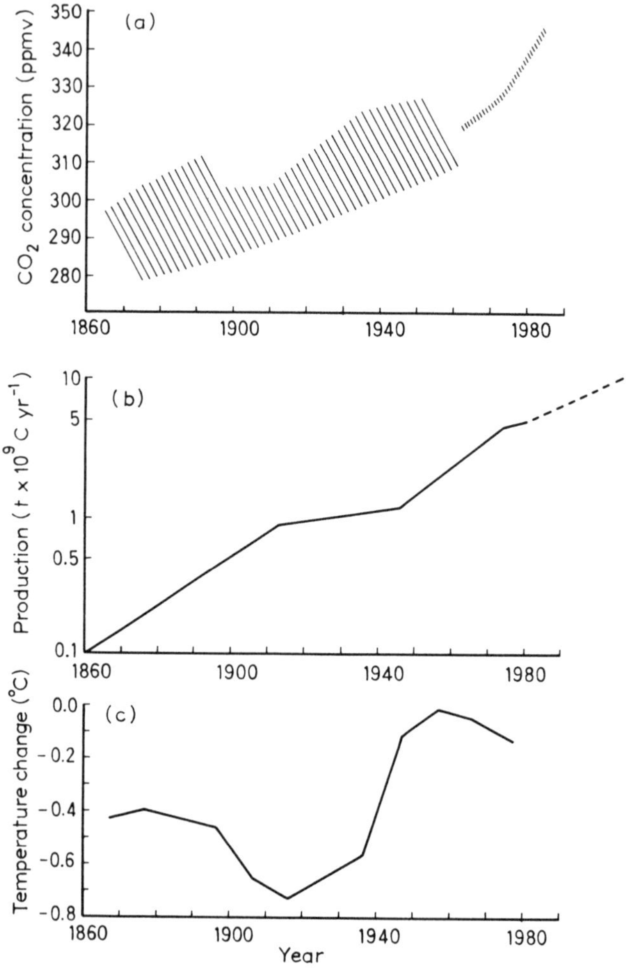

Figure 15.1 Global trends since 1860 AD in (a) atmospheric CO_2 concentration (the shaded area encompasses the degree of variability in the measurements), (b) annual CO_2 production (log scale) from fossil fuels and cement (the broken line is the projected trend to 2050 AD); (c) decadal average of temperature at the earth's surface (as departures from 1951–60 AD); (a) and (b) simplified from Liss and Crane [27]; (c) re-drawn from Folland and Kates [31].

7. Isolation of plant communities from human impact does not guarantee stability, as the communities remain susceptible to other forms of environmental change.

The logical conclusion from these premises is that management policies for plant communities should not strive to maintain the existing *status quo*, which relates to an arbitrary point in time in a changing environment, and may not be an equilibrium state, but should seek rather to minimize avoidable (i.e., human-induced) environmental damage, and to maintain a habitat capable of supporting a diversified plant cover (habitat conservation). As environments continue to change, so species that are currently rare in a particular community may increase in abundance, and *vice versa*; given this continual flux in plant populations it would seem dangerous to frame management policies that deliberately favour one species and disadvantage another. Perhaps to be avoided is the rather arbitrary categorization of species as 'native' or 'alien', and the subsequent decision often made that they are thus desirable or undesirable. A number of species that are designated as undesirable aliens in the British flora today (notably *Picea abies* and *Rhododendron ponticum*) undoubtedly occurred in Britain in the past as *bona fide* natives. Even a species that has clearly been introduced should not necessarily be regarded as undesirable. A successful alien such as sycamore (*Acer pseudoplatanus*) in Britain may be better equipped to cope with new environments in the future than native species which are evolutionarily stagnant. Analogies can perhaps be drawn with the European beech (*Fagus sylvatica*), which has been a highly successful tree over a range of altitudes and soil types in Europe in the later Holocene, but was rare in earlier interglacials [26], and with the redwoods, formerly widespread in northern hemisphere forests, and now very localized in range. It is clear where the future lies with these two examples, but for 'new' species such as sycamore where a palaeoecological perspective is not available, we should perhaps be more ready to refrain from trying to influence the course of change to our preconceived ideas. In a world that is going to change dramatically in the next century (see next section), we should aim to maximize the ability of the global plant cover to respond.

15.4 THE 21st CENTURY ENVIRONMENT

The likelihood of dramatic changes in the global environment in the next century is now widely recognized. Discussion has centred around the increasing concentrations of certain gases in the earth's atmosphere, and the probability of consequent climatic change. The concentration of carbon dioxide, in particular, is seen to be increasing. Although there is valid reason to question the accuracy of many of the measurements made in the 100 years prior to 1960 AD [27], the steady upward trend over the last three decades is undeniable: global concentrations of CO_2 have risen from <300 ppmv in 1860 to 320 ppmv in 1965, and 345 ppmv in 1985 [28] (Figure 15.1). The major source of this additional CO_2 in the atmosphere (currently increasing at a rate of $>2 \times 10^9$ tonnes of carbon per yr [27]) is the combustion of fossil fuels (coal, oil, gas and petrol). Here too there is an undeniable upward trend (Figure 15.1), with estimated values by the middle of the 21st century 5–10 times greater than at present [29]. This source of CO_2 input forms one part of a global carbon cycle which ultimately controls the concentration of CO_2 in the atmosphere. In addition to the fossil fuel input, CO_2 is also added naturally to the earth's atmosphere by the respiration of living plants and animals (including decomposer organisms), and is removed from it by absorption into the world's oceans and by the photosynthesis of land and water plants. More permanent removal of some part of the atmospheric CO_2 is achieved by incorporation into carbon compounds in the woody tissues of long-lived trees, in long-persistent soil organic matter, and in deep-ocean sediments. These natural processes of uptake and removal should maintain a nearly constant level of CO_2 in the atmosphere. However, under human influence the natural processes are continually being modified; forest clearance, for example, not only changes the short-term pattern of CO_2 exchange with the atmosphere by the living plant cover, but also reduces the longer-term carbon reservoirs in woody biomass and soil organic matter [30]. There have been various attempts to model carbon flux on a global scale, and in particular to study the responses of the oceanic reservoir to increasing atmospheric CO_2 concentrations, and there is rather general agreement between the models that a doubling of pre-baseline (i.e., 1860 AD) levels of atmospheric CO_2 (to 550–600 ppmv) is likely to be reached by, or before, the middle of the 21st century. There is also some consensus as to the consequences of these rapidly rising CO_2 levels. Gaseous carbon dioxide absorbs longwave radiation very efficiently, and even small concentrations in the atmosphere act as a heat shield preventing the emission of thermal energy (infrared radiation) from the earth's surface out into space. The balance between solar radiation absorbed by the earth's surface and infrared radiation emitted from it determines global temperature, and is inevitably altered as CO_2 builds up in the lower atmosphere. That this '**greenhouse effect**' is a real effect is borne out by modelling studies of the theoretical responses of factors believed to control global climate to increased concentrations of atmospheric CO_2 [27], and by measurements of the CO_2 content of air bubbles trapped in ice deposits of different ages. The models predict that a doubling of atmospheric CO_2 levels will result in increased global temperatures (though estimates of the increase range

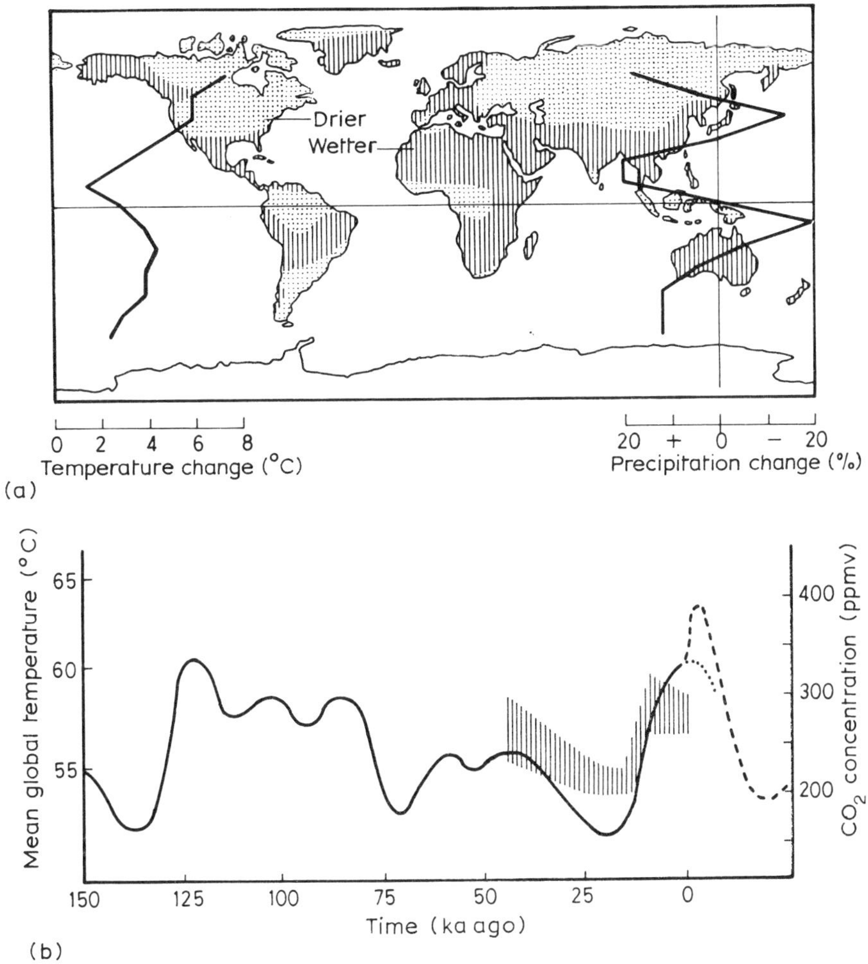

Figure 15.2 (a) The world in the middle of the 21st century, showing areas predicted to be drier ⠿⠿⠿ or wetter |||||||| than at present; superimposed (bold lines) are calculated temperature (left) and precipitation (right) deviations from present, for 10° belts of latitude in a 'doubled-CO_2 situation' (data of Perry and Perry [34]). (b) Generalized temperature curve for the last 150 ka, projected into the future for situations with ---- and without .... elevated levels of atmospheric CO_2; the shaded area superimposed shows the range of CO_2 concentrations recorded in air bubbles trapped in polar ice deposits formed over the last 40 ka. Re-drawn from Imbrie and Imbrie [33], Kellogg and Schware [29] and Liss and Crane [27].

from 1.1 to 4.5°C [28]), whilst 'fossil' CO_2 levels in ice clearly parallel deduced temperature levels over the last 40 ka (Figure 15.2). What is not yet clear is whether the temperature changes in the past were the cause or the result of the CO_2 changes. The models also predict that the rise in atmospheric CO_2 concentrations that has already occurred over the last 120 years should have raised global temperatures by 0.6–2.4°C [28]. A rise in global temperatures since 1900 AD of about 0.6°C is now indeed established (Figure 15.1), but it is not clear how much of this change should be attributed to the 'greenhouse effect' and how much to the natural recovery of climate since the Little Ice Age. A rise in temperature into the 21st century, however, seems inevitable, and will be augmented by the additive infrared-absorptive effects of other trace gases that are also increasing in concentration in the lower atmosphere: halocarbons, methane, nitrous oxide and ozone [32]. One prediction [33] is that 'consumption of the bulk of the world's known fossil fuel reserves would plunge the planet into a "super-interglacial age", unlike anything experienced in the last million years', and that this super-interglacial would probably endure for up to 2000 years before the 'greenhouse effect' ceased and the 'neoglacial' cooling trend was resumed (Figure 15.2). Even if we cannot yet establish the exact magnitude of this forthcoming temperature increase, we should accept its inevitability and attempt to plan for it – remembering that 'in the first half of the next century the world could well have twice as many people, consume three times as much food, and burn four times as much energy. Thus, impacts of climatic change must be superimposed on this backdrop' [29].

Possible scenarios for this 'super-interglacial' of the 21st century draw on three types of evidence: global climate models (GCMs); global climate patterns in recent years that have deviated greatly from the 20th-century temperature norm; and reconstructed vegetation patterns for the altithermal period and the last interglacial period. The main predictions of the GCMs for a doubled-CO_2 situation (likely to be reached between 2020 and 2030 AD [35]) are a warmer world that is also wetter and more humid, with the main warming effects at high latitudes; increased summer dryness in continental interiors at mid-latitudes; the low-latitude cyclone belt extending 200–400 km further towards the poles; and sea levels 20–140 cm higher as a result of thermal expansion of the water in the oceans [36]. One such scenario based on these predictions, and on the temperature and vegetation evidence, is shown in Figure 15.2. It must be regarded as a first approximation only, as the various GCMs currently in use give predictions that differ significantly in detail – either in the overall temperature increase to be expected for a given increase in atmospheric CO_2 concentrations [37], or in the geographical distribution of changes in temperature and precipitation. Table 15.1 shows the differences between the predictions of five

Table 15.1 Predictions of five climatic models for a doubled-CO_2 situation [39]

	Mean summer temperature change (°C)		Mean summer rainfall change (%)	
	World	United Kingdom	World	United Kingdom
Model 1	+3.3	+3.3	+ 7	+10
Model 2	+4.1	+5.9	+ 9	−17
Model 3	+4.3	+3.9	+11	−12
Model 4	+3.0	+3.1	+ 8	− 3
Model 5	+4.2	+5.0	+15	+38

models for global and United Kingdom temperature and precipitation changes with a doubled-CO_2 situation. Until some consensus of opinion over the likely climatic changes is achieved, any accompanying plant responses can be predicted only in very general terms. Displacements of plant ranges of 100–150 km polewards for each 1°C rise in temperature are nevertheless to be expected [38].

Parallels can be drawn with conditions during the altithermal period of 8–5 ka ago. However, it is unlikely that the vegetation patterns of that period will be restored, as the 21st-century climatic regime will probably differ in significant respects. More importantly, the plant assemblages available to respond to the new environmental regimes have been drastically changed since the altithermal period. In Europe, for example, elm (*Ulmus* spp.) and lime (*Tilia* spp.), major components of the altithermal forests, have declined markedly over the last 5 ka, and it is unlikely that their modern populations could recover quickly. Prediction of the patterns of plant response in the 21st century is now an urgent challenge that needs to be faced jointly by climatologists, palaeoecologists and conservationists. On these predictions may depend the quality of life in the 21st century.

REFERENCES

1. Tallis, J. H. (1983) Changes in wetland communities, in *Ecosystems of the World 4A. Mires: Swamp, Bog, Fen and Moor* (ed. A. J. P. Gore), Elsevier, Amsterdam, pp. 311–47.
2. Frenzel, B. (1983) Mires – repositories of climatic information or self-perpetuating ecosystems? in *Ecosystems of the World 4A. Mires: Swamp, Bog, Fen and Moor* (ed. A. J. P. Gore), Elsevier, Amsterdam, pp. 35–65.
3. Walker, D. and Walker, P. M. (1961) Stratigraphic evidence of regeneration in some Irish bogs. *J. Ecol.*, **49**, 169–85.
4. Heinselman, M. L. (1963) Forest sites, bog processes, and peatland types in the glacial Lake Agassiz region, Minnesota. *Ecol. Monogr.*, **33**, 327–74.

5. Heinselman, M. L. (1970) Landscape evolution, peatland types, and the environment in the Lake Agassiz Peatlands Natural Area, Minnesota. *Ecol. Monogr.*, **40**, 235–61.
6. Barber, K. E. (1981) *Peat Stratigraphy and Climatic Change*, Balkema, Rotterdam.
7. Tallis, J. H. (1965) Studies on southern Pennine peats. IV. Evidence of recent erosion. *J. Ecol.*, **53**, 509–20.
8. Phillips, J., Yalden, D. and Tallis, J. (1981) *Peak District Moorland Erosion Study. Phase 1 Report*. Peak Park Planning Board, Bakewell, Derbyshire.
9. Tallis, J. H. (1987) Fire and flood at Holme Moss: erosion processes in an upland blanket mire. *J. Ecol.*, **75**, 1099–129.
10. Taylor, J. A. (1983) The peatlands of Great Britain and Ireland, in *Ecosystems of the World 4B. Mires: Swamp, Bog, Fen and Moor* (ed. A. J. P. Gore), Elsevier, Amsterdam, pp. 1–46.
11. Bradshaw, R. and McGee, E. (1988) The extent and time-course of mountain blanket peat erosion in Ireland. *New Phytol.*, **108**, 219–24.
12. Jones, V. J., Stevenson, A. C. and Battarbee, R. W. (1987) *A Palaeolimnological Evaluation of Peatland Erosion*. Research Paper 28, Dept of Geography, University College, London.
13. Carling, P. A. (1986) Peat slides in Teesdale and Weardale, Northern Pennines, July 1983: description and failure mechanisms. *Earth Surface Proc. Landforms*, **11**, 193–206.
14. Tallis, J. H. (1981) Uncontrolled fires, in *Peak District Moorland Erosion Study, Phase 1 Report* (eds J. Phillips, D. Yalden and J. Tallis), Peak Park Planning Board, Bakewell, Derbyshire, pp. 176–82.
15. Walker, D. (1982) The development of resilience in burned vegetation, in *The Plant Community as a Working Mechanism* (ed. E. I. Newman), Blackwell, Oxford, pp. 27–43.
16. Bormann, F. H. and Likens, G. E. (1979) *Pattern and Process in a Forested Ecosystem*, Springer Verlag, New York.
17. Botkin, D. B., Janak, J. F. and Wallis, J. R. (1972) Some ecological consequences of a computer model of forest growth. *J. Ecol.*, **60**, 849–72.
18. Henry, J. D. and Swan, J. M. A. (1974) Reconstructing forest history from live and dead plant material – an approach to the study of plant succession in southwest New Hampshire. *Ecology*, **55**, 772–83.
19. Davis, M. B. and Botkin, D. B. (1985) Sensitivity of the fossil pollen record to sudden climatic change. *Quat. Res.*, **23**, 327–40.
20. Axelrod, D. I. (1975) Evolution and biogeography of Madrean–Tethyan sclerophyll vegetation. *Ann. Miss. Bot. Gard.*, **62**, 280–334.
21. Lems, K. (1960) Floristic botany of the Canary Islands. *Sarracenia*, **5**, 1–94.
22. Morat, Ph., Veillon, J.-R. and MacKee, H. S. (1984) Floristic relationships of New Caledonian rain forest Phanerogams, in *Biogeography of the Tropical Pacific* (eds F. J. Radovsky, P. H. Raven and S. H. Schmer), Lawrence, Kansas, pp. 71–128.
23. Raven, P. H. and Axelrod, D. I. (1972) Plate tectonics and Australasian paleobiogeography. *Science*, **176**, 1379–86.
24. Raven, P. H. and Axelrod, D. I. (1977) Origin and relationships of the California flora. *Univ. Calif. Publs Bot.*, **72**, 1–134.

25. Whitmore, T. C. and Prance, G. T. (1987) *Biogeography and Quaternary History in Tropical Latin America*, Oxford University Press, New York.
26. Huntley, B. and Birks, H. J. B. (1983) *An Atlas of Past and Present Pollen Maps for Europe: 0–13,000 Years Ago*, Cambridge University Press, Cambridge.
27. Liss, P. S. and Crane, A. J. (1983) *Man-made Carbon Dioxide and Climatic Change: a Review of Scientific Problems*, Geo Books, Norwich.
28. Ramanathan, V. (1988) The greenhouse theory of climatic change: a test by an inadvertent global experiment. *Science*, **240**, 293–9.
29. Kellogg, W. W. and Schware, R. (1981) *Climatic change and Society*, Westview Press, Boulder, CO.
30. Delcourt, H. R. and Harris, W. F. (1980) Carbon budget of the southeastern US biota: analysis of historical changes in trend from source to sink. *Science*, **210**, 321–3.
31. Folland, C. and Kates, F. (1984) Changes in decadally averaged sea surface temperature over the world 1861–1980, in *Milankovitch and Climate: understanding the Response to Astronomical Forcing* (ed. A. L. Berger), Reidel, Dordrecht, pp. 721–7.
32. Dickinson, R. E. and Cicerone, R. J. (1986) Future global warming from atmospheric trace gases. *Nature*, **319**, 109–15.
33. Imbrie, J. and Imbrie, K. P. (1979) *Ice Ages. Solving the Mystery*, Macmillan, London.
34. Perry, A. and Perry, V. (1986) *Climate and Society*, Bell and Hyman, London.
35. Wigley, T. M. L. (1988) Future CFC concentrations under the Montreal Protocol and their greenhouse-effect implications. *Nature*, **335**, 333–5.
36. Pittock, A. B. (1988) The greenhouse effect and future climatic change, in *Recent Climatic Change. A Regional Approach* (ed. S. Gregory), Belhaven Press, London, pp. 306–15.
37. Cess, R. D. and Potter, G. L. (1988) A methodology for understanding and intercomparing atmospheric climate feedback processes in general circulation models. *J. Geophys. Res.*, **93**, 8305–14.
38. Kerr, R. A. (1988) Report urges greenhouse action now. *Science*, **241**, 23–4.
39. Hulme, M. (1989) Generalists in hot pursuit. *Guardian*, 17 February 1989, p. 36.

Species index

Subject index